ELEMENTS OF HYDRODYNAMIC PROPULSION

Mechanics of fluids and transport processes
editor: R.J. Moreau

Books published in this series are:

J. Happel and H. Brenner, Low Reynolds number hydrodynamics. 1983.
ISBN 90-247-2877-0.
S. Zahorski, Mechanics of viscoelastic fluids. 1982. ISBN 90-247-2687-5.

Elements of hydrodynamic propulsion

J.A. Sparenberg
Department of Mathematics
University of Groningen
The Netherlands

1984 **MARTINUS NIJHOFF PUBLISHERS**
a member of the KLUWER ACADEMIC PUBLISHERS GROUP
THE HAGUE / BOSTON / LANCASTER

Distributors

for the United States and Canada: Kluwer Boston, Inc., 190 Old Derby Street, Hingham, MA 02043, USA
for all other countries: Kluwer Academic Publishers Group, Distribution Center, P.O.Box 322, 3300 AH Dordrecht, The Netherlands

Library of Congress Cataloging in Publication Data

Sparenberg, J. A.
Elements of hydrodynamic propulsion.

Includes index.
1. Ship propulsion. I. Title.
VM751.S7 1983 623.8'7 83-13127

ISBN-13: 978-94-009-6088-6 e-ISBN-13: 978-94-009-6086-2
DOI: 10.1007/978-94-009-6086-2

Softcover reprint of the hardcover 1st edition 1984

Preface

This is a treatment of a number of aspects of the theory of hydrodynamic propulsion. It has been written with in mind technical propulsion systems generally based on lift producing profiles.

We assume the fluid, which is admitted in conventional hydrodynamics, to be incompressible. Further we assume the occurring Reynolds numbers to be sufficiently high such that the inertia forces dominate by far the viscous forces, therefore we take the fluid to be inviscid. Ofcourse it must be realized that viscosity plays an important part in a number of phenomena displayed in real flows, such as flow separation at the nose of a profile and the entrainment of fluid by a ship's hull. Another approximation which will be used in general is that the problems are linearized. In other words it is assumed that the induced disturbance velocities are sufficiently small, such that their squares can be neglected with respect to these velocities themselves. Hence it is necessary to evaluate the domain of validity of the results with respect to these two a priori assumptions. Anyhow it seems advisable to have first a good understanding of the linearized non-viscous theory before embarking on complicated theories which describe more or less realistic situations. For elaborations of the theory to realistic situations we will refer to current literature.

In low Reynolds number flow, singular external forces and moments are very useful. It is one of the objectives of this book to promote the use of external force fields also in the case of incompressible and inviscid fluids as an expedient to generate velocity fields. Although in most text books external force fields appear in the equations of motion, usually it is assumed that they have an impulsive character or that they are the gradient of a potential function. In the latter case they have lost, in relation to incompressible fluids the ability of inducing velocities, they only change the pressure field. An interesting feature of non conservative external force fields is that they can generate vorticity in an inviscid fluid. By this we have no need in a discussion about the origin of vorticity, to make use of a slight viscosity which afterwards is abandoned again. Using external force fields the concepts of for instance pressure dipole and actuator disk, arise in a natural way from the integration of the equations of motion.

Another objective of the book is to describe a linearized optimization theory for propellers or more generally for systems of lifting surfaces. The theory applies ro rather general types of force actions, for instance to steady and to unsteady propulsion. It is assumed however that the lifting surfaces form angles with the direction of the desired force, which are not small. An exception is as we will show, the calculation of the optimum thrust of the sails of a yacht sailing close to wind. This problem can be reformulated as a problem of energy extraction and in this way it comes under the theory described here.

We mentioned already that in this treatment viscosity has been neglected and that we have to be careful with the interpretation of the results. This especially holds with regard to optimization theory, for this we refer to the introduction to chapter 5.

We also discuss the existence or non existence of optimum propulsion systems for a number of types. We do this mainly for the case of unsteady propulsion. It turns out that in some classes of admitted propellers, optimum propellers do exist and in other classes they do not. In the latter case it does not mean that the admitted class can not contain propellers with a high efficiency. On the contrary, a non-existence proof can be based on the fact that it is possible to construct a minimizing sequence in the considered class of propellers for which the loss of energy per unit of time theoretically tends to zero. However this will occur in general at the cost of wilder and wilder motions so that no acceptable propeller comes out in the limit procedure. Hence the non existence of an optimum propeller means only that we cannot construct an algorithm to find within the considered class a propeller with least energy losses.

For the proof of the existence of optimum propellers it seems that the abstract methods of functional analysis are unavoidable. The reason is that it has to be proved that the lost kinetic energy per unit of time of the propeller is a functional with some desired properties on some given set of motions, such that this functional assumes its minimum at one of the motions of the set.

The choice of the subjects and examples in this book, reflects the field of research of the author and his collaborators, it is not claimed that a complete survey of hydrodynamic propulsion theory is given. For instance slender body propulsion which is of importance in the biological sciences is not treated here. Also cavitation which is often (not always) an undesired phenomenon in propulsion is not considered.

We assume in this monograph the reader to be familiar with a number of basic concepts of hydrodynamics and with their application to wing theory. We mention the velocity potential, the streamfunction, Bernoulli's theorem, the concept of linearization, the law of Biot and Savart, the lifting line and the Kutta condition. Subjects which are fundamental for some types of propulsion, such as the unsteady suction force at the leading edge of a profile are discussed. Also the well-known trailing vorticity of a lifting surface is treated as an illustration of the use of external force fields.

Groningen J.A. SPARENBERG

Contents

Appendices

Acknowledgement

I am grateful to my wife Paula who typed the first version of this book and who spent a lot of time unriddling the manuscript and to Marga Wagenaar who skillfully typed the second version.

Also I am indebted to Mr. P.W. de Heer, from the Shipbuilding Laboratory of the Technical University Delft, who made the drawings.

Parts of the book have been read by some colleagues, of whom I mention Paul Urbach and Willem Potze, and by some students. My thanks are due to them for tracking down mistakes, and for the remaining ones the author has to be blamed.

Finally I am indebted to Martinus Nijhoff Publishers for offering me the opportunity to write this monograph.

J.A. Sparenberg

to Paula

1. External force actions

A body moving through a fluid, which we assume to be incompressible and inviscid, will induce velocities and pressures in this fluid. Hence the body will experience forces and moments caused by the integrated action of the pressures on its boundary. Inversely by the law, action equals reaction, the body will exert forces on the fluid. Sometimes these force actions are accompanied by the shedding of vorticity as in the case of a lifting surface of finite span, sometimes there is no vortex shedding as in the case of the accelerated motion of a sphere, where in both cases we assume that no flow separation occurs. In the first part of this chapter we will consider this vorticity shedding of a body more closely.

Next we consider the pressures and velocities induced by an external force field acting directly on the fluid, hence without the intermediary of a body. These considerations are mainly based on a linearized theory. It is discussed that force fields which are conservative are not of much interest in propulsion theory, these fields induce only pressures and no velocities. We calculate the work done per unit of time by an external force field. This gives for instance a possibility to calculate the induced resistance of a lifting surface.

A special case of an external force field is the singular force moving in one way or another through the fluid. The velocity field of a singular force yields the kernel function for a number of problems in lifting surface and actuator surface theory. We will determine this velocity field by means of limit considerations. For a mathematical discussion of the highly singular velocity field the theory of distributions should be used. This however complicates the reasoning to a large extent while it is less easy to recognize the simple physics behind it. It is shown by Urbach [64] that our results agree with those of the theory of distributions.

The vorticity induced by external force fields can be divided into bound vorticity and free vorticity. We will demonstrate that such a denomination is often subject to arbitrariness.

We conclude the chapter with the discussion of the suction force at the leading edge of a profile without thickness. This force is of importance for the calculation of the thrust delivered by a profile carrying out a small amplitude motion.

1.1. Hydrodynamic forces on a moving body

We will discuss here some general results which describe the forces and moments exerted by a fluid on a moving body. Our discussion will be restricted to the derivation of some basic formulas needed in the next section. For a more elaborate treatment of forces on rigid bodies we refer to [3] and [41].

Consider a body B of finite extent moving in an inviscid and incompressible fluid. In this fluid we have a Cartesian coordinate system (x, y, z) with respect to which the fluid at infinity is at rest. During its motion the body is allowed to change its shape and volume. We assume that no vorticity is shed into the fluid. Hence the fluid motion is irrotational and its velocity field $\boldsymbol{v}$ with components v_x, v_y and v_z in the x, y and z direction, can be derived from a potential function $\Phi = \Phi(x, y, z, t)$ at all points of space outside the body

$$\boldsymbol{v} = \operatorname{grad} \Phi. \tag{1.1.1}$$

Because the velocity field is free of divergence we have

$$\operatorname{div} \boldsymbol{v} = \Delta\Phi = \left(\frac{\partial^2}{\partial x^2} + \frac{\partial^2}{\partial y^2} + \frac{\partial^2}{\partial z^2}\right)\Phi = 0. \tag{1.1.2}$$

Consider around B a control surface $\tilde{H}$ which is coupled to the fluid particles, hence it floats with the fluid. To the fixed amount of fluid in the region $\tilde{\Omega}$ between $\tilde{H}$ and the boundary ∂B of B we can apply the theorem of momentum. This states: the resultant force exerted on an

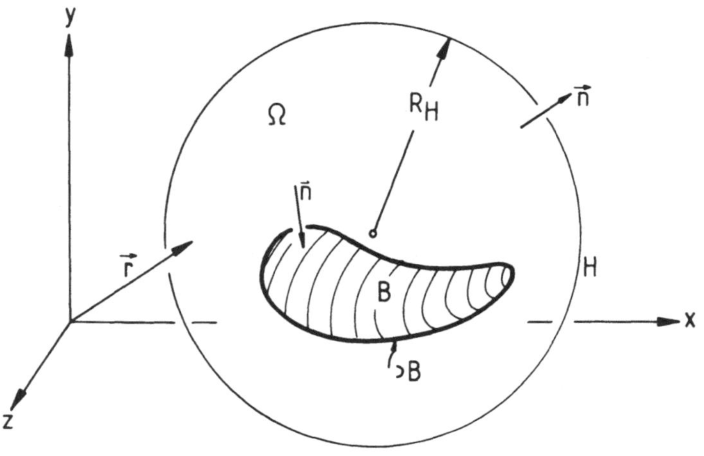

Fig. 1.1.1. Body with control surface H.

amount of fluid equals the change of its momentum per unit of time. For the formulation of the problem however it is more easy to replace $\tilde{H}$ by a surface H fixed in space and to consider the region Ω bounded by H and ∂B. But then we have to add to the rate of change of momentum of the fluid in Ω, the flux of momentum leaving Ω through H and to subtract the incoming flux. For H we take a sphere with sufficiently large radius R_H and with its centre in the neighbourhood of B.

The momentum $\boldsymbol{I} = \boldsymbol{I}(t)$ of the fluid in the region Ω is

$$\boldsymbol{I} = \mu \int_{\Omega} \operatorname{grad} \Phi \, \mathrm{dVol} = \mu \int_{\partial B + H} \Phi \boldsymbol{n} \, \mathrm{d}S, \tag{1.1.3}$$

where μ is the density of the fluid and the unit normal $\boldsymbol{n} = (n_x, n_y, n_z)$ points out of the region Ω. We want to calculate the force $\boldsymbol{F} = \boldsymbol{F}(\mathrm{t})$ exerted by the fluid on the body B during its motion. We also introduce the force $\boldsymbol{F}_H = \boldsymbol{F}_H(t)$ exerted by the fluid outside H at the fluid inside H. Then we can write the balance of momentum as

$$\boldsymbol{F}_H - \boldsymbol{F} = \mu \frac{\mathrm{d}}{\mathrm{d}t} \int_{\partial B + H} \Phi \boldsymbol{n} \, \mathrm{d}S + \mu \int_H \boldsymbol{v}(\boldsymbol{v} \cdot \boldsymbol{n}) \, \mathrm{d}S, \tag{1.1.4}$$

where the last term is the mentioned momentum flux through H.

The force $\boldsymbol{F}_H$ can be written as

$$\boldsymbol{F}_H = -\int_H p \boldsymbol{n} \, \mathrm{d}S. \tag{1.1.5}$$

Using Bernoulli's equation for unsteady motion

$$p + \tfrac{1}{2}\mu |\boldsymbol{v}|^2 + \mu \frac{\partial \Phi}{\partial t} = p_\infty + \tfrac{1}{2}\mu |\boldsymbol{v}_\infty|^2, \tag{1.1.6}$$

where p_∞ is the pressure at infinity and $\boldsymbol{v}_\infty$ the velocity at infinity, we write (1.1.5) as

$$\boldsymbol{F}_H = \mu \int_H \left\{ \frac{\partial \Phi}{\partial t} + \tfrac{1}{2} |\boldsymbol{v}|^2 \right\} \boldsymbol{n} \, \mathrm{d}S. \tag{1.1.7}$$

In (1.1.7) we used the fact that for any sufficiently smooth closed surface H

$$\int_H \boldsymbol{n} \, \mathrm{d}S = 0. \tag{1.1.8}$$

This vector equality follows from the repeated application of the scalar equality

$$\int_H \{ f_1(y,z)n_x + f_2(x,z)n_y + f_3(x,y)n_z \} \, \mathrm{d}S = 0, \tag{1.1.9}$$

which holds for "arbitrary" functions f_i $(i = 1, 2, 3)$ of the indicated arguments.

Because H is at rest with respect to our coordinate system we have

$$\int_H \frac{\partial \Phi}{\partial t} \boldsymbol{n} \, \mathrm{d}S = \frac{\mathrm{d}}{\mathrm{d}t} \int_H \Phi \boldsymbol{n} \, \mathrm{d}S. \tag{1.1.10}$$

Substitution of (1.1.7) and (1.1.10) into (1.1.4) yields

$$\boldsymbol{F} = -\mu \frac{\mathrm{d}}{\mathrm{d}t} \int_{\partial B} \Phi \boldsymbol{n} \, \mathrm{d}S + \mu \int_H \{ \tfrac{1}{2} |\boldsymbol{v}|^2 \boldsymbol{n} - \boldsymbol{v}(\boldsymbol{v} \cdot \boldsymbol{n}) \} \, \mathrm{d}S. \tag{1.1.11}$$

Next we consider the limit of (1.1.11) when the radius R_H of H tends to infinity. At large distances the velocities induced by B tend to zero as R_H^{-2}. This happens when B changes its volume, otherwise the induced velocities tend to zero more quickly. Anyhow the contribution of the integral over H in (1.1.11) tends to zero for $R_H \to \infty$. This means that

$$\boldsymbol{F} = -\mu \frac{\mathrm{d}}{\mathrm{d}t} \int_{\partial B} \Phi \boldsymbol{n} \, \mathrm{d}S. \tag{1.1.12}$$

In an analogous way we can derive a result for the moment $\boldsymbol{M} = \boldsymbol{M}(t)$, caused by the hydrodynamic pressures at ∂B. This moment will be calculated with respect to the origin O. We apply the theorem: the sum of the moments about O of external forces acting at an amount of fluid equals the change of the moment of momentum about O per unit of time of that amount of fluid.

The moment of momentum about O of the fluid in Ω is

$$\mu \int_\Omega (\boldsymbol{r} * \boldsymbol{v}) \, \mathrm{dVol} = \mu \int_{\partial B + H} \Phi \cdot (\boldsymbol{r} * \boldsymbol{n}) \, \mathrm{d}S, \tag{1.1.13}$$

where $*$ denotes the vector product and the equality follows by partial integration. When we denote by $\boldsymbol{M}_H = \boldsymbol{M}_H(t)$ the moment about O of the hydrodynamic forces at H exerted by the fluid outside H, the just mentioned theorem assumes the form

$$\boldsymbol{M}_H - \boldsymbol{M} = \mu \frac{\mathrm{d}}{\mathrm{d}t} \int_{\partial B + H} \Phi \cdot (\boldsymbol{r} * \boldsymbol{n}) \, \mathrm{d}S + \mu \int_H (\boldsymbol{r} * \boldsymbol{v})(\boldsymbol{v} \cdot \boldsymbol{n}) \, \mathrm{d}S, \tag{1.1.14}$$

where the last term is the moment of momentum flux through the fixed surface H.

The moment $\boldsymbol{M}_H$ can, by using (1.1.6) and (1.1.9), be written as

$$\boldsymbol{M}_H = \mu \int_H \left\{ \frac{\partial \Phi}{\partial t} + \tfrac{1}{2}|\boldsymbol{v}|^2 \right\} (\boldsymbol{r} * \boldsymbol{n}) \, \mathrm{d}S. \tag{1.1.15}$$

Now we substitute (1.1.15) into (1.1.14). Using an analogous formula as (1.1.10) and taking the limit $R_H \to \infty$ we obtain

$$\boldsymbol{M} = -\mu \frac{\mathrm{d}}{\mathrm{d}t} \int_{\partial B} \Phi \cdot (\boldsymbol{r} * \boldsymbol{n}) \, \mathrm{d}S. \tag{1.1.16}$$

Here the integral over the surface H tends to zero because its integrand tends to zero as R_H^{-3} or more quickly when B does not change its volume.

Next we consider the case of an infinitely long cylinder B with generators parallel to the z axis. This cylinder is allowed to move arbitrarily and to change its shape, but such that its generators remain parallel to the z axis. Then the induced fluid flow depends only on the two coordinates x and y.

Again we assume the fluid to be at rest with respect to our coordinate system, at large distances from B. We surround the cylinder B by a circular cylindrical control surface H with radius R_H, fixed in space. Because our consideration will be given for a slab of space of unit width in the z direction, we consider H and the boundary ∂B of B as lines in the (x, y) plane (fig. 1.1.2) and Ω is the two dimensional region bounded by them. An essential difference with the previous three dimensional case is that here Ω is doubly connected.

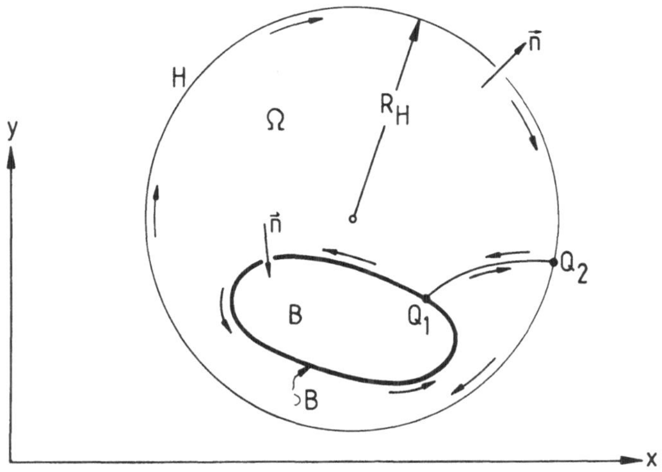

Fig. 1.1.2. Cross section of cylinder B and control surface H.

The momentum $\boldsymbol{I} = (I_x(t), I_y(t))$ of the fluid in the region Ω is

$$\boldsymbol{I} = (I_x, I_y) = \mu \iint_{\Omega} \left(\frac{\partial \Phi}{\partial x}, \frac{\partial \Phi}{\partial y} \right) \mathrm{d}x\, \mathrm{d}y, \tag{1.1.17}$$

where $\Phi = \Phi(x, y, t)$ is again the velocity potential. Because Ω is doubly connected, the function Φ can be "multivalued", we make Ω simply connected by introducing a cut (Q_1, Q_2) from ∂B, towards H.

By a partial integration we can change the double integrals (1.1.17) into integrals along the line L consisting of H, ∂B and (Q_1, Q_2) in the indicated directions. We find

$$(I_x, I_y) = -\mu \left\{ \int_L \Phi\, \mathrm{d}y, -\int_L \Phi\, \mathrm{d}x \right\} = \mu \int_L (y, -x)\, \mathrm{d}\Phi. \tag{1.1.18}$$

When there is circulation around B the function Φ will assume different values at both sides of (Q_1, Q_2). This circulation has to be independent of time, otherwise vorticity would be shed into the fluid which here just as in the three dimensional case, is assumed not to happen. Because the difference of Φ at both sides of the cut is a constant, the contributions to (1.1.18) from the two sides of (Q_1, Q_2) cancel. When i is the imaginary unit and $\zeta = x + iy$ we can write (1.1.18) as

$$I = (I_x + iI_y) = -i\mu \int_{\partial B + H} \zeta\, \mathrm{d}\Phi. \tag{1.1.19}$$

The resultant hydrodynamic force on ∂B is denoted by $F = (F_x(t) + iF_y(t))$, the resultant force on H by the outside fluid is

$$F_H = -i \int_H p\, \mathrm{d}\zeta. \tag{1.1.20}$$

Now we apply again the theorem of momentum. Then we have as in the previous case, to consider the momentum flux through H. Analogous to the reasoning for the three dimensional case it is easily seen that the contribution of this to the force F tends to zero when the radius R_H tends to infinity. Hence we write

$$F_H - F \approx -i\mu \frac{\mathrm{d}}{\mathrm{d}t} \int_{\partial B + H} \zeta\, \mathrm{d}\Phi, \tag{1.1.21}$$

where the symbol $\approx$ means that the momentum flux through H has been left out of consideration.

Because H is fixed we have

$$\frac{\mathrm{d}}{\mathrm{d}t} \int_H \zeta\, \mathrm{d}\Phi = \int_H \zeta\, \mathrm{d}\frac{\partial \Phi}{\partial t} = -\int_H \frac{\partial \Phi}{\partial t}\, \mathrm{d}\zeta. \tag{1.1.22}$$

The latter equality is based on the following consideration. The potential Φ can change by a certain amount by encircling the body B. This amount however is independent of time because we supposed the circulation around B to be constant, hence $\partial\Phi/\partial t$ assumes the same value after encircling B. From this (1.1.22) follows.

Next we write (1.1.20) by substitution of (1.1.6) and substitute the result together with (1.1.22) into (1.1.21). Then we find for the limit $R_H \to \infty$

$$F = \left(F_x + iF_y\right) = i\mu \frac{\mathrm{d}}{\mathrm{d}t} \int_{\partial B} \zeta \,\mathrm{d}\Phi. \tag{1.1.23}$$

1.2. Force actions and shed vorticity

Consider a body B of finite extent moving through an incompressible and inviscid fluid which is again at rest at infinity with respect to our Cartesian coordinate system (x, y, z). The body B will move with a mean velocity U in the positive x direction and repeats its velocities after each time period τ or after each covered distance

$$b = U\tau, \tag{1.2.1}$$

while also the neighbouring field of flow has the same periodicity. When we assume that no vorticity is shed by B we prove that no mean force can be exerted by B on the fluid.

In this case (1.1.12) is valid. The mean value of $\boldsymbol{F}(t)$ over one period τ of time becomes

$$\frac{1}{\tau}\int_t^{t+\tau} \boldsymbol{F}(t)\,\mathrm{d}t = -\frac{\mu}{\tau}\int_{\partial B} [\Phi(x+b, y, z, t+\tau)\boldsymbol{n}(x+b, y, z, t+\tau) - \Phi(x, y, z, t)\boldsymbol{n}(x, y, z, t)]\,\mathrm{d}S. \tag{1.2.2}$$

The velocities of the fluid at times t and $t+\tau$ are the same for the points (x, y, z) and $(x+b, y, z)$. Hence the difference of the potential at corresponding points and times can only be a constant c, then

$$\frac{1}{\tau}\int_t^{t+\tau} \boldsymbol{F}(t)\,\mathrm{d}t = -\frac{\mu c}{\tau}\int_{\partial B} \boldsymbol{n}\,\mathrm{d}S = 0. \tag{1.2.3}$$

From (1.2.3) we find that a body of finite extent, moving periodically in the way as we described, cannot experience a force with a non zero mean value without shedding vorticity. Inversely, by the principle action equals reaction, such a body cannot exert a mean force on the fluid without leaving behind vorticity. It cannot act as a lift producing wing or a thrust producing propeller. When vorticity is shed periodically the function Φ is not defined in the whole space and the velocities do not tend to zero at infinity in the way as was needed for the derivation of (1.1.12). Hence the foregoing argument does not hold. Because the velocity field belonging to the shed vorticity represents kinetic energy of the fluid, we can state; when a periodically moving body of finite extent inducing a periodic neighbouring field of flow, exerts a mean force on the fluid this has to be accompanied by energy losses.

Next we consider the moment with respect to 0, of the fluid pressures exerted at the body B. Now we use (1.1.16) which is valid when no vorticity is shed. Hence in that case we find for the mean value of the moment around the x axis

$$\frac{\boldsymbol{e}_x}{\tau}\cdot\int_0^\tau \boldsymbol{M}(t)\,\mathrm{d}t = -\frac{\mu}{\tau}\boldsymbol{e}_x\cdot\int_{\partial B}\big[\Phi(x+b,y,z,t+\tau)$$
$$\times\{(\boldsymbol{r}(x,y,z,t)+b\boldsymbol{e}_x)*\boldsymbol{n}(x+b,y,z,t+\tau)\}$$
$$-\Phi(x,y,z,t)\{\boldsymbol{r}(x,y,z,t)*\boldsymbol{n}(x,y,z,t)\}\big]\,\mathrm{d}S, \tag{1.2.4}$$

where $\boldsymbol{e}_x$ is the unit vector in the x direction. Because

$$\boldsymbol{n}(x+b,y,z,t+\tau)=\boldsymbol{n}(x,y,z,t) \tag{1.2.5}$$

and again the difference of the potential at corresponding points and times can be only a constant c we find

$$\frac{\boldsymbol{e}_x}{\tau}\cdot\int_0^\tau \boldsymbol{M}\,\mathrm{d}t = -\frac{\mu c}{\tau}\int_{\partial B}(yn_z - zn_y)\,\mathrm{d}S = 0. \tag{1.2.6}$$

The last equality follows from (1.1.9).

From (1.2.6) it follows that a periodically moving body of finite extent which induces a periodic field of flow and which exerts at the fluid a non zero mean moment around a line parallel to its mean direction of motion, has to shed vorticity.

We can also consider moments around lines l_1 and l_2, parallel to the y-axis and z-axis respectively, which translate in the positive x direction with the velocity U. It can be seen, that in the case of no vorticity shedding, the moments around these lines need not to have zero mean values. A simple example is a flat wing of zero thickness of large aspect ratio having constant chordlength. The wing has a nonzero angle of incidence while its circulation is zero at every section along its span. Then the flow around it will not satisfy the Kutta condition, which states that the pressure jump over the profile tends to zero at the trailing edge. However this is not an objection in the case of our inviscid and incompressible fluid. Such a wing will have a moment for instance around a line through the midpoints of its chords ([3], page 437).

The two dimensional case is different from the three dimensional one, as follows also from our considerations in section 1.1. In fact it can be considered as three dimensional, while the velocities are independent of the z coordinate, hence they do not tend to zero at infinity. For instance a two dimensional wing can have a lift force per unit of span without shedding vorticity.

For the time dependent force F per unit of span exerted by the fluid on the possibly flexible contour of the profile, we use (1.1.23). Then the mean value of $F(t)$ over one period τ of time can be written as

$$\frac{1}{\tau}\int_t^{t+\tau} F(t)\,\mathrm{d}t = \frac{1}{\tau}\int_t^{t+\tau}\left(F_x(t) + iF_y(t)\right)\mathrm{d}t$$

$$= \frac{i\mu}{\tau}\int_{\partial B}\{(\zeta + b)\,\mathrm{d}\Phi(x+b, y, t+\tau) - \zeta\,\mathrm{d}\Phi(x, y, t)\}$$

$$= \frac{i\mu b}{\tau}\int_{\partial B}\mathrm{d}\Phi(x, y, t) = \frac{i\mu b}{\tau}\Gamma, \tag{1.2.7}$$

where Γ is the circulation around the profile, which is a real number. From (1.2.7) we find

$$\frac{1}{\tau}\int_t^{t+\tau} F_x(t)\,\mathrm{d}t = 0, \qquad \frac{1}{\tau}\int_t^{t+\tau} F_y(t)\,\mathrm{d}t = \frac{\mu b\Gamma}{\tau} = \mu U\Gamma. \tag{1.2.8}$$

Hence we have the result; when a periodically moving two dimensional body which induces a periodic neighbouring field of flow does not leave behind free vorticity, it cannot exert a mean force in the mean direction of its motion.

1.3. External force fields and vorticity

We consider the equations which describe the motion of an inviscid and incompressible fluid with respect to an inertial Cartesian frame of reference (x, y, z),

$$\frac{\mathrm{d}\boldsymbol{v}}{\mathrm{d}t} = \frac{\partial \boldsymbol{v}}{\partial t} + (\boldsymbol{v}\cdot\operatorname{grad})\boldsymbol{v} = -\frac{1}{\mu}\operatorname{grad} p + \frac{1}{\mu}\boldsymbol{F}, \tag{1.3.1}$$

$$\operatorname{div}\boldsymbol{v} = 0, \tag{1.3.2}$$

where $\boldsymbol{F} = \boldsymbol{F}(x, y, z, t)$ is an external force field per unit of volume, acting on the fluid. The way in which this force field is created physically is not a question of issue at this stage, later on we will show how the action of for instance a wing can be represented by it. Here and in the following we assume that the field of flow belonging to a force field, exists and is uniquely determined by suitably chosen initial and boundary conditions.

First consider a force field which satisfies rot $\boldsymbol{F} = 0$ and which comes into action at $t = 0$. Such a field can be represented by

$$\boldsymbol{F} = \text{grad}\, \psi(x, y, z, t), \qquad t > 0, \tag{1.3.3}$$

where ψ is some sufficiently smooth scalar function. Assume that for $t \leq 0$ the fluid is at rest hence $\boldsymbol{v}(x, y, z, t) = 0$, $t \leq 0$. Then we can satisfy (1.3.1) and (1.3.2) by

$$p = \psi(x, y, z, t), \qquad \boldsymbol{v} \equiv 0, \qquad t > 0. \tag{1.3.4}$$

Hence this field does not induce any motion in the fluid. In general only force fields will be of interest for which rot $\boldsymbol{F} \neq 0$ for some region of space.

Next consider two force fields $\boldsymbol{F}_1(x, y, z, t)$ and $\boldsymbol{F}_2(x, y, z, t)$ to which belong the pressure fields and velocity fields $(p_1, \boldsymbol{v}_1)$ and $(p_2, \boldsymbol{v}_2)$ respectively. The velocity fields are assumed to satisfy the same initial conditions at $t = 0$. The question can be posed, when do we have

$$\boldsymbol{v}_1(x, y, z, t) = \boldsymbol{v}_2(x, y, z, t), \qquad t > 0. \tag{1.3.5}$$

The answer is of course closely related to (1.3.3) and (1.3.4), in fact this will happen when

$$\boldsymbol{F}_1(x, y, z, t) - \boldsymbol{F}_2(x, y, z, t) = \text{grad}\, \psi(x, y, z, t). \tag{1.3.6}$$

Then by

$$p_2 = p_1 - \psi, \tag{1.3.7}$$

the right hand sides of (1.3.1) in both cases will be the same and hence (1.3.5) is valid. The equivalence relation (1.3.6) between two external force fields is important in the theory of actuator surfaces (section 2.4).

External force fields are useful with respect to theories where vorticity is created in an inviscid and incompressible fluid. Indeed in domains where the rotation of the force field is non zero, in a natural way rotation of the motion of the fluid, hence vorticity is induced. We write (1.3.1) in the form

$$\frac{\partial \boldsymbol{v}}{\partial t} - \boldsymbol{v} * \boldsymbol{\omega} = -\frac{1}{\mu}\operatorname{grad}\left(p + \tfrac{1}{2}\mu|\boldsymbol{v}|^2\right) + \frac{1}{\mu}\boldsymbol{F}, \tag{1.3.8}$$

where $\boldsymbol{\omega} = \operatorname{rot}\boldsymbol{v}$. Application of the operation rot to both sides, yields

$$\frac{\partial \boldsymbol{\omega}}{\partial t} - \operatorname{rot}(\boldsymbol{v} * \boldsymbol{\omega}) = \frac{1}{\mu}\operatorname{rot}\boldsymbol{F}. \tag{1.3.9}$$

From this equation it follows that when $\operatorname{rot}\boldsymbol{F} \neq 0$ then $\boldsymbol{\omega} \neq 0$.

It is easily seen that a force field which is confined to a finite region of space and which has a non zero resultant or which has a non zero moment about some point, cannot be the gradient of a potential function. This means that its rotation is non zero, hence it induces vorticity.

Consider a closed contour C in the fluid, coupled to the fluid particles, hence it is transported by the velocity field. At some moment of time we calculate the circulation $\Gamma = \Gamma(t)$ of $\tilde{C}$ defined by

$$\Gamma = \int_{\tilde{C}} \boldsymbol{v} \cdot \mathrm{d}\boldsymbol{s}, \tag{1.3.10}$$

where the integration is taken along $\tilde{C}$ and $\mathrm{d}\boldsymbol{s}$ is a directed line element of length $\mathrm{d}s$. We determine

$$\frac{\mathrm{d}\Gamma}{\mathrm{d}t} = \int_{\tilde{C}} \frac{\mathrm{d}\boldsymbol{v}}{\mathrm{d}t} \cdot \mathrm{d}\boldsymbol{s} + \int_{\tilde{C}} \boldsymbol{v} \cdot \frac{\mathrm{d}}{\mathrm{d}t}(\mathrm{d}\boldsymbol{s}). \tag{1.3.11}$$

The second integral can be written as

$$\int_{\tilde{C}} \boldsymbol{v} \cdot \frac{\mathrm{d}}{\mathrm{d}s}\boldsymbol{v}\,\mathrm{d}s = \frac{1}{2}\int_{\tilde{C}} \frac{\mathrm{d}}{\mathrm{d}s}|\boldsymbol{v}|^2\,\mathrm{d}s = 0 \tag{1.3.12}$$

and by (1.3.1) the first integral as

$$\frac{1}{\mu}\int_{\tilde{C}} (-\operatorname{grad} p + \boldsymbol{F}) \cdot \mathrm{d}\boldsymbol{s} = \frac{1}{\mu}\int_{\tilde{C}} \boldsymbol{F} \cdot \mathrm{d}\boldsymbol{s}. \tag{1.3.13}$$

Hence we have the result

$$\frac{\mathrm{d}\Gamma}{\mathrm{d}t} = \frac{1}{\mu}\int_{\tilde{C}} \boldsymbol{F} \cdot \mathrm{d}\boldsymbol{s}, \tag{1.3.14}$$

which is zero for each contour $\tilde{C}$, when $\operatorname{rot}\boldsymbol{F} = 0$ in the whole space.

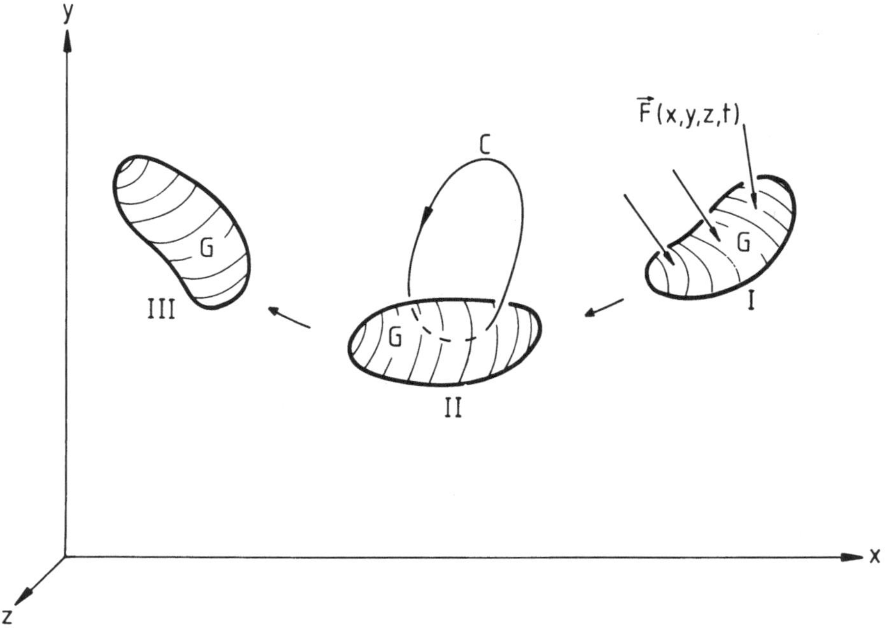

Fig. 1.3.1. Moving force field and probing contour C.

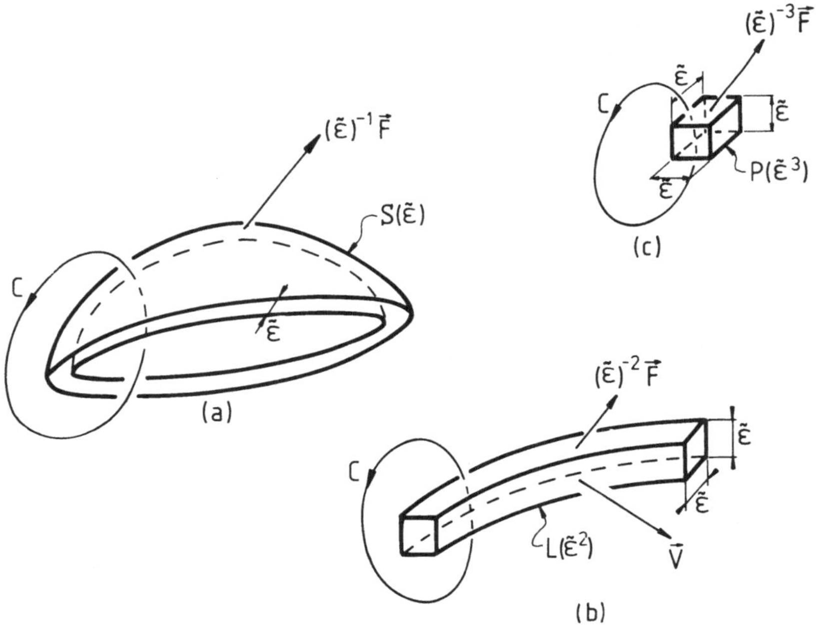

Fig. 1.3.2. Probing contour C with force field, (a) concentrated on a surface, (b) concentrated on a line, (c) concentrated in a point.

We now suppose that the external force field $\boldsymbol{F}$ is small of $O(\varepsilon)$, where ε is a small parameter with respect to which we linearize our equations of motion. We assume that in this case also the velocity field $\boldsymbol{v}$, its derivatives and the pressure field p induced by $\boldsymbol{F}$ are $O(\varepsilon)$. Then it follows that in (1.3.1) the term $(\boldsymbol{v} \cdot \operatorname{grad} \boldsymbol{v})$ is of $O(\varepsilon^2)$, and will be neglected for ε sufficiently small. Hence the linearized version of (1.3.1) becomes

$$\frac{\partial \boldsymbol{v}}{\partial t} = -\frac{1}{\mu} \operatorname{grad} p + \frac{1}{\mu} \boldsymbol{F}, \tag{1.3.15}$$

which is correct up to and including $O(\varepsilon)$. When we consider instead of a floating contour $\tilde{C}$ a contour C fixed in space, it is allowed to simplify (1.3.11) to

$$\frac{\mathrm{d}\Gamma}{\mathrm{d}t} = \int_C \frac{\partial \boldsymbol{v}}{\partial t} \cdot \mathrm{d}\boldsymbol{s}. \tag{1.3.16}$$

Substitution of (1.3.15) leads to

$$\frac{\mathrm{d}\Gamma}{\mathrm{d}t} = \frac{1}{\mu} \int_C \boldsymbol{F} \cdot \mathrm{d}\boldsymbol{s}. \tag{1.3.17}$$

This equation is correct up to and including $O(\varepsilon)$, which is sufficient for a linearized theory.

Equation (1.3.17) can be used to investigate in a linearized theory whether vorticity is left behind in the fluid by a force field $\boldsymbol{F}(x, y, z, t)$, confined to a moving finite region G (figure 1.3.1). Suppose that the circulation of the "probing" contour C is zero when the force field is in position I$(t = t_0)$. Then in general it becomes non zero when G in position II, is cut by C (1.3.17). When G moves on to position III $(t = t_e)$, it has no longer contact with C, hence the circulation Γ remains constant and has the value

$$\Gamma = \frac{1}{\mu} \int_{t_0}^{t_e} \left(\int_C \boldsymbol{F} \cdot \mathrm{d}\boldsymbol{s} \right) \mathrm{d}t. \tag{1.3.18}$$

In this way we can determine by using different kinds of contours C where and of what strength, vorticity has been shed in the fluid by $\boldsymbol{F}(x, y, z, t)$.

Now consider a small, sufficiently smooth and flat contour C of area $\Delta\sigma$, with circulation $\Delta\Gamma$. The unit normal $\boldsymbol{n}$ at C is coupled with a right hand screw to the direction of s along C. Then, as is well known, the component of the vorticity of the fluid at that place in the direction of $\boldsymbol{n}$, has approximately the value

$$\boldsymbol{n} \cdot \operatorname{rot} \boldsymbol{v} \approx \frac{\Delta \Gamma}{\Delta \sigma}. \tag{1.3.19}$$

Taking a series of contours for which $\Delta\sigma \to 0$, (1.3.19) becomes an equality.

We conclude this section with some remarks about the calculation of (1.3.17) or (1.3.18) along a smooth contour C which cuts a force field confined to a region of dimension less than three. First suppose that the force region is a surface S in space and that the field has the intensity $\boldsymbol{F}$ per unit of area. Then we consider a narrow three-dimensional region $S(\tilde{\varepsilon})$ of width $\tilde{\varepsilon}$, in the neighbourhood of S (figure 1.3.2a). This $\tilde{\varepsilon}$ has no relation with our linearization parameter.

We replace the force field $\boldsymbol{F}$ by a spatial field $\tilde{\varepsilon}^{-1}\boldsymbol{F}$ spread evenly over the region $S(\tilde{\varepsilon})$. Then the integration in (1.3.17) can be performed, we take the limit $\tilde{\varepsilon} \to 0$ and obtain

$$\int_C \boldsymbol{F} \cdot \mathrm{d}\boldsymbol{s} = \boldsymbol{F} \cdot \boldsymbol{l}, \tag{1.3.20}$$

where $\boldsymbol{l}$ is the unit tangent vector to C at S. Also the integration with respect to t in (1.3.18) can now be carried out.

Next we consider a force field $\boldsymbol{F}$ per unit of length concentrated on a line L. We suppose that this line has a velocity $\boldsymbol{V}$ relatively to the smooth contour C, which we assumed to be at rest. Now consider a narrow three-dimensional region of space $L(\tilde{\varepsilon}^2)$ with a square cross section, which sides $\tilde{\varepsilon}$ in the neighbourhood of L (figure 1.3.2b). In this case we replace the force field $\boldsymbol{F}$ by a spatial field of intensity $\tilde{\varepsilon}^{-2}\boldsymbol{F}$ spread evenly over the region $L(\tilde{\varepsilon}^2)$. Now we cannot give a meaning to (1.3.17) because in the limit $\tilde{\varepsilon} \to 0$ the intensity of $\tilde{\varepsilon}^{-2}\boldsymbol{F}$ becomes too high, however we can give a meaning to (1.3.18), where t_0 is a time before $L(\tilde{\varepsilon}^2)$ has contact with C and t_e is a time where the contact is finished. Then we perform the integration in (1.3.18) and take the limit $\tilde{\varepsilon} \to 0$, we obtain

$$\int_{t_0}^{t_e} \left(\int_C \boldsymbol{F} \cdot \mathrm{d}\boldsymbol{s} \right) \mathrm{d}t = \frac{\boldsymbol{F} \cdot \boldsymbol{l}}{V_n}, \tag{1.3.21}$$

where $\boldsymbol{l}$ is again the unit tangent vector to C at L and V_n is the component of $\boldsymbol{V}$ normal at C and L, at the point of their intersection.

Such a meaning cannot be given to an integration along C with respect to the force concentrated at a point P (figure 1.3.2c). Following our procedure we have to consider a cube $P(\tilde{\varepsilon}^3)$ with edges of length $\tilde{\varepsilon}$. The singular force is then replaced by a field of intensity $\tilde{\varepsilon}^{-3}\boldsymbol{F}$ inside the cube. The intensity of this field for $\tilde{\varepsilon} \to 0$ is too large to yield a simple limit for anyone of the two integrals (1.3.17) and (1.3.18). Still this procedure can give information as we will discuss in section 1.7.

As we mentioned already it is possible to show [64] that the results we obtained here are independent of the way in which we distributed the concentrated force fields.

1.4. Solution of linearized equation of motion

We start from the linear equation (1.3.15), application of the operation div to both sides of it yields

$$\Delta p = \operatorname{div} \boldsymbol{F}. \tag{1.4.1}$$

When we assume that the pressure at infinity is zero, we find by the Poisson integral ([12] page 261), the following unique solution of (1.4.1)

$$p(x, y, z, t) = -\frac{1}{4\pi} \iiint_{G(t)} \frac{\operatorname{div} \boldsymbol{F}(\xi, \eta, \zeta, t)}{R} \, \mathrm{d}\xi \, \mathrm{d}\eta \, \mathrm{d}\zeta, \tag{1.4.2}$$

where

$$R = |\boldsymbol{R}|, \qquad \boldsymbol{R} = (x - \xi, y - \eta, z - \zeta), \tag{1.4.3}$$

and $G(t)$ is the possibly time dependent finite closed region of space to which $\boldsymbol{F}$ is confined. Assuming $\boldsymbol{F}$ to be sufficiently smooth we find by partial integration

$$p = \frac{1}{4\pi} \iiint_{G(t)} \frac{\boldsymbol{F} \cdot \boldsymbol{R}}{R^3} \, \mathrm{d}\xi \, \mathrm{d}\eta \, \mathrm{d}\zeta. \tag{1.4.4}$$

Application of the operation rot to both sides of (1.3.15) yields

$$\operatorname{rot} \frac{\partial \boldsymbol{v}}{\partial t} = \frac{1}{\mu} \operatorname{rot} \boldsymbol{F}. \tag{1.4.5}$$

The unique solution of this equation that satisfies $\operatorname{div} \dfrac{\partial \boldsymbol{v}}{\partial t} = 0$ and vanishes at infinity is given by ([3] page 86)

$$\frac{\partial \boldsymbol{v}}{\partial t} = \frac{1}{4\pi\mu} \operatorname{rot} \iiint_{G(t)} \frac{\operatorname{rot} \boldsymbol{F}(\xi, \eta, \zeta, t)}{R} \, \mathrm{d}\xi \, \mathrm{d}\eta \, \mathrm{d}\zeta. \tag{1.4.6}$$

We now consider points (x, y, z) which have a non zero distance to the region A travelled through by $G(t)$. The region A is of finite extent and is the union of the regions G, hence

$$A = \bigcup_{t' \in [t_0, t]} G(t'). \tag{1.4.7}$$

Then, because $R \neq 0$, the integrand in (1.4.6) has no singularity. It follows from (1.4.6) by partial integration with respect to ξ, η and ζ and by integration with respect to time that

$$\boldsymbol{v}(x, y, z, t) = -\frac{1}{4\pi\mu}\int_{t_0}^{t}\left[\iiint_{G(t)}\left\{\frac{\boldsymbol{F}}{R^3} - \frac{3\boldsymbol{R}\cdot(\boldsymbol{F}\cdot\boldsymbol{R})}{R^5}\right\} \mathrm{d}\xi\, \mathrm{d}\eta\, \mathrm{d}\zeta\right] \mathrm{d}t,$$

$$(x, y, z) \notin A, \tag{1.4.8}$$

where we assumed the fluid to be at rest for $t < t_0$ and the force field to be "switched on" at $t = t_0$. From (1.4.8) it follows that outside A the velocity field is free of rotation. Hence outside A exists a potential function $\Phi = \Phi(x \cdot y \cdot z \cdot t)$ with

$$\boldsymbol{v} = \operatorname{grad} \Phi. \tag{1.4.9}$$

By (1.3.15) and (1.4.4) we find

$$\Phi(x, y, z, t) = -\frac{1}{4\pi\mu}\int_{t_0}^{t}\left[\iiint_{G(t)} \frac{\boldsymbol{F}\cdot\boldsymbol{R}}{R^3}\, \mathrm{d}\xi\, \mathrm{d}\eta\, \mathrm{d}\zeta\right] \mathrm{d}t,$$

$$(x, y, z) \notin A. \tag{1.4.10}$$

From the foregoing the following conclusions can be made. The pressure field (1.4.4) depends instantaneously on the force field, while the velocity field (1.4.8) or (1.4.10) depends on the behaviour of the force field in the past. When the force field is "switched off" it follows that the pressure becomes zero everywhere and that the velocity field becomes independent of time. Of course these results are a consequence of the neglect of viscosity and the linearization of the equation of motion.

In potential theory the field of a dipole or doublet is given by

$$-\frac{\boldsymbol{k}\cdot\boldsymbol{R}}{4\pi R^3}, \tag{1.4.11}$$

where $|\boldsymbol{k}|$ is its strength and its axis is by convention in the direction of $\boldsymbol{k}$. It is seen from (1.4.4) that the pressure induced by $\boldsymbol{F}$ is at each moment a superposition of fields of pressure dipoles distributed over the region G, with their axes opposite to the local force direction at that time (force direction is from low pressures to high pressures). The potential Φ (1.4.10) is a superposition of the potentials of source doublets or velocity dipoles with their axes (from sink to source, $\boldsymbol{v} = \text{grad} - \boldsymbol{k} \cdot \boldsymbol{R}/4\eta R^3$) in the local force direction.

As we mentioned already at the end of the previous section we can make several specializations of the force field. When we concentrate the field on a finite part of a surface H given by

$$H(x, y, z, t) = 0, \tag{1.4.12}$$

we have the linearized theory of what will be called an actuator surface. These surfaces will be discussed in chapter 2. When it happens that $\boldsymbol{F}$ is perpendicular to H and if the induced velocities are tangent to H, we have the linearized theory of a possibly deformable wing in an incompressible and inviscid fluid. Apparently a wing is a special case of an actuator surface.

We now will consider in detail the case that the force field is concentrated in a point, hence the field is degenerated to a singular force. The point of application $Q = Q(t)$ of the force will move in a general, but sufficiently smooth, way through space. Its path L (figure 1.4.1) is given by

$$L = (\xi(t), \eta(t), \zeta(t)), \qquad t \geq t_0, \tag{1.4.13}$$

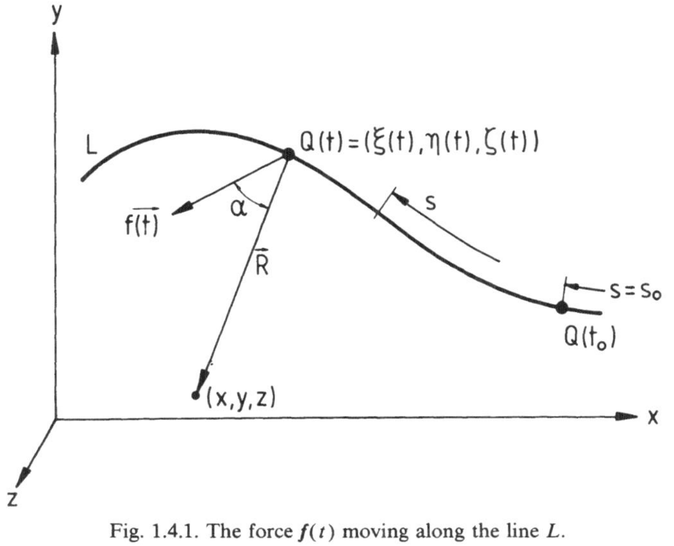

Fig. 1.4.1. The force $\boldsymbol{f}(t)$ moving along the line L.

and we put

$$\boldsymbol{F}(x, y, z, t) = \boldsymbol{f}(t)\delta(x - \xi(t))\delta(y - \eta(t))\delta(z - \zeta(t)), \qquad (1.4.14)$$

where δ is the delta function of Dirac.

By (1.4.14) our previous results (1.4.4), (1.4.8) and (1.4.10) change into

$$p = \frac{1}{4\pi} \frac{\boldsymbol{f} \cdot \boldsymbol{R}}{R^3}, \qquad (1.4.15)$$

$$\boldsymbol{v} = -\frac{1}{4\pi\mu} \int_{t_0}^{t} \left\{ \frac{\boldsymbol{f}}{R^3} - \frac{3\boldsymbol{R} \cdot (\boldsymbol{f} \cdot \boldsymbol{R})}{R^5} \right\} \mathrm{d}t, \qquad (x, y, z) \notin L \qquad (1.4.16)$$

and

$$\Phi = -\frac{1}{4\pi\mu} \int_{t_0}^{t} \frac{\boldsymbol{f} \cdot \boldsymbol{R}}{R^3} \mathrm{d}t, \qquad (x, y, z) \notin L. \qquad (1.4.17)$$

Hence the pressure is, as could be expected, the field of one pressure dipole at Q in the direction of $-\boldsymbol{f}$ and of strength $f = |\boldsymbol{f}|$.

A simple case appears when Q and $\boldsymbol{f}$ are independent of time. Then the velocity potential has the form

$$\Phi = -\frac{1}{4\pi\mu} \frac{\boldsymbol{f} \cdot \boldsymbol{R}}{R^3}(t - t_0), \qquad t \geq t_0. \qquad (1.4.18)$$

Hence in the linearized theory, the velocity field induced by the constant force $\boldsymbol{f}$ which starts at $t = t_0$ is a source doublet of which the strength increases linearly with time.

We continue with the more general case that Q and $\boldsymbol{f}$ are time dependent and assume that the velocity $\boldsymbol{V} = \boldsymbol{V}(t)$ of the point Q satisfies

$$V = |\boldsymbol{V}| = (\dot{\xi}^2 + \dot{\eta}^2 + \dot{\zeta}^2)^{1/2} \geq \tilde{V} > 0, \qquad (1.4.19)$$

where $\tilde{V}$ is some positive number. Along L we introduce the length parameter s. By (1.4.19) it is possible to write for the value of s at the point Q

$$s = s(t), \qquad s_0 = s(t_0), \qquad (1.4.20)$$

hence we can use in the following either the time t or the length s along L, as the variable on which quantities depend. When there is no risk of confusion we will use the notation $g(s)$ or $g(t)$ which denotes that some quantity g is thought of to depend on s or on t. It does not mean that the dependence of g as a function of s or of t is the same.

Introducing s as the independent variable we can write (1.4.16) and (1.4.17) as

$$\boldsymbol{v} = -\frac{1}{4\pi\mu}\int_{s_0}^{s(t)}\left\{\frac{\boldsymbol{f}}{VR^3} - \frac{3\boldsymbol{R}\cdot(\boldsymbol{f}\cdot\boldsymbol{R})}{VR^5}\right\}\mathrm{d}s, \qquad (x, y, z) \notin L, \quad (1.4.21)$$

and

$$\Phi = -\frac{1}{4\pi\mu}\int_{s_0}^{s(t)}\frac{\boldsymbol{f}\cdot\boldsymbol{R}}{VR^3}\,\mathrm{d}s, \qquad (x, y, z) \notin L, \quad (1.4.22)$$

where integrations are along L and the functions in the integrands depend on s.

Of course the linearized theory is not valid in the neighbourhood of Q, where the induced velocity field assumes arbitrarily high values. However the results can be considered as kernel functions from which by integration solutions can be derived, for which the linearized theory holds.

1.5. Vortex representation of solution

Before giving an interpretation of the solution described in the previous section by means of vortices, we recapitulate some general results ([3] section 2.4). We prescribe from a flow of an incompressible and inviscid fluid in an unbounded space its divergence $\theta = \theta(x, y, z, t)$ and its vorticity $\boldsymbol{\omega} = \boldsymbol{\omega}(x, y, z, t)$, which for simplicity are confined to a finite region of space. We assume that the velocity field is zero at infinity, then we have the well known unique solution

$$\boldsymbol{v}(x, y, z, t) = \operatorname{grad}\Theta(x, y, z, t) + \operatorname{rot}\boldsymbol{\psi}(x, y, z, t) \quad (1.5.1)$$

where

$$\Theta = -\frac{1}{4\pi}\int\frac{\theta}{R}\,\mathrm{dVol}, \qquad \boldsymbol{\psi} = \frac{1}{4\pi}\int\frac{\boldsymbol{\omega}}{R}\,\mathrm{dVol}. \quad (1.5.2)$$

In the following we suppose $\theta = 0$ and $\boldsymbol{\omega}$ to be concentrated on a closed line J (figure 1.5.1) along which we have a length parameter s. In fact we consider the limit of a narrow tube around J, outside of which $\boldsymbol{\omega} = 0$ and inside of which $\boldsymbol{\omega} \neq 0$, homogeneously distributed and "tangent" to J. When σ is the area of the cross section of this tube, we denote

$$\lim_{\sigma\to 0}\omega\sigma = \Gamma. \quad (1.5.3)$$

From

$$\text{div rot } \boldsymbol{v} = \text{div } \boldsymbol{\omega} = 0, \tag{1.5.4}$$

we obtain $\Gamma = |\boldsymbol{\Gamma}|$ is constant along J. Because dVol $= \sigma$ ds we find from (1.5.2)

$$\psi = \frac{1}{4\pi} \int_J \frac{\boldsymbol{\Gamma}}{R} \, \mathrm{d}s. \tag{1.5.5}$$

Application of the operation rot to both sides of (1.5.5) with $(x, y, z) \notin J$, yields

$$\boldsymbol{v} = \frac{1}{4\pi} \int_J \frac{\boldsymbol{\Gamma} * \boldsymbol{R}}{R^3} \, \mathrm{d}s = \frac{\Gamma}{4\pi} \int_J \frac{\boldsymbol{k} * \boldsymbol{R}}{R^3} \, \mathrm{d}s, \tag{1.5.6}$$

where $\boldsymbol{k} = (\mathrm{d}\xi/\mathrm{d}s, \mathrm{d}\eta/\mathrm{d}s, \mathrm{d}\zeta/\mathrm{d}s)$ is the unit vector tangent to J. Formula (1.5.6) is the law of Biot and Savart for a closed vortex line. It follows that the direction of $\boldsymbol{\Gamma}$ is coupled with a right hand screw to its locally induced velocities.

Outside the vortex line J the fluid flow is free of rotation. When we make space outside J simply connected by introducing a sufficiently smooth surface S bounded by J, we can introduce a velocity potential $\Phi = \Phi(x, y, z)$, where we assumed J and $\boldsymbol{\Gamma}$ to be independent of time, otherwise Φ depends also on t. We now derive a formula for Φ. Consider for instance, the x component v_x of $\boldsymbol{v}$, which by (1.5.6) and the expression for $\boldsymbol{k}$, can be written as

$$v_x = \frac{\Gamma}{4\pi} \int_J \left\{ \frac{\partial}{\partial \zeta}\left(\frac{1}{R}\right) \mathrm{d}\eta - \frac{\partial}{\partial \eta}\left(\frac{1}{R}\right) \mathrm{d}\zeta \right\} . \tag{1.5.7}$$

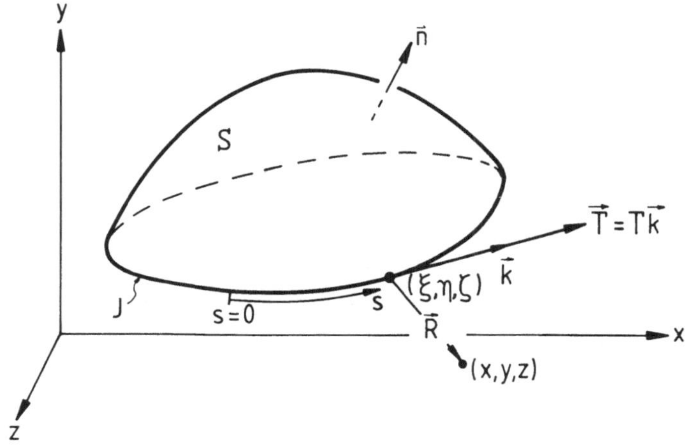

Fig. 1.5.1. Vortex line J.

We use Stokes' formula

$$\int_J \boldsymbol{g}(\xi, \eta, \zeta) \cdot \mathrm{d}\boldsymbol{s} = \int_S (\operatorname{rot} \boldsymbol{g}) \cdot \boldsymbol{n} \, \mathrm{d}S, \tag{1.5.8}$$

where $\boldsymbol{g} = (g_x, g_y, g_z)$ is some vector field and $\boldsymbol{n}$ is the unit normal on S. The normal $\boldsymbol{n}$ is coupled to the increase of s along J by a right hand screw (figure 1.5.1). When we take in (1.5.8)

$$g_x = 0, \qquad g_y = \frac{\Gamma}{4\pi} \frac{\partial}{\partial \zeta}\left(\frac{1}{R}\right), \qquad g_z = -\frac{\Gamma}{4\pi} \frac{\partial}{\partial \eta}\left(\frac{1}{R}\right), \tag{1.5.9}$$

we can write (1.5.7) as

$$v_x = +\frac{\Gamma}{4\pi} \frac{\partial}{\partial x} \int_S \left(\operatorname{grad} \frac{1}{R}\right) \cdot \boldsymbol{n} \, \mathrm{d}S = -\frac{\Gamma}{4\pi} \frac{\partial}{\partial x} \int_S \frac{\boldsymbol{R} \cdot \boldsymbol{n}}{R^3} \, \mathrm{d}S, \tag{1.5.10}$$

where grad is with respect to x, y, z and we used

$$\Delta \frac{1}{R} = 0, \quad (x, y, z) \neq (\xi, \eta, \zeta); \quad \frac{\partial}{\partial x} \frac{1}{R} = -\frac{\partial}{\partial \xi} \frac{1}{R}, \quad \text{etc..} \tag{1.5.11}$$

The expression

$$\frac{\boldsymbol{R} \cdot \boldsymbol{n} \, \mathrm{d}S}{R^3} = \frac{\cos(n, R) \, \mathrm{d}S}{R^2}, \tag{1.5.12}$$

at the right hand side of (1.5.10) is however the solid angle $\mathrm{d}\Lambda$ under which we see the surface element $\mathrm{d}S$, from the point (x, y, z). Hence when Λ is the total solid angle at (x, y, z) subtended by J we find

$$v_x = -\frac{\Gamma}{4\pi} \frac{\partial}{\partial x} \Lambda. \tag{1.5.13}$$

Analogously the other components v_y and v_z of $\boldsymbol{v}$ can be treated. Then we find for the potential

$$\Phi = -\frac{\Gamma}{4\pi} \Lambda \tag{1.5.14}$$

which is discontinuous over S.

We now return to the result (1.4.22) of the previous section and give a simple interpretation of it in terms of vorticity. Consider a small flat vortex ring of area $\mathrm{d}S$ and strength Γ at some point $Q(t_1)$, $t \geq t_1 \geq t_0$ (figure 1.5.2). The smallness of the ring is with respect to the other dimensions of the problem, namely the length of the line L (figure 1.4.1) and its radii of curvature. We erect at the centre of this ring the unit normal $\boldsymbol{n}$, related to Γ by a right hand screw. The potential $\mathrm{d}\tilde{\Phi}$ of this vortex has by (1.5.14) and (1.5.12) the value

$$\mathrm{d}\tilde{\Phi} = -\frac{\Gamma}{4\pi}\frac{\boldsymbol{n}\cdot\boldsymbol{R}}{R^3}\,\mathrm{d}S = -\frac{\Gamma}{4\pi}\frac{\cos\alpha}{R^2}\,\mathrm{d}S, \tag{1.5.15}$$

where α is the angle between $\boldsymbol{n}$ and $\boldsymbol{R}$. It is seen from (1.4.11) and (1.5.15) that the small ring vortex is equivalent to a source doublet, in this case of strength $\Gamma\mathrm{d}S$, with its axis in the $\boldsymbol{n}$ direction.

From (1.5.15) it follows that we can consider the potential Φ in (1.4.22) as a superposition of potentials of small flat ring vortices of area $\mathrm{d}S$ and perpendicular to $\boldsymbol{f}(s)$. The vorticity Γ of such a ring is connected to the direction of $\boldsymbol{f}(s)$ by a right hand screw and has the magnitude

$$\Gamma = \frac{1}{\mu}\frac{f(s)}{V(s)}\frac{\mathrm{d}s}{\mathrm{d}S}. \tag{1.5.16}$$

Somewhat more precisely we have to consider the limits $\mathrm{d}s \to 0$ and $\mathrm{d}S \to 0$ for the potential of these ring vortices. We remark that the shape of the ring vortices does not appear in the theory, it can be chosen for instance rectangular or circular.

In order to elaborate this vortex model further we split the singular force $\boldsymbol{f}(t)$ into two parts

$$\boldsymbol{f}(t) = \boldsymbol{h}(t) + \boldsymbol{g}(t), \tag{1.5.17}$$

where $\boldsymbol{h}(t)$ is perpendicular to L and $\boldsymbol{g}(t)$ is tangent to L. These

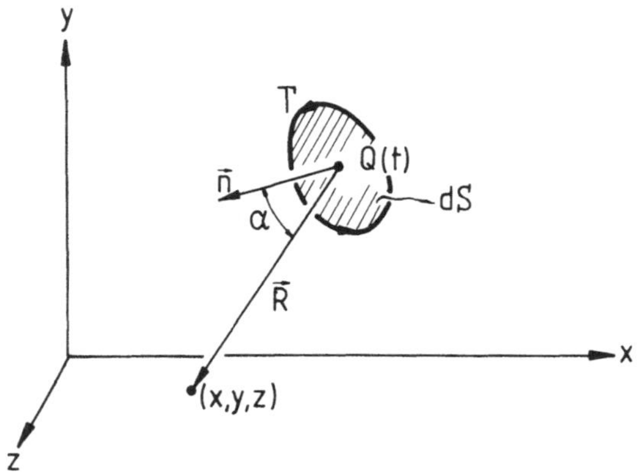

Fig. 1.5.2. Small ring vortex.

components are uniquely determined and because our theory is linear we can add the velocities induced by each of these to obtain the velocity field induced by $\boldsymbol{f}$. In the next two sections we discuss $\boldsymbol{h}(t)$ and $\boldsymbol{g}(t)$ separately.

1.6. Singular force perpendicular to its velocity

In this section we discuss the picture of the vorticity left behind by a singular force perpendicular to its velocity. We divide the line L from the point $s = s_0$ where the force $\boldsymbol{h}$ is switched on to the point $s = s_e$ where the force has arrived, in N equal intervals of length $\Delta s = (s_e - s_0)/N$. Then we consider the points σ_n, $n = 1, \ldots, N$, which are the midpoints of the intervals. At such a point $s = \sigma_n$ (figure 1.6.1), we erect three mutual orthogonal unit vectors. The vector $\boldsymbol{i}$ tangent to L in the direction of increasing s, the vector $\boldsymbol{j}$ along $\boldsymbol{h}(\sigma_n)$ and the vector $\boldsymbol{k}$ perpendicular to both $\boldsymbol{i}$ and $\boldsymbol{j}$, so that $\boldsymbol{i}$, $\boldsymbol{j}$ and $\boldsymbol{k}$ form a right handed system.

We replace the continuous action of the force by singular forces $\boldsymbol{h}(\sigma_n)$ acting during the time interval

$$\int_{s_0}^{s_n} \frac{\mathrm{d}s}{V(s)} \leq t \leq \int_{s_0}^{s_{n+1}} \frac{\mathrm{d}s}{V(s)}, \tag{1.6.1}$$

which has approximately the length $\Delta t = \Delta s / V(\sigma_n)$. Outside this time interval the force at σ_n is zero. During this time interval the force gives a contribution to the potential of the magnitude (1.4.18)

$$\Delta\Phi = -\frac{1}{4\pi\mu} \frac{\boldsymbol{h} \cdot \boldsymbol{R}}{R^3} \frac{\Delta s}{V(\sigma_n)}. \tag{1.6.2}$$

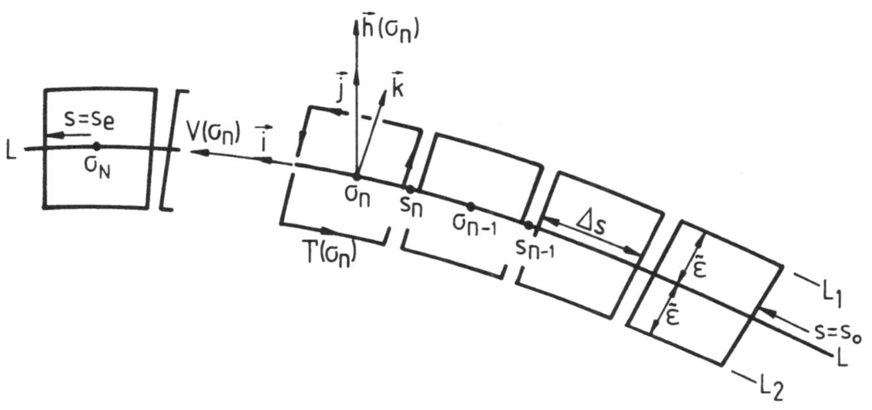

Fig. 1.6.1. Vorticity induced by $\boldsymbol{h}$.

This contribution can also be delivered by a little flat rectangular closed vortex, perpendicular to $\boldsymbol{h}(\sigma_n)$, of length Δs in the $\boldsymbol{i}$ direction and of width $2\tilde{\varepsilon}$ in the $\boldsymbol{k}$ direction and of which the midpoint coincides with $s=\sigma_n$. The strength of this vortex follows from (1.5.16)

$$\Gamma=\frac{1}{\mu}\frac{h(\sigma_n)}{V(\sigma_n)}\frac{1}{2\tilde{\varepsilon}}, \tag{1.6.3}$$

where $h=|\boldsymbol{h}|$. The direction of Γ of the little closed vortex is coupled with a right hand screw to $\boldsymbol{h}$. In this way we obtain the vortex representation drawn in figure 1.6.1. We introduce in the neighbourhood of L two lines L_1 and L_2 which have the representation

$$(\xi(s),\eta(s),\zeta(s)\pm\tilde{\varepsilon}\boldsymbol{k}(s)), \tag{1.6.4}$$

where the $+(-)$ sign belongs to $L_1(L_2)$. By the length parameter s on L we have also a parameter s on L_1 and L_2, however this one does not measure exactly lengths along L_1 and L_2.

By taking the limit $\Delta s\to 0$ we find the following vortex system which consists of five parts:

1) At $s=s_0$ a concentrated vortex in the direction of $\boldsymbol{k}$, of length $2\tilde{\varepsilon}$ and of strength

$$\frac{1}{2\tilde{\varepsilon}\mu}\frac{h(s_0)}{V(s_0)}, \tag{1.6.5}$$

this is the "starting vortex".

2) At $s=s_e$ a concentrated vortex in the direction of $-\boldsymbol{k}$, of length $2\tilde{\varepsilon}$ and of strength

$$\frac{1}{2\tilde{\varepsilon}\mu}\frac{h(s_e)}{V(s_e)}, \tag{1.6.6}$$

this is the bound vorticity of the "lifting line" which represents the force $\boldsymbol{h}$. Herewith we have the relation between the force of magnitude h and the vorticity at the point of application of the force. In essence this is the well known theorem of Joukowski. When the force is switched off at $s=s_e$ this vortex element continues to exist at $s=s_e$ and can be called the "ending vortex".

3) Along L_1, we have a concentrated "tip vortex" of strength

$$\frac{1}{2\tilde{\varepsilon}\mu}\frac{h(s)}{V(s)} \tag{1.6.7}$$

with a right hand screw coupled to the $+s$ direction because $\boldsymbol{h}$ and $\boldsymbol{j}$ have the same direction.

4) Along L_2 we have a concentrated tip vortex of strength (1.6.7) however in the opposite direction.
5) Distributed vorticity between L_1 and L_2 in the $\boldsymbol{k}$ direction, of length $2\tilde{\varepsilon}$ and of strength

$$\frac{1}{2\tilde{\varepsilon}\mu}\left(\frac{\mathrm{d}}{\mathrm{d}s}\frac{h(s)}{V(s)}\right), \tag{1.6.8}$$

per unit of length along L in the s direction. This vorticity arises from the neighbouring sides of the rectangles (figure 1.6.1).

We can now calculate by means of the law of Biot and Savart the velocity field induced by the vortex system 1)... 5). Because a vortex field is free of divergence hence can be considered to be composed of "closed vortex lines", it follows that we can use the differential form of (1.5.6)

$$\mathrm{d}\boldsymbol{v}=\frac{\Gamma}{4\pi}\left(\frac{\boldsymbol{k} * \boldsymbol{R}}{R^3}\right)\mathrm{d}s, \tag{1.6.9}$$

where the meaning of the symbols is the same as in (1.5.6).

The velocity induced by the starting vortex and the bound vortex representing the force becomes

$$-\frac{1}{4\pi}\frac{1}{2\tilde{\varepsilon}\mu}\frac{h(s)}{V(s)}\boldsymbol{k} * \frac{\boldsymbol{R}}{R^3}2\tilde{\varepsilon}\Big|_{s_0}^{s_e} = -\frac{1}{4\pi}\frac{h(s)}{\mu V(s)}\boldsymbol{k} * \frac{\boldsymbol{R}}{R^3}\Big|_{s_0}^{s_e}. \tag{1.6.10}$$

The unit vectors along L_1 and L_2 have the form

$$\left(\boldsymbol{i}\pm\tilde{\varepsilon}\frac{\mathrm{d}\boldsymbol{k}}{\mathrm{d}s}\right)\Big/\left|\boldsymbol{i}\pm\tilde{\varepsilon}\frac{\mathrm{d}\boldsymbol{k}}{\mathrm{d}s}\right|, \tag{1.6.11}$$

here and in the next formulas upper signs refer to L_1 and lower ones to L_2. Then the induced velocity by the vorticity along L_1 and L_2 becomes

$$\pm\frac{1}{4\pi\mu}\frac{1}{2\tilde{\varepsilon}}\int_{s_0}^{s_e}\frac{h(s)}{V(s)}\left\{\frac{\left(\boldsymbol{i}\pm\tilde{\varepsilon}\frac{\mathrm{d}\boldsymbol{k}}{\mathrm{d}s}\right)}{\left|\boldsymbol{i}\pm\tilde{\varepsilon}\frac{\mathrm{d}\boldsymbol{k}}{\mathrm{d}s}\right|} * \frac{(\boldsymbol{R}\mp\tilde{\varepsilon}\boldsymbol{k})}{|\boldsymbol{R}\mp\tilde{\varepsilon}\boldsymbol{k}|^3}\right\}\left|\boldsymbol{i}\pm\tilde{\varepsilon}\frac{\mathrm{d}\boldsymbol{k}}{\mathrm{d}s}\right|\mathrm{d}s. \tag{1.6.12}$$

Adding these and taking the limit $\tilde{\varepsilon}\to 0$ we obtain

$$\frac{1}{4\pi\mu}\int_{s_0}^{s_e}\frac{h(s)}{V(s)}\frac{\mathrm{d}}{\mathrm{d}\lambda}\left\{\left(\boldsymbol{i}+\lambda\frac{\mathrm{d}\boldsymbol{k}}{\mathrm{d}s}\right) * \frac{(\boldsymbol{R}-\lambda\boldsymbol{k})}{|\boldsymbol{R}-\lambda\boldsymbol{k}|^3}\right\}\Bigg|_{\lambda=0}\mathrm{d}s$$

$$=\frac{1}{4\pi\mu}\int_{s_0}^{s_e}\frac{h(s)}{V(s)}\left[3\frac{(\boldsymbol{R}\cdot\boldsymbol{k})}{R^5}\cdot\boldsymbol{i} * \boldsymbol{R}+\frac{1}{R^3}\frac{\mathrm{d}\boldsymbol{k}}{\mathrm{d}s} * \boldsymbol{R}-\frac{1}{R^3}\boldsymbol{i} * \boldsymbol{k}\right]\mathrm{d}s. \tag{1.6.13}$$

Finally the velocity induced by the distributed vorticity (1.6.8) becomes

$$\frac{1}{4\pi\mu}\int_{s_0}^{s_e}\frac{\mathrm{d}}{\mathrm{d}s}\left(\frac{h(s)}{V(s)}\right)\boldsymbol{k}*\frac{\boldsymbol{R}}{R^3}\,\mathrm{d}s. \tag{1.6.14}$$

It can be shown that by adding the three parts (1.6.10), (1.6.13) and (1.6.14) we find again the velocity field (1.4.21). To this end a partial integration has to be carried out in (1.6.13), in order to get rid of the derivative of $\boldsymbol{k}$ with respect to s.

We now give a simple example how we can determine the velocity fields belonging to more complicated singular external actions, by means of the velocity field induced by an external force as discussed in this section. Consider two forces parallel to the y axis (figure 1.6.2) and moving one after the other with a constant velocity V in the negative x direction. The distance between the two forces is a and their strength is a^{-1} and $-a^{-1}$. In the limit $a \to 0$ we obtain a singular moment of unit strength, acting at the fluid, with its vector pointing in the positive z direction (righthand screw). It follows by comparing this situation with that of figure 1.6.1, that now only a small rectangular vortex remains, the strength of it has by (1.6.5)-(1.6.7) the value

$$\Gamma=\frac{1}{2\mu V\tilde{\varepsilon}a}. \tag{1.6.15}$$

In the limit the flow field induced by this vortex is the same as the flow field of a source doublet (below (1.5.15)), of strength

$$3\tilde{\varepsilon}a\Gamma=\frac{1}{\mu V}. \tag{1.6.16}$$

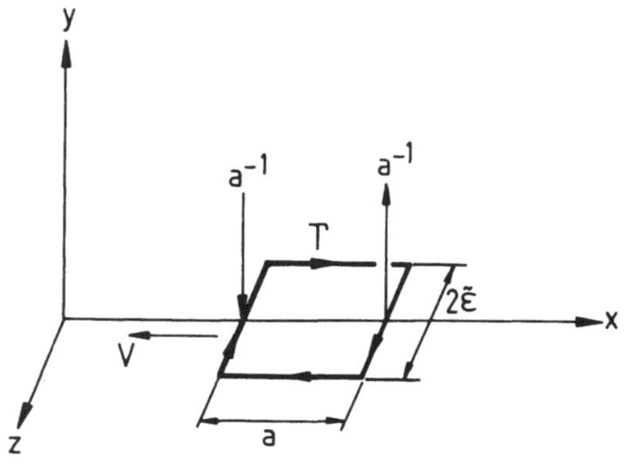

Fig. 1.6.2. Singular unit moment.

We have to be careful with the choice of the limit procedure used for the construction of a singular moment. For instance we have drawn in figure 1.6.3 two possibilities for creating a moment of unit strength acting at the fluid at the point of application $Q=(-Vt, 0, 0)$. The moment vector is pointing in both cases in the negative x direction. We first remark that the external force action $\boldsymbol{F}(x, y, z, t)$ in figure 1.6.3b in the limit case $a \to 0$ is axisymmetric with respect to the x axis. It can be written as

$$\boldsymbol{F}(x,y,z,t)=\{0, \tfrac{1}{2}\delta(x-Vt)\delta(y)\delta'(z), -\tfrac{1}{2}\delta(x-Vt)\delta'(y)\delta(z)\}$$

$$=\text{rot}\ \tfrac{1}{2}\{\delta(x-Vt)\delta(y)\delta(z), 0, 0\}. \qquad (1.6.17)$$

As is known the distribution $\delta(y)\delta(z)$ in the (y, z) plane is a radial distribution hence it can be considered as a generalization of a function $f(r)$, $r^2=y^2+z^2$. However then $\boldsymbol{F}(x, y, z, t)$ originates from the vector $(\delta(x-Vt)\delta(y)\delta(z), 0, 0)$ which is axisymmetric with respect to the x axis, by applying the operation rot to that vector. It follows that $\boldsymbol{F}$ can be considered as a generalization of a force field which is axisymmetric around the x axis.

The velocity fields induced by the unit moments of figure 1.6.3a and of figure 1.6.3b are different, because the flow field belonging to figure 1.6.3a is not axisymmetric, while the flow field belonging to figure 1.6.3b has to be axisymmetric as follows from $\boldsymbol{F}(x, y, z, t)$ having that property. Besides the latter flow field outside the x axis is zero. These properties are easily checked by calculating the pressure field from (1.4.15).

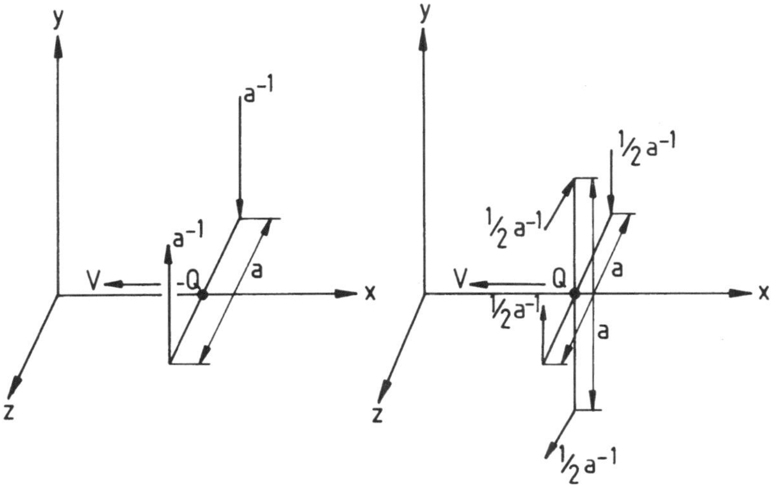

Fig. 1.6.3. Two unit singular moments, moment vector in negative x direction.

We do not elaborate the problems connected to singular moments, because they seem to be not of much interest with respect to propulsion theory in inviscid and incompressible fluids. We note that singular moments are important in the theory of Stokes' flow for highly viscous fluids [24], where the representation of a moment given in figure 1.6.3b is called a "rotlet". There the velocity field belonging to such a moment is not zero outside the x axis and is indeed axisymmetric around it.

1.7. Singular force in the direction of its velocity

In this case the force $\boldsymbol{g}$ is tangent to L at each point of L. Because we are interested in propulsion it seems natural to consider forces $\boldsymbol{g}$ which act on the fluid in the negative s direction. Then by action equals reaction, the force on the system which causes $\boldsymbol{g}$, is in the positive s direction, hence in the direction of the velocity V of the system. In other words this reaction is a propulsive force. We assume again as in the previous section that the force $\boldsymbol{g}$ is switched on at $s = s_0$ and has arrived at $s = s_e$. Then we can write (1.4.22) as

$$\Phi = \frac{1}{4\pi\mu} \int_{s_0}^{s_e} \frac{g(s)}{V(s)} \frac{\mathrm{d}}{\mathrm{d}s}\left(\frac{1}{R}\right) \mathrm{d}s, \qquad (x, y, z) \notin L, \tag{1.7.1}$$

where $g = |\boldsymbol{g}|$.

By partial integration we obtain

$$\Phi = \frac{1}{4\pi\mu} \left\{ \frac{g}{V} \frac{1}{R} \Big|_{s_0}^{s_e} - \int_{s_0}^{s_e} \frac{1}{R} \frac{\mathrm{d}}{\mathrm{d}s}\left(\frac{g(s)}{V(s)} \right) \mathrm{d}s \right\}, \qquad (x, y, z) \notin L. \tag{1.7.2}$$

This formula has a simple interpretation. It is known ([3] page 88) that a source placed at (ξ, η, ζ), which produces a unit volume of fluid per unit of time, has the velocity potential

$$-\frac{1}{4\pi} \frac{1}{R}. \tag{1.7.3}$$

Hence (1.7.2) can be considered as a source system which consists of three parts:

1) At $s = s_0$ we have a "starting source" of strength

$$\frac{g(s_0)}{\mu V(s_0)}. \tag{1.7.4}$$

2) At $s = s_e$ we have a source of strength

$$\frac{-g(s_e)}{\mu V(s_e)}, \tag{1.7.5}$$

hence at the place of the propulsive force we have a sink. When the propulsive force is switched off at $s = s_e$, this sink remains as an "ending sink".
3) Along L, from $s = s_0$ towards $s = s_e$, we have a source distribution of strength

$$\frac{1}{\mu}\frac{\mathrm{d}}{\mathrm{d}s}\left(\frac{g(s)}{V(s)}\right) \tag{1.7.6}$$

per unit of length.
The velocity field follows from (1.7.2) as

$$\boldsymbol{v} = \frac{1}{4\pi\mu}\left\{\frac{-g\boldsymbol{R}}{VR^3}\Big|_{s_0}^{s_e} + \int_{s_0}^{s_e}\frac{\boldsymbol{R}}{R^3}\frac{\mathrm{d}}{\mathrm{d}s}\left(\frac{g(s)}{V(s)}\right)\mathrm{d}s\right\}, \qquad (x, y, z) \notin L. \tag{1.7.7}$$

By the representation of the potential Φ as given in (1.7.2), hence as a combination of potentials of sources and a source distribution, we do not satisfy the equation of mass conservation (1.3.2). However (1.7.2) holds for $(x, y, z) \notin L$. The only way to repair this imperfection of the theory, is to admit at the line L itself a jet of zero cross-section area and with finite volume transport per unit of time, hence the fluid velocity inside the jet has to be "infinite". This transport, in order to save mass conservation, has to have the value

$$\frac{g(s)}{\mu V(s)}, \qquad s_0 \le s \le s_e, \tag{1.7.8}$$

in the direction of decreasing values of s.

This transport along L is in agreement with the description by means of vortices given in section 1.5, which belong to the force field $\boldsymbol{g}$. There we found that around L we have a system of small closed vortices, which for instance can be chosen circular with their centre at L (figure 1.7.1). Their planes have to be perpendicular to L in this case. When we take $\tilde{\varepsilon}$ as their radius, we find from (1.5.16) for their strength $\gamma = \gamma(s)$ per unit of length along L

$$\gamma(s) = \frac{1}{\pi\mu\tilde{\varepsilon}^2}\frac{g(s)}{V(s)}. \tag{1.7.9}$$

When for instance L is straight and g and V are constant it is easily seen

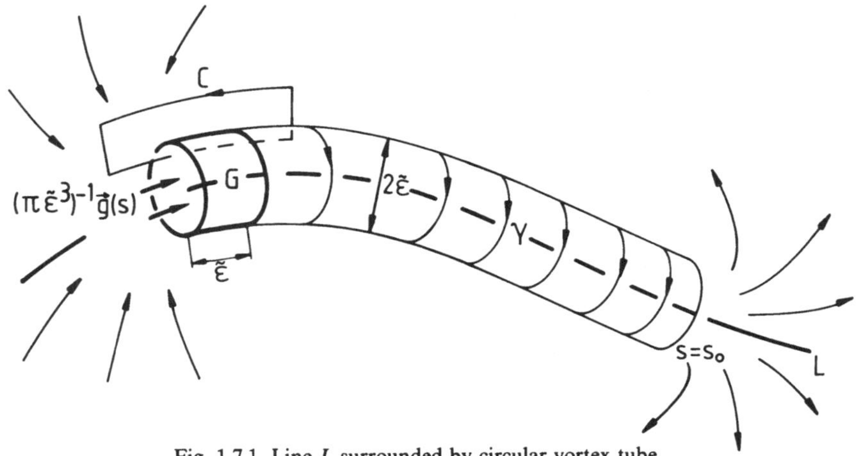

Fig. 1.7.1. Line L surrounded by circular vortex tube.

that in the limit $\tilde{\varepsilon} \to 0$, the volume transport by this vortex system, with cross section area $\pi\tilde{\varepsilon}^2$, equals (1.7.8).

A direct way to calculate the vorticity surrounding the line L is by replacing the singular force $\mathbf{g}(s)$ by a force field of strength $(\pi\tilde{\varepsilon}^3)^{-1}\mathbf{g}(s)$ confined to a small cylinder G of radius $\tilde{\varepsilon}$ and length $\tilde{\varepsilon}$. Using a probing contour C as drawn in figure 1.7.1 the vorticity $\gamma(s)$ of (1.7.9) is recovered. Here we used the contour C in the sense of figure 1.3.2c.

With the respect to the relevance of limit procedures in relation with combinations of singular forces of the type we discussed in this section, we refer to the end of the previous section.

We conclude these sections about the linearized theory of the velocities induced by a singular force with a remark about the non linear theory. In [3] (page 205) is given the exact solution found by Landau, for the non linear theory of the velocity field induced by a singular force acting on a viscous fluid. When we take in this result the limit of vanishing viscosity, then the velocity at each fixed point at a finite distance from the point of application of the force disappears. This is a rather extra-ordinary result in comparison with the linearized theory. It shows among others that the influence of a propeller at a large distance in non linear potential theory, cannot be represented by a singular force.

1.8. Work done by external force field and moving body

We consider an incompressible and inviscid fluid which is at rest at infinity with respect to a Cartesian coordinate system (x, y, z). First we derive an expression for the change of the kinetic energy E of the fluid when an external force field $\boldsymbol{F}(x, y, z, t)$ and a moving body $B(t)$ is present.

We start from the nonlinear equations of motion (1.3.1) and from the equation of mass conservation (1.3.2), hence from

$$\frac{\mathrm{d}\boldsymbol{v}}{\mathrm{d}t}=\frac{\partial \boldsymbol{v}}{\partial t}+(\boldsymbol{v}\cdot\operatorname{grad})\boldsymbol{v}=-\frac{1}{\mu}\operatorname{grad}p+\frac{1}{\mu}\boldsymbol{F} \tag{1.8.1}$$

and

$$\operatorname{div}\boldsymbol{v}=0. \tag{1.8.2}$$

The possibly time dependent region $G(t)$ for which $\boldsymbol{F}\neq 0$ is finite and is surrounded by a large spherical control surface H of radius R_H, fixed in space. Inside H is also the body $B(t)$ which is allowed to change its shape when moving. We assume that the region $G(t)$ and the body stay away from each other. The region inside H with the exception of B is called Ω. From (1.8.1) it follows

$$\mu\int_\Omega \boldsymbol{v}\cdot\left(\frac{\partial \boldsymbol{v}}{\partial t}+v_x\frac{\partial \boldsymbol{v}}{\partial x}+v_y\frac{\partial \boldsymbol{v}}{\partial y}+v_z\frac{\partial \boldsymbol{v}}{\partial z}\right)\mathrm{dVol}$$

$$=-\int_\Omega \boldsymbol{v}\cdot\operatorname{grad}p\,\mathrm{dVol}+\int_G \boldsymbol{v}\cdot\boldsymbol{F}\,\mathrm{dVol}. \tag{1.8.3}$$

The lefthand side of this equation can, by using (1.8.2), be reduced to

$$\frac{\mu}{2}\int_\Omega\left\{\frac{\partial \boldsymbol{v}^2}{\partial t}+\operatorname{div}(\boldsymbol{v}^2\boldsymbol{v})\right\}\mathrm{dVol}. \tag{1.8.4}$$

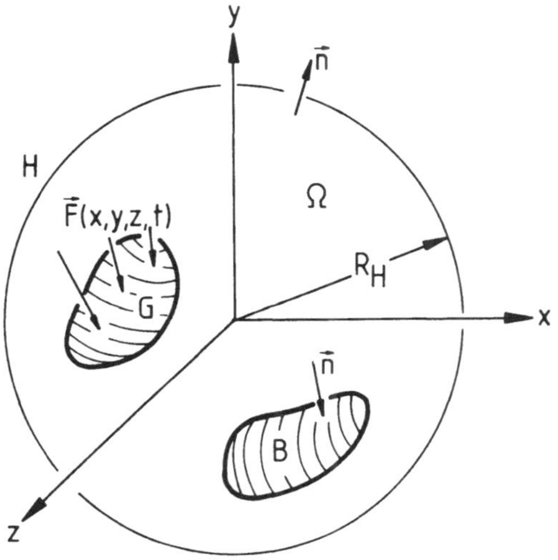

Fig. 1.8.1. Force field $\boldsymbol{F}(x,y,z,t)$ and body B.

The first part of this integral is rewritten as

$$\frac{\mu}{2}\int_{\Omega}\frac{\partial \boldsymbol{v}^2}{\partial t}\,\mathrm{dVol}=\frac{\mu}{2}\frac{\mathrm{d}}{\mathrm{d}t}\int_{\Omega}\boldsymbol{v}^2\,\mathrm{dVol}-\frac{\mu}{2}\int_{\partial B+H}\boldsymbol{v}^2(\boldsymbol{v}\cdot\boldsymbol{n})\,\mathrm{d}S$$

$$=\frac{\mathrm{d}E}{\mathrm{d}t}-\frac{\mu}{2}\int_{\partial B+H}\boldsymbol{v}^2(\boldsymbol{v}\cdot\boldsymbol{n})\,\mathrm{d}S, \qquad (1.8.5)$$

where E is the kinetic energy of the fluid inside H and $\boldsymbol{n}$ is the unit normal at the boundary ∂B or H, pointing out of Ω. The second part of the integral (1.8.4) can be written as

$$\frac{\mu}{2}\int_{\Omega}\mathrm{div}(\boldsymbol{v}^2\cdot\boldsymbol{v})\,\mathrm{dVol}=\frac{\mu}{2}\int_{\partial B+H}\boldsymbol{v}^2(\boldsymbol{v}\cdot\boldsymbol{n})\,\mathrm{d}S. \qquad (1.8.6)$$

Hence (1.8.4) becomes

$$\frac{\mathrm{d}E}{\mathrm{d}t}. \qquad (1.8.7)$$

The first term at the right hand side of (1.8.3) can be brought into the form

$$-\int_{\Omega}\mathrm{div}\,p\boldsymbol{v}\,\mathrm{dVol}=-\int_{H}p(\boldsymbol{v}\cdot\boldsymbol{n})\,\mathrm{d}S-\int_{\partial B}p(\boldsymbol{v}\cdot\boldsymbol{n})\,\mathrm{d}S. \qquad (1.8.8)$$

It is easily seen that the integrations over H tend to zero for $R_H\to\infty$. Hence by (1.8.7) and (1.8.8) we find from (1.8.3)

$$\frac{\mathrm{d}E}{\mathrm{d}t}=\int_{G}\boldsymbol{v}\cdot\boldsymbol{F}\,\mathrm{dVol}-\int_{\partial B}p(\boldsymbol{v}\cdot\boldsymbol{n})\,\mathrm{d}S. \qquad (1.8.9)$$

The first term at the right hand side can be interpreted as the work done by the force field, it is the integral of the scalar product of the "infinitesimal" forces $\boldsymbol{F}$ dVol with the velocity $\boldsymbol{v}$ at their point of application. The second term at the right hand side is the work done by the moving body B. The work of the force field or of the body can be positive or negative, although the total amount has to be non negative. Formula (1.8.9) is of importance for the concept of thrust deduction and will be used in section 2.3.

In case that $\boldsymbol{F}$ is the gradient of a scalar function ψ and when the body B is absent, (1.8.9) changes into

$$\frac{\mathrm{d}E}{\mathrm{d}t}=\int_{\Omega}\boldsymbol{v}\cdot\mathrm{grad}\,\psi\,\mathrm{dVol}=\int_{\Omega}\mathrm{div}(\psi\boldsymbol{v})\,\mathrm{dVol}=\int_{H}\psi\cdot(\boldsymbol{v}\cdot\boldsymbol{n})\,\mathrm{d}S. \qquad (1.8.10)$$

Because the force field is supposed to be zero outside of the finite region

$G(t)$ we have $\psi = \text{const.}$ at H for R_H sufficiently large, and because the fluid is incompressible and div $\boldsymbol{v} = 0$, we find from (1.8.10)

$$\frac{\mathrm{d}E(t)}{\mathrm{d}t} = 0. \tag{1.8.11}$$

This is in agreement with the fact that a conservative force field does not induce a velocity field (1.3.4).

The above result, derived from the non linear equation of motion, is easily seen to hold also for the linearized equation (1.3.15).

We can also consider the case of a periodically moving external force field $\boldsymbol{F}(x, y, z, t)$, with a mean velocity U in the direction of the positive x-axis. The length period of the motion will be h. We assume the linearized equation of motion to be valid. Then within the accuracy of the theory, the created free vorticity $\boldsymbol{\gamma}$ stays where it has been formed. Hence in the linearized theory we have behind the force field a periodic system of vortices up to $x = -\infty$. It is clear that when we assume that the force field has been switched on not infinitely long ago but very long ago, the velocity field in the neighbourhood of the force region is nearly periodic. Then the increase of kinetic energy during one period of motion is nearly the kinetic energy between two planes S_1 and S_2 (figure 1.8.2) parallel to the (y, z) plane, at a distance h apart and situated half way the most left point of the shed vorticity and the force field. However in this case our previous reasoning is valid, it means that this increase equals the work done by $\boldsymbol{F}$.

From this it follows that in the purely periodic case, in which the force field is switched on at $t = -\infty$ hence at $x = -\infty$, the work done by $\boldsymbol{F}$ per period equals the kinetic energy between two planes S_1 and S_2 a period apart and situated infinitely far behind $\boldsymbol{F}$.

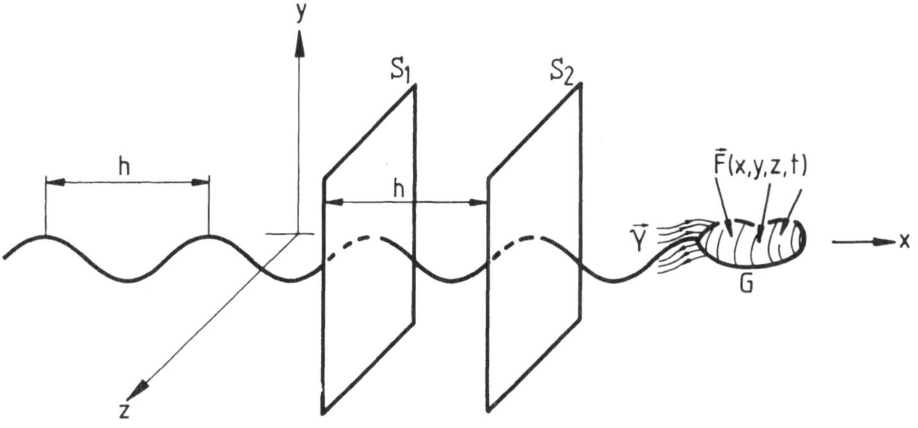

Fig. 1.8.2. Periodically moving force field $\boldsymbol{F}(x, y, z, t)$.

1.9. Vorticity of a lifting surface, linearized theory

In the following we apply some of the previous considerations to the description of the vorticity of a flat wing of finite span. The results we arrive at belong to the basic knowledge treated in many text books on hydrodynamics, however the method we use is different.

We use linearized theory and consider a wing without thickness which is nearly parallel to a plane $y = \text{const.}$ (figure 1.9.1) and of which the tips are rounded off. This wing moves with a velocity U in the positive x direction and started infinitely long ago at $x = -\infty$.

We prescribe for simplicity a constant pressure difference $[p]_+^-$ between the upper $(-)$ and lower $(+)$ side of the wing, $[p]_+^- = p^- - p^+ < 0$. In linearized theory it is allowed to replace the wing by its planform W which is flat and parallel to the (x, z) plane. At this planform we place a distributed external force field $\boldsymbol{F}(x - Ut, z) = [p]_+^- \boldsymbol{e}_y$ acting at the fluid, where $\boldsymbol{e}_y$ is the unit vector in the y direction and $\boldsymbol{F}$ has the magnitude

$$F = [p]_+^- < 0 \tag{1.9.1}$$

This force field is "downwards" when the wing has a lift force upwards in the positive y direction. We now check by means of the probing contours A and B, both fixed in space, the vorticity which belongs to this wing.

First we consider the rectangular contour A of which the sides are parallel to the y and z axis. We assume the position of A such that when W moves on, it is cut by the side $A_2 A_3$ of A along the line $W_1 W_2$ of length

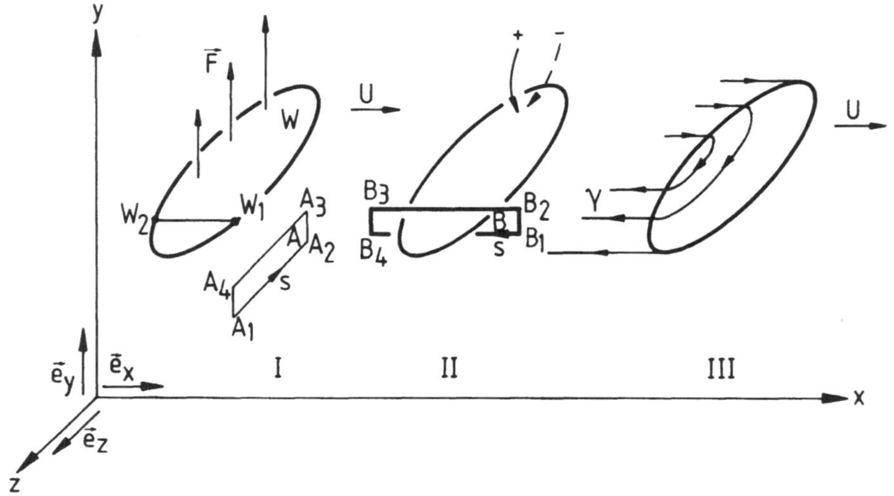

Fig. 1.9.1. Wing with probing contours A and B.

$|W_1W_2|$. It follows from (1.3.14) and (1.3.20) that as long as the wing is cut by A_2A_3, the circulation Γ_A round A satisfies

$$\frac{d\Gamma_A}{dt} = \frac{F}{\mu}. \tag{1.9.2}$$

Because $\Gamma_A = 0$ when W is in position I and because the interval of time it takes for W_1W_2 to pass A_2A_3, equals $|W_1W_2|/U$, we find for Γ_A when W is in position III

$$\Gamma_A = \frac{|W_1W_2|}{\mu U} F < 0, \tag{1.9.3}$$

which remains unaltered when the wing moves on. It follows, that behind the wing we have trailing vorticity γ which within A amounts to (1.9.3) and of which the direction, coupled with a right hand screw to the locally induced velocities, is denoted in figure (1.9.1).

Next we consider the rectangular contour B with sides parallel to the x and y axis. Its position is such that the side B_2B_3 cuts the wing also along the line W_1W_2, while we take $|B_1B_2| > |W_1W_2|$. Again when W is in position I the circulation $\Gamma_B = \Gamma_B(t)$ round B is zero. When W is in position II the circulation Γ_B has been calculated by means of B_2B_3 cutting W in the same way as Γ_A has been calculated by means of A_2A_3 cutting W, hence Γ_B equals Γ_A (1.9.3). When W is in position III, $\Gamma_B = 0$ again, because when B_1B_4 cuts W the force field is in the opposite direction with respect to s as it was on B_2B_3. This means that no free vorticity is left behind by the wing which has a component perpendicular to contour B. When the force field F is time dependent this will not be true in general.

The vorticity detected by contour B in case the wing is in position II is sometimes called the bound vorticity of profile W_1W_2 of W. It is said to be able to sustain a pressure difference between upper and lower side of the wing. We can also say, there exists equilibrium between the external force field and the pressure jump created by it. When we take B_1B_4 very short it is seen that this bound vorticity is continuously distributed over the chord length with a strength Γ given by

$$\Gamma = -\frac{F}{\mu U}, \tag{1.9.4}$$

per unit of length in the x direction, where Γ is coupled with a right hand screw to the positive z direction. This formula is nothing but the well known law of force production by bound vorticity, or Joukowski's theorem.

We can also take A_1A_2 very short and such that both A_2A_3 and A_4A_1 cut W, in order to describe more locally the vorticity component perpendicular to the plane of A. This vorticity we call, free vorticity, also at the wing itself.

We assumed in the foregoing the loading of the wing to be known. The question arises what are the profiles needed to induce this loading. We do not discuss this problem here but refer to chapter 3, where the analogous problem is treated for the blades of the ship screw.

We can determine the induced resistance R_i of the wing by calculating the work done by the force field $\boldsymbol{F}$ per unit of time. We allow here that the pressure jump is a function of position at the wing hence $F = F(x, z)$, which we assume for simplicity to be a continuous function at the closure of the planform of the wing, hence including the edge. From (1.8.9) we find for this power which equals the rate of change of the kinetic energy in the fluid

$$\frac{\mathrm{d}E}{\mathrm{d}t} = \int\int_W v_y(x, y = \text{const.}, z) F(x, z)\, \mathrm{d}x\, \mathrm{d}z = UR_i, \tag{1.9.6}$$

where $R_i > 0$ is the force we have to exert in the $+x$ direction to keep the wing moving. When F is $O(\varepsilon)$ then also v_y is $O(\varepsilon)$ hence R_i is $O(\varepsilon^2)$.

Another way of looking at (1.9.6) is as follows. We divide both sides of the second equality by U, then we can interpret $-v_y/U$ as the local angle of incidence of the wing. The reaction forces $-\boldsymbol{F}$ acting at the wing are perpendicular to its surface. Hence the integral now represents the resultant component R_i of $-\boldsymbol{F}$ in the negative x direction.

As is well known (1.9.6) is not valid when $F(x, z)$ has a square root singularity at the leading edge of W, because then we have to take into account a suction force of $O(\varepsilon^2)$ at the leading edge. We return to this in section 1.11.

Next we discuss the concept of bound vorticity somewhat more closely by means of a linearized two-dimensional flow. We consider in the neighbourhood of the x axis a profile (a, b) without thickness of chord length c and moving with a constant velocity U in the positive x direction (figure 1.9.2). When the profile has a lift force in the positive y direction it exerts a force field $\boldsymbol{F} = F\boldsymbol{e}_y$ with $F < 0$ at the fluid. Then analogous to (1.9.4) it follows that we have bound vorticity along the profile of strength

$$\Gamma(s) = -\frac{F(s)}{\mu U}, \tag{1.9.7}$$

where s is a length parameter along the projection of the profile at the x axis $0 \le s \le c$. We call $\Gamma(s)\,\mathrm{d}s$ an amount of vorticity bound to the local force $\boldsymbol{F}(s)\,\mathrm{d}s$. This denomination is coupled to the possibility of giving a meaning to "a force $\boldsymbol{F}(s)\,\mathrm{d}s$ acting at the fluid for a given value of s and moving with a velocity U". When however we consider for instance a profile of time dependent length $c = c(t)$ and of which the forces are time dependent $\boldsymbol{F}(s, t)\,\mathrm{d}s$, then this connection is lost. The velocity of a point of the profile or of the elementary force $\boldsymbol{F}(s, t)\,\mathrm{d}s$ can be chosen arbitrary.

This holds even for the simple case of figure 1.9.2 and we will show the ambiguity of the concept bound vorticity for this case. For simplicity we assume a constant pressure jump over the profile hence

$$\Gamma = -\frac{F}{\mu U} = \text{const.}. \tag{1.9.8}$$

We imagine a row of concentrated elementary forces placed along the entire x axis $(-\infty < x < +\infty)$ at intervals of length $\mathrm{d}s$. These forces are parallel to the y axis and move with a velocity $V > U$ in the positive x direction. Behind the profile the strength of these forces is zero. When they catch up with the profile, they are switched on at the trailing edge, to the desired strength $F\,\mathrm{d}s$. Then their concentrated bound vorticity $\tilde{\Gamma}$ has the strength

$$\tilde{\Gamma}\,\mathrm{d}s = -\frac{F\,\mathrm{d}s}{\mu V}, \tag{1.9.9}$$

while the starting vortices of strength $-\tilde{\Gamma}\,\mathrm{d}s$ remain at the place where they are formed. Behind the profile these latter vortices give rise to a free vortex sheet of strength

$$-\tilde{\Gamma}\,\mathrm{d}s\left(\frac{V-U}{U\,\mathrm{d}s}\right) = \frac{F(V-U)}{\mu V U}, \tag{1.9.10}$$

per unit of length in the x direction. This follows from the scheme given

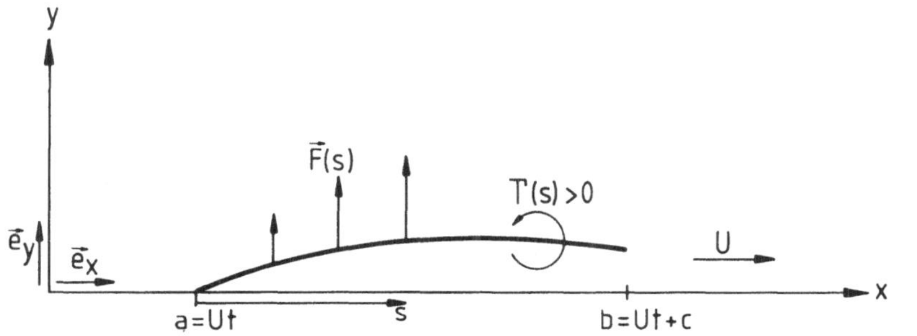

Fig. 1.9.2. Stationary motion of profile $a-b$.

in figure 1.9.3 where it is the intention that the two rows of elementary forces and the two profiles are all at the x axis.

Then the elementary forces move along the profile to the leading edge and create a bound vortex layer of strength

$$\tilde{\Gamma} = -\frac{F}{\mu V}, \qquad (1.9.11)$$

which differs from (1.9.8).

Next the elementary forces reach the leading edge and when they pass it are swtiched off and leave behind their ending vortex at the place where it was formed. This creates a free vortex layer behind the leading edge of strength per unit of length

$$\tilde{\Gamma}\,\mathrm{d}s\left(\frac{V-U}{U\,\mathrm{d}s}\right) = \frac{-F(V-U)}{\mu VU}. \qquad (1.9.12)$$

Hence at the profile we have two vortex layers of strengths (1.9.11) and (1.9.12), which together form a layer of strength

$$-\frac{F}{\mu V} - \frac{F(V-U)}{\mu VU} = -\frac{F}{\mu U}, \qquad (1.9.13)$$

which equals the strength of the bound vorticity in the first approach (1.9.8). Behin the trailing edge we have the two layers of strength (1.9.10) and (1.9.12) which cancel each other, as it has to be in comparison with our first approach.

From this it follows that although the bound vorticity differs in both approaches, their flow fields are the same because the law of Biot and Savart makes no distinction between bound and free vorticity. When it is obvious we will use the words "bound vorticity" of for instance a screw propeller blade, in the sense as was used in the first approach (1.9.7).

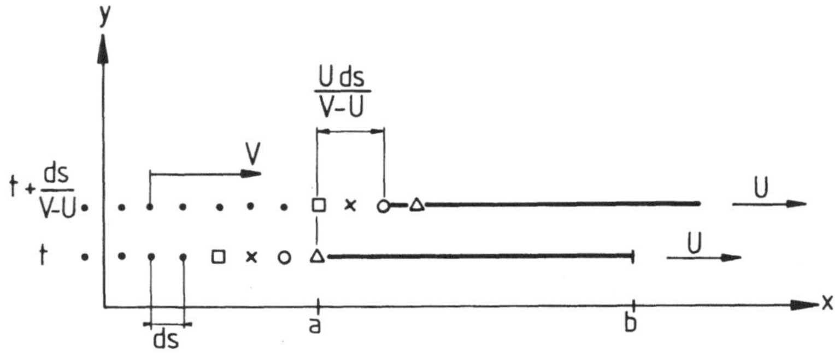

Fig. 1.9.3. Profile overtaken by a row of elementary forces.

The foregoing can be generalized to arbitrarily moving flexible lifting surfaces which are allowed to expand or to contract. The velocities of its points are of no interest and can be chosen within certain limits at will. The same holds for the velocities of the time dependent elementary forces which represent the lifting surfaces, only the vorticity created by them is of importance.

The description given here will be useful when we consider the optimization of flexible wings.

1.10. Bound vortex "ending" at a plate

Sometimes the statement is heard; a bound vortex can end at a rigid plate. This kind of configuration is rather important, for instance in ship propulsion where we have the ducted propeller. The tips of the screw blades with their bound vorticity move closely to the inner side of the duct.

In order to simplify the problem without losing its essence, we consider the case of a flat and infinitely thin plate of finite extent. The plate H (figure 1.10.1) coincides with part of the (x, y) plane. The fluid in which it is embedded has a velocity U in the positive x direction. Concentrated at the z axis for $z > a$ we have a force field $\boldsymbol{F} = F\boldsymbol{e}_y$, where $\boldsymbol{e}_y$ is the unit vector in the y direction and $F = \text{const.} > 0$. The vorticity which belongs to it can be tested by probing contours such as A and B which float with velocity U in the positive x direction. We consider for

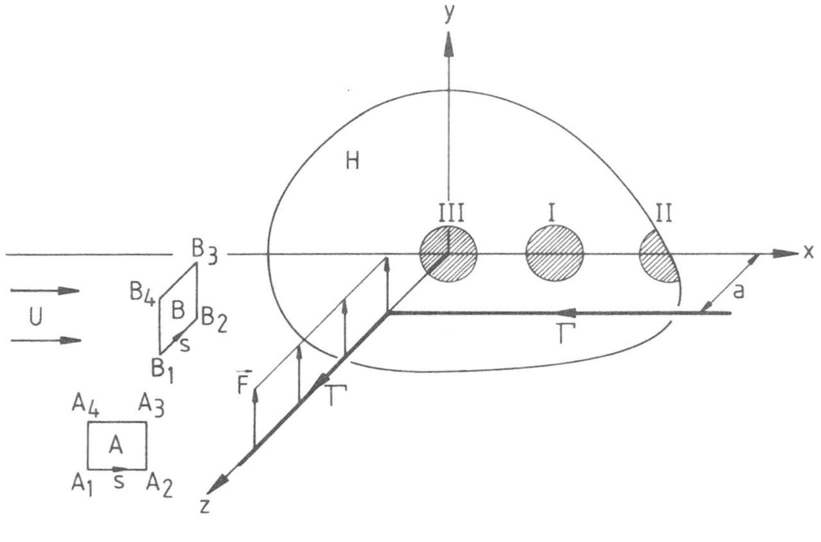

Fig. 1.10.1. Bound vortex ending closely to a rigid plate.

instance contour B with sides parallel to the y and z axis which is positioned so that its centre moves along the line $y = 0$, $z = a$. This means that the sides B_1B_4 and B_2B_3 will cut the z axis for values $z > a$ and $z < a$, respectively. Then it follows from (1.3.18) and (1.3.21) that the circulation Γ_B of B, when B has drifted to positive values of x, assumes the value

$$\Gamma_B = \frac{1}{\mu}\frac{F\boldsymbol{e}_y \cdot -\boldsymbol{e}_y}{U} = -\frac{F}{\mu U} < 0. \tag{1.10.1}$$

This circulation does not change anymore when B floats downstream, hence a free vortex exists along the half infinite line $x > 0$, $y = 0$, $z = a$. Analogously the bound vortex along the z axis is detected by contour A. The strength of both vortices is equal and has the value

$$\Gamma = \frac{F}{\mu U}, \tag{1.10.2}$$

where the indicated directions in figure 1.10.1 are coupled with a right hand screw to the locally induced velocities.

We now raise the question; what happens when $a \to 0$, that is when the free vortex tends to the positive x axis and hence will coincide partly with the plate. We consider this limit separately for three small fixed regions I, II, and III of the plate which are all cut by the x axis. Region I is somewhat in between 0 and the trailing edge, region II is partly bounded by the trailing edge and region III is a neighbourhood of the origin.

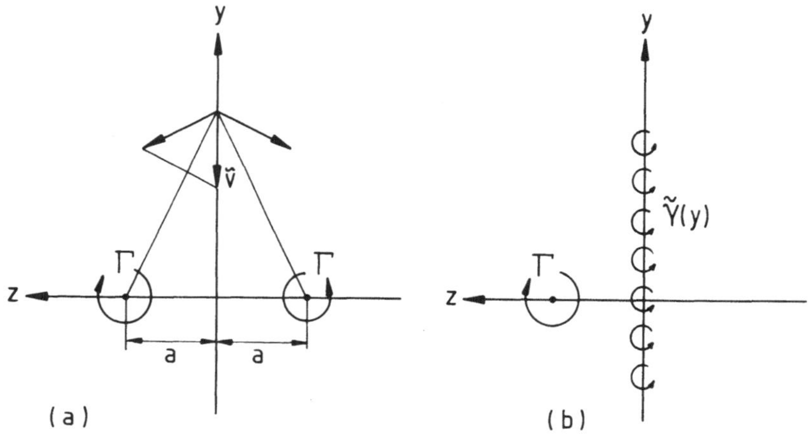

Fig. 1.10.2. Cross section of region I, $a \to 0$.

First region I. When a is sufficiently small we have predominantly the situation of a two sided infinite free vortex of strength Γ in the neighbourhood of the unbounded (x, y) plane. The occurring velocity field is simple. At points with $z > 0$ the influence of the plate is the same as the influence of a vortex which is the image of the free vortex under discussion (figure 1.10.2a). By this the tangential velocity $\tilde{v} = \tilde{v}(y)$ is known, $\tilde{v} = -\Gamma a/\pi(a^2 + y^2)$. Behind the unbounded plate $(z < 0)$ the velocity induced by Γ is zero. Hence by the definition of a vortex layer, the vorticity $\tilde{\gamma} = \tilde{\gamma}(y)$ which can represent the plate is

$$\tilde{\gamma} = \frac{\Gamma a}{\pi\left(a^2 + y^2\right)}, \tag{1.10.3}$$

which is parallel to the x axis and coupled with a right hand screw to the positive x direction (figure 1.10.2b). We note that the total amount of vorticity $\tilde{\gamma}$ has the value

$$\int_{-\infty}^{+\infty} \tilde{\gamma}(y)\, \mathrm{d}y = \frac{\Gamma a}{\pi} \int_{-\infty}^{+\infty} \frac{\mathrm{d}y}{\left(a^2 + y^2\right)} = \Gamma. \tag{1.10.4}$$

From Bernoulli's equation (1.1.6) it follows that in our linearized theory the pressure in the fluid equals the pressure p_∞ at infinity because the disturbance velocities $(0, v_y, v_z)$ are perpendicular to the main flow $(U, 0, 0)$. We neglected disturbance velocities, due to the vortex Γ along the positive z axis and the vorticity at H induced by it.

Second we consider region II, which for sufficiently small values of a behaves for the dominant vorticity as a half infinite plate with an oblique trailing edge. The solution to this problem is simple, the vorticity is the same as we found for region I. We only have to take away the part of the

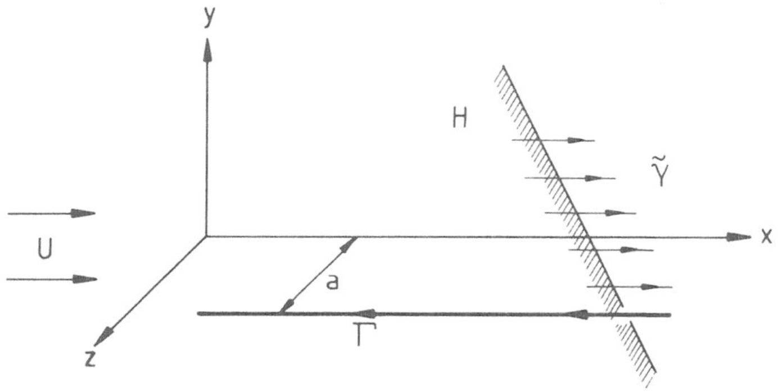

Fig. 1.10.3. Region II with trailing edge.

unbounded plate connected with region I, which is downstream of the trailing edge $\tilde{l}$ (figure 1.10.3), however we let intact the vortex system. This is correct because we satisfy upstream of the trailing edge the boundary condition of vanishing normal component of the velocity and we satisfy at $\tilde{l}$ the Kutta condition, because as we already remarked all "dominant" pressures in the whole space are equal to p_∞, hence we have no pressure jump at the trailing edge.

For the two regions I and II we consider the limit $a \to 0$. It follows from (1.10.3) that $\tilde{\gamma}(y) \to 0$ for each fixed $y \neq 0$, however as follows from (1.10.4) the total strength of $\tilde{\gamma}$ remains equal to Γ. Hence for $a \to 0$ the free vortex Γ is annihilated by the vorticity $\tilde{\gamma}(y)$, at H as well as behind the trailing edge.

Our real problem is more complicated. We have a finite plate H and also the influence of the bound vortex along the positive z axis. These differences however cannot cause infinite induced velocities at regions I and II and hence will not give rise to concentrated free vorticity. We conclude from this that also in the real problem when $a \to 0$ the concentrated free vortex disappears, only distributed free vorticity flows from the trailing edge. Because the vector field of vorticity is free of divergence, the total strength of this free vorticity equals the strength Γ of the bound vorticity ending at the plate.

It is also easy to describe the limit $a = 0$ for region III, hence the dominant vorticity in the neighbourhood of the origin O where the bound vorticity meets the plate H. Then we neglect the influence of the boundary and consider a vortex ending at an unbounded plate. By r we denote the distance from a point of the plate to O. In this case the solution is as follows. At the plate we have a radially converging vortex system of strength $\Gamma/2\pi r$ per unit of length along a circle with radius r (figure 1.10.4). Then it is seen that the total three dimensional vorticity

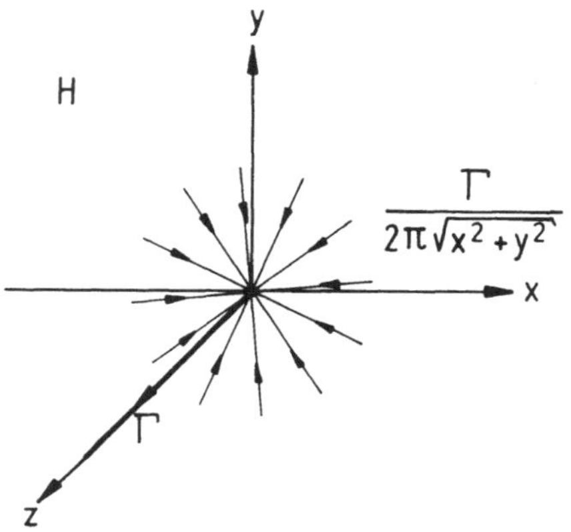

Fig. 1.10.4. The dominant vorticity at region III, $a = 0$.

field is without divergence and it can be proved simply by symmetry considerations that the component of the induced velocity normal to the plate is zero. The reader can check that behind the plate, the velocity of the half infinite bound vortex Γ is opposite to the velocity induced by the radial vorticity at the plate, hence behind the plate the velocity is zero.

When we consider the situation of a finite endplate (figure 1.10.1) for the case $a = 0$, it is plausible from the foregoing that the spreading of the vorticity is qualitatively as given in figure 1.10.5. Because we have to satisfy the Kutta condition at the trailing edge, hence no pressure jump is allowed, the vorticity at the plate has to meet the trailing edge, parallel to the direction of the incoming main flow, hence parallel to the x axis.

At last we make a remark concerning the induced resistance of the configuration drawn in figure (1.10.1) ($a \neq 0$). Then we have parallel to the x axis, a concentrated half infinite free vortex, shed by the force distribution of constant strength along the z axis. Such a vortex has per unit of length an infinite amount of kinetic energy around it. Hence when instead of an incoming flow we consider the force system $\boldsymbol{F}$ moving in the negative x direction with a velocity U, the kinetic energy in the fluid increases per unit of time, by an infinite amount. This means that the induced resistance of the force system ($a \neq 0$) is infinite. However when $a = 0$ the bound vortex connected to the force system ends at the plate and the shed vorticity starting at the trailing edge of the plate has no concentrated part, hence the energy losses are finite. It follows that end plates can have a favourable influence, anyhow from the point of view of potential theory, on the induced resistance of wings or screw propeller blades with loaded tips.

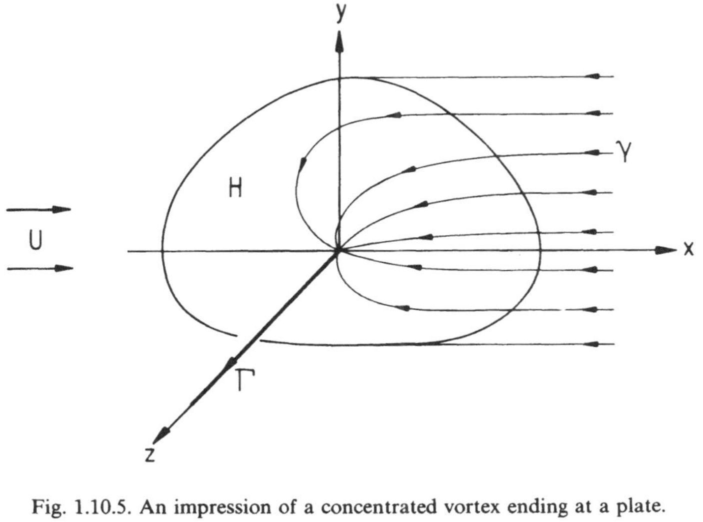

Fig. 1.10.5. An impression of a concentrated vortex ending at a plate.

1.11. Suction force at leading edge

We consider a thin profile under an angle of incidence. When no flow separation occurs the flow resembles the pattern drawn in figure 1.11.1a. The strong curvature of the flow at the nose is caused by a low pressure in the region between A and B. This low pressure results in a force in the x direction at that part of the nose. The fluid particles behave like mass points which rotate around the nose kept in their orbit by a string on which they exert a "centrifugal force". When the profile becomes thinner the frontal area of the nose diminishes, however the flow velocity and curvature at that place increase hence the pressure drop becomes stronger. In the limit when the profile has become an infinitely thin plate (figure 1.11.1b), the frontal area is zero and the pressure has become infinitely low in such a way that theoretically still a suction force is exerted at the leading edge in the positive x direction. In the following we calculate the limit value of this suction force for the case of a two dimensional wing. Thrust production by means of suction forces does not seem to be very reliable in practice because flow separation in the neighbourhood of the leading edge can disturb this phenomenon rather easily.

It is clear that the suction force has to be caused by infinite velocities at the leading edge of the plate, hence we can neglect for the calculation of it the vorticity at a "finite distance" of the leading edge. For this reason we consider the following problem. Along the x axis we assume vorticity of strength

$$\begin{aligned} &\Gamma(x,t)=\frac{\Gamma_l(t)}{(-x)^{1/2}}, \qquad -a\le x\le 0; \\ &\Gamma(x,t)=0, \qquad x<-a \quad \text{or} \quad x>0; \qquad a>0, \end{aligned} \tag{1.11.1}$$

which is the characteristic singular behaviour of the vorticity in the neighbourhood of the nose, represented by O, of an infinitely thin profile

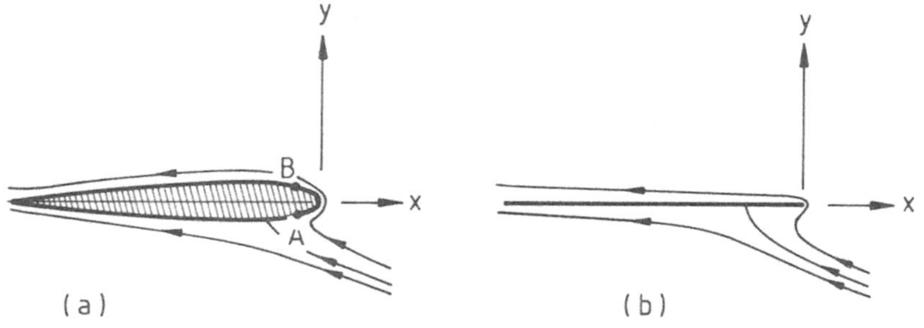

Fig. 1.11.1. Flow around nose of a profile.

(section 4.3 or [3] page 468). The strength of this singular behaviouor determined by $\Gamma_l(t)$ is allowed to be time dependent.

First we state the velocities and the potential belonging to a two sided infinitely long straight vortex. The vortex is perpendicular to the (x, y) plane (figure 1.11.2), cuts this plane at $(\xi, 0)$ and is of strength $\tilde{\Gamma}$. Its velocity field is given by

$$v_x = -\frac{\tilde{\Gamma}}{2\pi}\frac{y}{\left((x-\xi)^2+y^2\right)}, \qquad v_y = \frac{\tilde{\Gamma}}{2\pi}\frac{(x-\xi)}{\left((x-\xi)^2+y^2\right)}, \tag{1.11.2}$$

and its potential by

$$\tilde{\Phi} = \frac{\tilde{\Gamma}\theta}{2\pi}, \tag{1.11.3}$$

where θ is the angle denoted in figure 1.11.2.

By superposition we find from (1.11.2) for the x component of the velocity induced by the vortex layer (1.11.1) at the point $x = \tilde{\varepsilon}\cos\psi$, $y = \tilde{\varepsilon}\sin\psi$,

$$v_x = -\frac{\Gamma_l(t)}{2\pi}\int_{-a}^{0}\frac{1}{(-\xi)^{1/2}}\frac{\tilde{\varepsilon}\sin\psi}{\left((-\xi+\tilde{\varepsilon}\cos\psi)^2+\tilde{\varepsilon}^2\sin^2\psi\right)}\,\mathrm{d}\xi. \tag{1.11.4}$$

Replacing the integration variable ξ by $-\tilde{\varepsilon}\eta$ we find

$$v_x = -\frac{\Gamma_l(t)\sin\psi}{2\pi\tilde{\varepsilon}^{1/2}}\int_0^{a/\tilde{\varepsilon}}\frac{1}{\eta^{1/2}(\eta^2+2\eta\cos\psi+1)}\,\mathrm{d}\eta. \tag{1.11.5}$$

Later on we want to take the limit $\tilde{\varepsilon}\to 0$, by this the upper bound of the

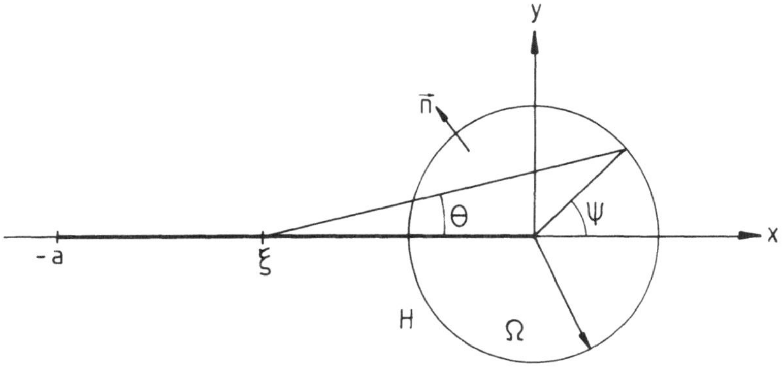

Fig. 1.11.2. Coordinate system at leading edge.

integral tends to ∞, the resulting integral is known ([21] II page 31). Because this integral is absolutely convergent for each fixed value of ψ with $-\pi < \psi < +\pi$, we can write (1.11.5) as

$$v_x = -\frac{\Gamma_l(t)}{2(2\tilde{\varepsilon})^{1/2}}\left(\frac{\sin\psi}{(1+\cos\psi)^{1/2}} + o(\tilde{\varepsilon})\right), \qquad -\pi < \psi < +\pi, \tag{1.11.6}$$

where $o(\tilde{\varepsilon})$ means a quantity with $o(\tilde{\varepsilon})/\tilde{\varepsilon} \to 0$ for $\tilde{\varepsilon} \to 0$. Analogously we find for the velocity component in the y direction

$$v_y = \frac{\Gamma_l(t)}{2(2\tilde{\varepsilon})^{1/2}}\left((1+\cos\psi)^{1/2} + o(\tilde{\varepsilon})\right), \qquad -\pi < \psi < +\pi. \tag{1.11.7}$$

From (1.11.6) and (1.11.7) we obtain the velocity component v_n in the direction of the normal at the control circle H with radius $\tilde{\varepsilon}$ around O,

$$v_n = (v_x \cos\psi + v_y \sin\psi) = \frac{\Gamma_l(t)}{2(2\tilde{\varepsilon})^{1/2}}\left(\frac{\sin\psi}{(1+\cos\psi)^{1/2}} + o(\tilde{\varepsilon})\right),$$

$$-\pi < \psi < +\pi. \tag{1.11.8}$$

It follows from (1.11.6), (1.11.7) and (1.11.8) that these results are easily continuously extended to the closed interval $-\pi \le \psi \le \pi$.

Next we apply the theorem of momentum to the small circular region Ω around the origin O. The total force acting on the fluid inside H equals the change of momentum of the fluid occupying the region Ω, to which we have to add the flux of momentum leaving Ω through H.

First we consider the force exerted at Ω by the fluid outside H. The pressure follows from Bernoulli's equation which reads (1.1.6)

$$p = p_\infty + \tfrac{1}{2}\mu|\boldsymbol{v}_\infty|^2 - \tfrac{1}{2}\mu|\boldsymbol{v}|^2 - \mu\frac{\partial\Phi}{\partial t}. \tag{1.11.9}$$

The constants at the right hand side of (1.11.9) cannot give a resultant force, hence we have to calculate $|\boldsymbol{v}|^2$ and $\partial\Phi/\partial t$. From (1.11.6) and (1.11.7) it follows

$$|\boldsymbol{v}|^2 = v_x^2 + v_y^2 \approx \frac{\Gamma_l^2(t)}{4\tilde{\varepsilon}} + \frac{o(\tilde{\varepsilon})}{\tilde{\varepsilon}}, \tag{1.11.10}$$

hence also this term does not yield a contribution to the resultant force at Ω for $\tilde{\varepsilon} \to 0$. The potential of the vortex layer has by (1.11.3) the value

$$\Phi = \frac{\Gamma_l(t)}{2\pi}\int_{-a}^{0}\frac{1}{(-\xi)^{1/2}}\operatorname{arctg}\frac{y}{(x-\xi)}\,d\xi, \tag{1.11.11}$$

which remains bounded for all values of (x, y) in the neighbourhood of the origin, hence also $\partial\Phi/\partial t$ remains bounded and also this term cannot contribute to the pressure forces exerted at Ω for $\tilde{\varepsilon} \to 0$.

Next we consider the flux of momentum through H. This becomes

$$\mu \int_0^{2\pi} v_x v_n \tilde{\varepsilon}\, \mathrm{d}\psi = -\frac{\mu\pi\Gamma_l^2(t)}{4} + \mathrm{o}(\tilde{\varepsilon}), \tag{1.11.12}$$

and

$$\mu \int_0^{2\pi} v_y v_n \tilde{\varepsilon}\, \mathrm{d}\psi = \mathrm{o}(\tilde{\varepsilon}), \tag{1.11.13}$$

in the x and y direction respectively.

At last we check the momentum of the fluid inside Ω. Because the velocities (1.11.6) and (1.11.7) are $\mathrm{O}((x^2+y^2)^{-1/4})$ for $x^2+y^2 \to 0$, and the area of Ω is $\mathrm{O}(\tilde{\varepsilon}^2)$ it follows that the momentum as well as its time derivative tend to zero with $\tilde{\varepsilon} \to 0$.

From the foregoing follows the desired result. On the fluid inside the small circle with radius $\tilde{\varepsilon}$ around O acts a force with x and y components given by (1.11.12) and (1.11.13) respectively.

Hence by the principle action is reaction, at the nose of the profile acts a suction force K in the positive x direction given by

$$K = \frac{\mu\pi}{4}\Gamma_l^2(t). \tag{1.11.14}$$

This force occurs when the vorticity of the profile in the neighbourhood of the leading edge has a singular behaviour as described by (1.11.1).

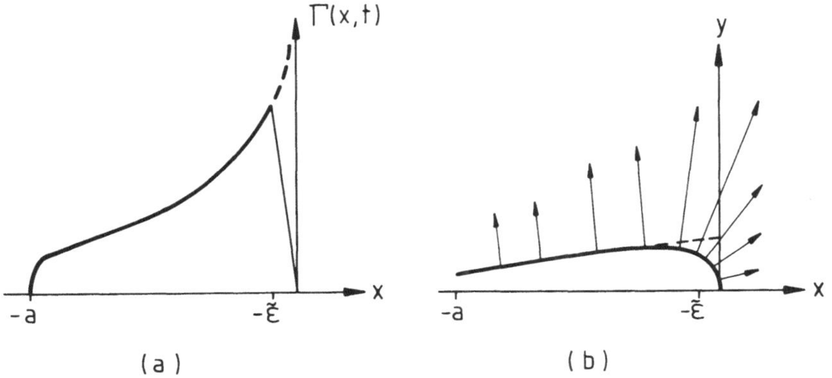

Fig. 1.11.3. Mutilated vorticity and profile belonging to it.

We conclude this section with an heuristic argument, which can be made rigorous, for the existence of a suction force. In (1.11.1) we considered vorticity with a square root singularity. We now mutilate this vorticity as is drawn in figure 1.11.3a. From $x = -\tilde{\varepsilon}$ to $x = 0$, the vorticity tends to zero, for instance linearly. Then it is not difficult to show that the profile which belongs to this vorticity, is curved very steeply downwards in the neighbourhood of the leading edge (figure 1.11.3b). Hence the force action of the fluid on the profile gets a horizontal component which changes, in the limit $\tilde{\varepsilon} \to 0$, into the suction force.

This indicates that the suction force is not a new concept for infinitely thin profiles, but has the same origin as the other horizontal components of the forces induced by the pressure difference. Only part of these have been shifted to the leading edge. This approach will be used with respect to formulas (7.2.12) and (7.2.13).

2. The actuator surface

An actuator surface can be defined as a two dimensional geometric region in a fluid at which a discontinuity in any flow property can occur [26]. We confine ourselves to a more restricted definition. We will use the term actuator surface for a two dimensional region in a fluid on which an external force field is concentrated. At this surface we have to admit a possible discontinuity of the pressure or of the tangential component of the velocity field or of both.

Following this definition a lifting surface is a special type of actuator surface. Perpendicular to it we have an external force field (section 1.9) which by its pressure dipole layer creates a pressure jump over the surface. The flow is tangential to it and has a discontinuity corresponding with the vorticity at the surface. This type of actuator surface will be treated in chapter 3, where the lifting surface theory of ship screw blades is discussed.

The actuator surfaces we discuss in this chapter are passed through by the fluid particles. For instance consider a circular region perpendicular to the main stream at which the force field represents the propulsive action of a screw propeller. This type of actuator surface is generally called an actuator disk. It can be used when the complicated detailed pressure field and flow field induced by a screw propeller is not so much of interest and only a global knowledge of these fields will be sufficient.

We start with a rather general linearized theory of an actuator surface moving in one way or another through the fluid. From this we find by specialization the flow field belonging to the actuator disk. As an application of this theory we discuss a simple aspect of the interaction of a propeller and a body, which sheds some light upon the complicated phenomenon called thrust deduction. Theoretically also the opposite effect can occur although this does not seem to be of practical importance.

We conclude the chapter with a discussion of the non linear theory of the heavily loaded actuator disk and try to obtain insight in the singular behaviour of the pressure and the flow field at its edge.

2.1. Linearized actuator surface theory

In this section we give a rather general description, with respect to hydrodynamics, of the concept actuator surface. We assume the force field at the surface to be $O(\varepsilon)$, where ε is a small parameter, hence we develop a linear theory.

We have a Cartesian coordinate system (x, y, z), embedded in an inviscid and incompressible fluid, which is assumed to be at rest at infinity. Consider an immaterial geometrical surface

$$G(x, y, z, t) = 0. \tag{2.1.1}$$

The normal component of the velocity of this surface amounts to

$$-\frac{1}{|\text{grad } G|}\frac{\partial G}{\partial t}, \tag{2.1.2}$$

in the direction of grad G. It is assumed that this velocity is $O(\varepsilon^0)$.

When the surface G has at a certain time a certain position in space, the force $\boldsymbol{K}$ per unit of area at each point (x, y, z) of G is given by its components

$$K_x = K_x(x, y, z), \quad K_y = K_y(x, y, z), \quad K_z = K_z(x, y, z), \tag{2.1.3}$$

which are sufficiently smooth prescribed functions of $O(\varepsilon)$ in a three dimensional region A of space.

The pressure field induced by the external force field is in connection with (1.4.15) the field of a layer of pressure dipoles in the direction of $-\boldsymbol{K}$ (1.4.11), strength $|\boldsymbol{K}|$ per unit of area, situated at $G = 0$. We note that this layer need not to be the usual pressure dipole layer, where it is generally assumed that the axes of the dipoles are perpendicular to the surface. In this case the axes of the dipoles can have also a tangential component, then by the integration process also a layer of pressure "poles" occurs.

At each point of the region A a direction is given by (2.1.3), we assume that there are no points where all three components of $\boldsymbol{K}$ vanish. Then we consider the lines $l(x = \tilde{x}(s),\ y = \tilde{y}(s),\ z = \tilde{z}(s))$ which satisfy the differential equations

$$\frac{\mathrm{d}\tilde{x}}{\mathrm{d}s} = \frac{K_x(x, y, z)}{|\boldsymbol{K}|}, \quad \frac{\mathrm{d}\tilde{y}}{\mathrm{d}s} = \frac{K_y(x, y, z)}{|\boldsymbol{K}|}, \quad \frac{\mathrm{d}\tilde{z}}{\mathrm{d}s} = \frac{K_z(x, y, z)}{|\boldsymbol{K}|}, \tag{2.1.4}$$

where s is a length parameter along l. We assume that for $s = s_1$

$$\tilde{x}(s_1) = x_1, \qquad \tilde{y}(s_1) = y_1, \qquad \tilde{z}(s_1) = z_1, \tag{2.1.5}$$

with

$$G(x_1, y_1, z_1, t_1) = 0. \tag{2.1.6}$$

Hence we have a two parameter family of lines l, which for $s = s_1$, pierce the surface

$$G(x, y, z, t_1) = 0. \tag{2.1.7}$$

We also assume that $\boldsymbol{K}$ is nowhere tangent to any surface $G(x, y, z, t) = 0$ for the considered interval of time. Then the velocity V with which the point of intersection of a line l with G moves along l becomes

$$V(t) = -\frac{|\boldsymbol{K}|}{|\boldsymbol{K} \cdot \text{grad } G|} \frac{\partial G}{\partial t}, \tag{2.1.8}$$

reckoned positive when it has a positive component in the direction of grad G.

Next we consider the two dimensional region $B(t_1)$, defined as the set of points of intersection of $G(x, y, z, t_1) = 0$ with the lines l (2.1.4). We divide this region into nearly rectangular small surface elements $\Delta S_n(t_1)$, $1 \leq n \leq N$. Then we approximate the loading of the actuator surface at $t = t_1$, by N elementary singular forces

$$\Delta \boldsymbol{K}_n(t_1) = \boldsymbol{K} \Delta S_n(t_1), \qquad 1 \leq n \leq N, \tag{2.1.9}$$

where $\boldsymbol{K}$ is given by (2.1.3) for some point $(x_n, y_n, z_n) \in \Delta S_n(t_1)$.

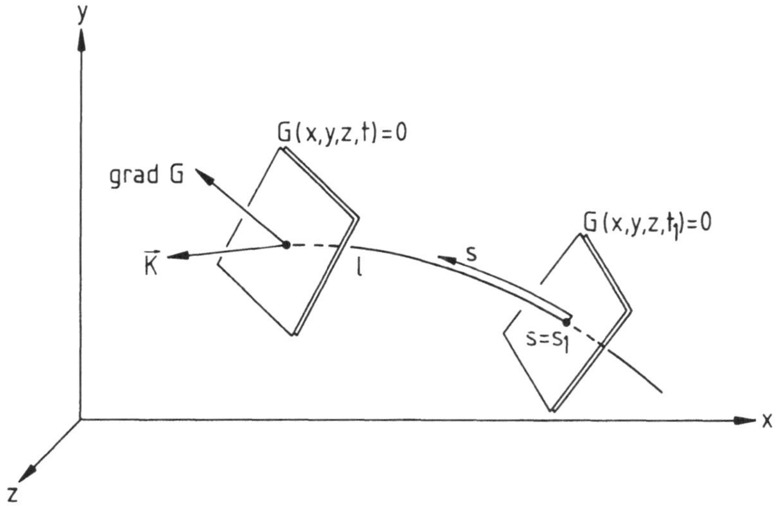

Fig. 2.1.1. Surface $G(x, y, z, t) = 0$ moving through space.

We also introduce tubes M_n formed by the lines l through the boundaries of the $\Delta S_n(t_1)$. In these tubes lie the lines l denoted by l_n, through the points (x_n, y_n, z_n). When time increases $t > t_1$, the points of intersection of $G(x, y, z, t) = 0$ with these lines l_n are the moving points of application of the elementary singular forces $\Delta \boldsymbol{K}_n$. The elementary area's $\Delta S_n(t)$, $t > t_1$ are the area's of the intersections of the tubes M_n with the surface $G(x, y, z, t) = 0$. These $\Delta S_n(t)$ form together the region $B(t)$ of $G(x, y, z, t) = 0$ where the external force field is acting at time t. The strength of the forces $\Delta K_n = \Delta K_n(t)$ is given by (2.1.9) but now for $t \neq t_1$.

In this way we have replaced the actuator surface by N singular forces which vary in a well defined way their velocity and strength, when moving along the lines l_n. These forces are in the direction of their motion because they are tangent to the l_n, hence we can apply the results of section 1.7.

It follows that distributed over each intersection of a tube M_n with $G = 0$ we have a source strength per unit of area of magnitude (1.7.5)

$$\frac{|\boldsymbol{K}|\Delta S_n(t)}{\mu V(t)} \frac{1}{\Delta S_n(t)} = -\frac{(\boldsymbol{K} \cdot \operatorname{grad} G)}{\mu \dfrac{\partial G}{\partial t}}. \tag{2.1.10}$$

We have to take into account that in section 1.7 we assumed the force to be positive when it acted in the negative s direction, which is by (2.1.4) opposite to the assumption made in this section. When in tube M_n at $t = t_0$ the force is switched on we have there, distributed over the intersection of $G(x, y, z, t_0) = 0$ with M_n, starting source strength per unit of area of magnitude (1.7.4)

$$\left.\frac{(\boldsymbol{K} \cdot \operatorname{grad} G)}{\mu \dfrac{\partial G}{\partial t}}\right|_{t=t_0} \tag{2.1.11}$$

Inside the tube M_n we have distributed a source strength per unit of volume of magnitude (1.7.6)

$$-\frac{1}{\mu} \frac{\mathrm{d}}{\mathrm{d}s}\left(\frac{|\boldsymbol{K}|\Delta S_n}{V}\right) \frac{1}{|\Delta S_n \cos \alpha|}, \tag{2.1.12}$$

where α is the angle between $\boldsymbol{K}$ and grad G. At last we have a fluid transport through the tube M_n of magnitude determined by (1.7.8).

Hence a velocity field inside M_n

$$\frac{\boldsymbol{K}\cdot\Delta S_n}{\mu V}\,\frac{1}{|\Delta S_n\cos\alpha|}=\frac{\boldsymbol{K}}{\mu V|\cos\alpha|}=-\frac{\boldsymbol{K}|\text{grad } G|}{\mu\dfrac{\partial G}{\partial t}}. \tag{2.1.13}$$

Herewith we can calculate the velocity field at an arbitrary point of space induced by the actuator surface. This velocity field outside the region passed through by the actuator surface is approximated by calculating the velocities induced by the mentioned source and sink distributions. Inside the region passed through by the actuator surface we have to add the velocity (2.1.13). In fact we have to take the limit $N \to \infty$ in order to obtain the exact solution.

It is not difficult to show by the use of probing contours (section 1.3) that only the component of $\boldsymbol{K}$ tangential to the actuator surface is responsible for concentrated vorticity at this surface.

When the actuator surface becomes a lifting surface the foregoing has to be changed because then, in the linearized theory, the surface has no finite velocity field of $O(\varepsilon^0)$ perpendicular to itself. This will be discussed in chapter 3 where the lifting surface theory of a screw propeller blade is treated.

It is remarked that instead of the lines l defined by (2.1.4) also an arbitrary two parameter family of lines could have been chosen, which covers the region A. Then the same reasoning could have been built up by means of other elementary area's ΔS_n. However then the elementary forces $\Delta \boldsymbol{K}_n$ (2.1.9) are no longer in the direction of their velocity. This means that also the vorticity, discussed in section 1.6, belonging to their component perpendicular to the velocity has to be added.

2.2. Actuator disk

We restrict ourselves here to a simple special case. The geometrical surface (2.1.1) is given by the plane

$$G(x, y, z, t) = x + Ut = 0. \tag{2.2.1}$$

The force function $\boldsymbol{K}$ is chosen to be equal to

$$K_x = f(y, z) = O(\varepsilon), \qquad K_y = K_z = 0, \tag{2.2.2}$$

where we assume for instance $f(y, z) > 0$, hence $\boldsymbol{K}$ acts in the positive x direction and the reaction is a propulsive force at the disk. The three dimensional region A of the previous section is the cylinder of finite

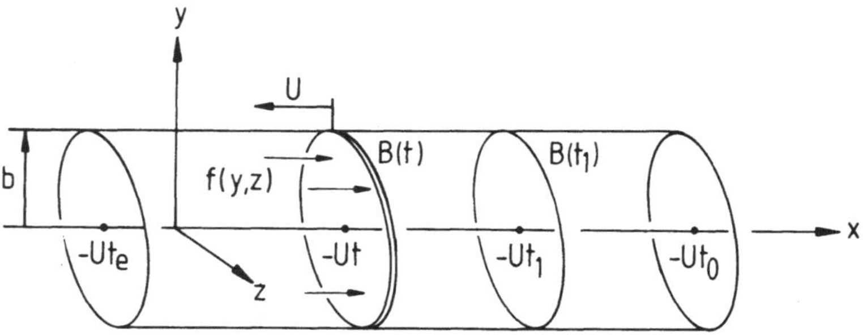

Fig. 2.2.1. Actuator disk, moving with velocity U in negative x direction.

length

$$A: -Ut_e \leq x \leq -Ut_0, \qquad y^2+z^2 \leq b^2, \tag{2.2.3}$$

hence the force field is switched on at $t=t_0$ and switched off at $t=t_e$. The lines l are parallel to the x axis and the two dimensional region $B(t_1)$ is the circular disk $x=-Ut_1$, $y^2+z^2 \leq b^2$ with $t_e \geq t_1 \geq t_0$.

The pressure field induced by the force field is the field of a pressure dipole layer at the disk, it becomes by (1.4.4)

$$p(x,y,z,t)=\frac{1}{4\pi}\int\int_{B(t)} \frac{f(\eta,\zeta)(x+Ut)\,\mathrm{d}\eta\,\mathrm{d}\zeta}{\left\{(x+Ut)^2+(y-\eta)^2+(z-\zeta)^2\right\}^{3/2}}. \tag{2.2.4}$$

The velocity field induced by the disk can now be represented by three parts. First, the velocity field induced by a layer of sinks of strength

$$\frac{f(y,z)}{\mu U} \tag{2.2.5}$$

at the disk. These sinks remain at $x=-Ut_e$ as ending sinks for $t>t_e$. Second, the velocity field induced by a layer of starting sources of strength (2.2.5) at $x=-Ut_0$, $y^2+z^2 \leq b^2$. Third, in the region

$$-Ut \leq x \leq -Ut_0, \qquad y^2+z^2 \leq b^2, \tag{2.2.6}$$

we have the velocity field (2.1.13)

$$\boldsymbol{v}=\left(\frac{f(y,z)}{\mu U}, 0, 0\right). \tag{2.2.7}$$

It is easy to check that the total velocity field at the disk and at the layer of starting sources is continuous. The jumps of the normal velocity components at the source and sink layer, are compensated by the jumps of (2.2.7) at those layers. Hence the divergence of the flow is zero everywhere. Also it is clear that no concentrated vorticity is present at the disk.

The only vorticity that occurs is inside region (2.2.6) and at the cylindrical part of its boundary. Another way of describing the induced velocity field is by applying the law of Biot and Savart to this vorticity field, because as has been said, no divergence is present in the total flow field.

From the rate of change of the region (2.2.6) of the third part of the velocity field follows the rate of change of the total impulse of the fluid in the x direction, which equals the resultant external force

$$\mu U \iint_{y^2+z^2 \leq b^2} \frac{f(y,z)}{\mu U}\, \mathrm{d}y\, \mathrm{d}z = T, \tag{2.2.8}$$

where T is the thrust exerted by the disk, reckoned positive in the negative x direction.

When the actuator disk started its action very long ago, hence when $t_0 = -\infty$, we can also calculate easily the rate of change of kinetic energy of the fluid. Each unit of time is added an amount of kinetic energy E_i which equals

$$\begin{aligned} E_i &= \tfrac{1}{2}\mu \int_0^U \iint_{y^2+z^2 \leq b^2} \left\{ \frac{f(y,z)}{\mu U} \right\}^2 \mathrm{d}y\, \mathrm{d}z\, \mathrm{d}x \\ &= \frac{1}{2\mu U} \iint_{y^2+z^2 \leq b^2} f^2(y,z)\, \mathrm{d}y\, \mathrm{d}z. \end{aligned} \tag{2.2.9}$$

From (2.2.9) follows the efficiency η_T of the actuator disk. The efficiency is defined as the quotient of the useful work UT and the total work $UT + E_i$,

$$\eta_T = \frac{UT}{UT + E_i} = (1 - \mathrm{O}(\varepsilon)). \tag{2.2.10}$$

The latter equality of (2.2.10) follows from the fact that $T = \mathrm{O}(\varepsilon)$ and $E_i = \mathrm{O}(\varepsilon^2)$. Note that by adapting the cross section of the cylinder A the foregoing also holds for non-circular actuator disks.

The free vorticity shed by the actuator disk of figure 2.2.1 consists of closed vortex lines which are parallel to the (y, z) plane. In case $t_0 = -\infty$ we have a half infinite tube of this vorticity. Hence it follows that the x components of the disturbance velocities at the disk are half the disturbance velocities at $x = \infty$, which are parallel to the x axis. The half infinite cylinder downstream of the disk is called the slipstream region.

The most simple special case of the actuator disk of figure 2.2.1, arises when the load of the disk is independent of y and z hence $f(y, z) = f =$ const. and when $t_0 = -\infty$. Instead of the coordinate system of figure 2.2.1 we use a system which moves with the disk. Then we have, in order to obtain the total velocity field, to add to the induced velocities the incoming velocity $(U, 0, 0)$ and we obtain the situation of figure 2.2.2a.

In this case the only vorticity occurs at the half infinite cylinder $x > 0$, $y^2 + z^2 = b^2$, it is of strength

$$\gamma = f/\mu U, \tag{2.2.11}$$

per unit of length in the x direction and is parallel to the (y, z) plane. Also here the disturbance velocity field can be described in two different ways. First, at the disk we have a layer of sinks of strength $f/\mu U$ (2.2.5) and behind the disk we have to add a parallel flow in the positive x direction of magnitude $f/\mu U$ (2.2.7). Second, by applying the law of Biot and Savart to the circular vorticity described above (2.2.11).

A cross section of the flow is given in figure 2.2.2b. It is clear that this picture is only correct up to and including $O(\varepsilon)$. The free vorticity which in the linearized theory is at the half infinite cylinder with radius b, in fact is transported by the fluid. Hence this vorticity lies in reality at the stream tube which passes through the edge of the disk.

The question arises if we can give a representation of the disk under consideration which is more related to a screw propeller. We introduce a cylindrical coordinate system (x, r, φ) as drawn in figure 2.2.3. Consider a straight bound vortex OA of length b. The endpoint O coincides with the origin of the coordinate system and the vortex rotates with angular

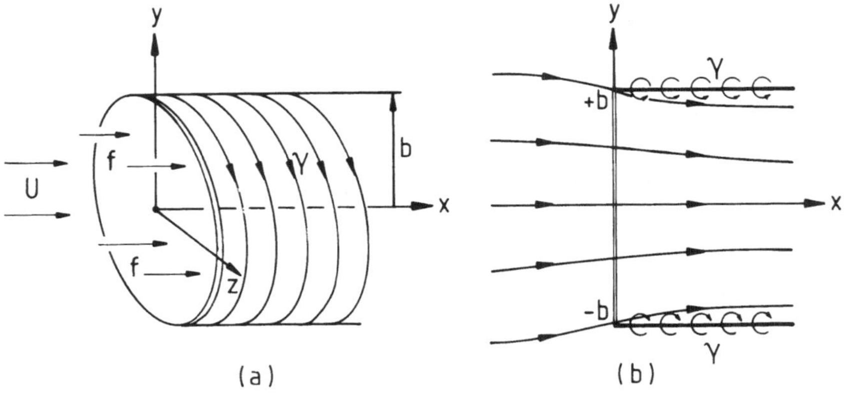

Fig. 2.2.2. Actuator disk with constant normal load in parallel flow.

velocity ω in the plane $x=0$. The strength Γ of this vortex is coupled with a right hand screw to the negative r direction. From the endpoint 0 starts a free vortex of strength Γ stretching along the positive x axis and from the endpoint A starts a free vortex of the same strength along the helicoidal line

$$\varphi-\omega t+ax=0, \qquad r=b, \tag{2.2.12}$$

where $a=\omega/U$. These two vortices are connected by the starting vortex O_2A_2, which was shed long ago at the beginning of the process and which makes the vortex field free of divergence.

The velocity of a point of OA, at a distance r of O, relatively to the fluid is $(-U, 0, +\omega r)$ where the components, which are the physical components, are in the x, r and φ direction (see below formula (B.1.9) of appendix B, in the following always physical components are used). Hence by the "theorem of Joukowski" (1.6.6) the force per unit of length acting at the fluid at that point is

$$\mu(\omega r\Gamma, 0, U\Gamma). \tag{2.2.13}$$

The opposite force acts at the rotating vortex, for $\Gamma>0$ the rotating vortex acts as a propeller.

Now suppose that ω increases indefinitely and Γ decreases so that $\omega\Gamma$ remains constant. Then several limits have to be considered. First, the vorticity at the disk, the free vortex along the x axis and the starting vortex disappear. Second, the helicoidal vortex changes into a layer of circular vortices of which the strength per unit of length in the x direction assumes the value

$$\Gamma\frac{\omega}{2\pi}\frac{1}{U}=\frac{\Gamma\omega}{2\pi U}, \tag{2.2.14}$$

coupled with a right hand screw to the positive φ direction. Third, the

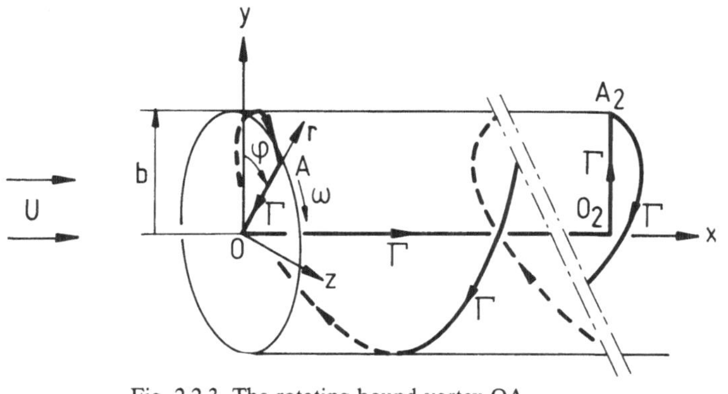

Fig. 2.2.3. The rotating bound vortex OA.

force action exerted at the fluid becomes perpendicular to the disk ($x=0, 0 \le r \le b$) and its mean value per unit of area at a radius r becomes, by using (2.2.13),

$$\frac{\mu\omega r\Gamma\,\mathrm{d}r}{2\pi r\,\mathrm{d}r}=\frac{\mu\omega\Gamma}{2\pi}, \tag{2.2.15}$$

which is independent of r. This force action is in the positive x direction. If we suppose

$$\lim_{\omega\to\infty}\omega\Gamma=2\pi f/\mu, \tag{2.2.16}$$

we obtain the actuator disk of figure 2.2.2.

The efficiency of the disk in this case, hence with $f(y,z)=f=$ const. and $t_0=-\infty$ follows from (2.2.10) it becomes

$$\eta_T=\left(1+\frac{T}{2\mu U^2 S}\right)^{-1}=1-\mathrm{O}(\varepsilon), \tag{2.2.17}$$

where $S=\pi b^2$ is the area of the disk. Later on (section 5.7) it will be shown that this efficiency is with respect to linearized theory, the lowest upperbound for the efficiency of all possible propellers acting in an inviscid and incompressible fluid, with thrust T, "working area" S and mean velocity of advance U. Therefore this special actuator disk is sometimes called an "ideal propeller".

We now discuss shortly an actuator disk with a more general type of loading. In case of a real screw propeller we have behind the propeller besides a jet directed backwards, also a rotation of the fluid in the jet around the axis of the propeller. This rotation can be induced by admitting at the disk besides forces parallel to the x axis, also tangential forces per unit of area. We assume in this case the disk region (figure 2.2.4) to be given by

$$x=0, \qquad 0<b_i\le r\le b, \qquad 0\le\varphi<2\pi, \tag{2.2.18}$$

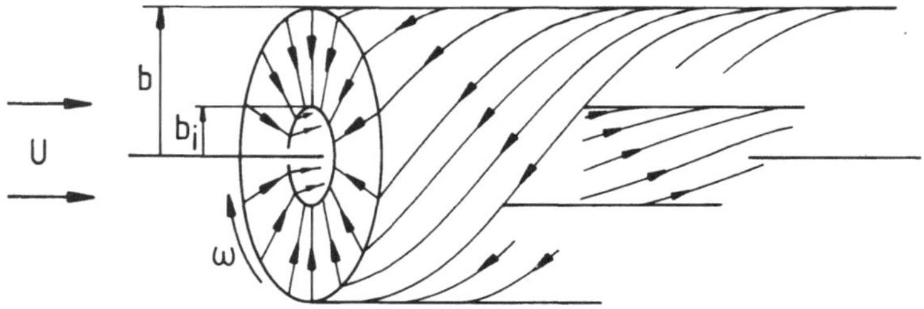

Fig. 2.2.4. Bound vortex system with finite rotational velocity.

and take the loading, (x, r, φ) components, per unit of area as

$$\left(f, 0, \frac{g}{r}\right), \tag{2.2.19}$$

where f and g are constants. Then it follows by using probing contours (section 1.3) floating with the incoming parallel flow U, that at the disk we have vorticity. In fact at the disk the fluid velocity shows a jump in its φ component. It follows from an adequate interpretation of (1.6.3) that we can represent at the disk the force field (2.2.19) by distributed bound vorticity in the negative r direction (right hand screw) of strength $\tilde{\Gamma}$ per unit of length in the φ direction

$$\tilde{\Gamma} = \frac{g}{\mu U r}, \tag{2.2.20}$$

which rotates with an angular velocity $\omega = fU/g$. It is seen that this is allowable vorticity because at the disk it is free of divergence. On the two half infinite cylinders $x \geq 0$, $r = b$ or $r = b_i$ there is helicoidal free vorticity.

When $b_i = 0$, we have a concentrated free vortex along the positive x axis. Such a vortex has an infinite amount of kinetic energy around it, hence in that case the efficiency of the disk is zero. This configuration is not realistic because cavitation would occur along the positive x axis by which the kinetic energy induced by the disk in the fluid per unit of time will be finite again.

When we consider actuator disks with axi-symmetric loading placed in a uniform parallel flow, the stream tube through the edge of the disk will be also axisymmetric. When the normal component of the loading of the disk is in the downstream direction of the incoming flow there will be a contraction of this streamtube. Suppose the radius of the disk is b and the radius of the stream tube infinitely far upstream at $x = -\infty$ is $b + \Delta b$, then the radius of the streamtube infinitely far downstream at $x = +\infty$ is $b - \Delta b$. This is caused by the fact that the axial component of the induced velocities at the disk is half the axial component of the induced velocities at $x = +\infty$. This will in general no longer be true for a heavily loaded disk, described by a non linear theory.

The problem of the actuator disk can be posed in a number of ways. The first one is, as we have partly discussed here, by prescribing explicitly the external force field at the disk. The second one is, by prescribing the bound vorticity rotating or moving in some way with velocities of $O(\varepsilon^0)$ in the plane of the disk. These two methods are in our linearized theory practically equivalent. From the forces the vorticity can be found explicitly and inversely. This is no longer true in the non linear theory where from the prescribed moving bound vorticity first by solving the problem the induced velocities have to be calculated, from which then follow the forces at the vorticity. The reason for this is that in a non linear theory the induced velocities and the velocities of the bound vorticity are of the same order of magnitude ($O(\varepsilon^0)$).

A third more complicated problem arises when at the disk a rotating homogeneously distributed material structure is defined, which represents for instance a large number of rotating blades of a screw propeller. Then the vorticity at the disk has to be determined such that the flow which arises satisfies certain conditions imposed by the moving structure. For such a problem we refer to Hess [25] who calculated the forces on a quickly rotating boomerang of prescribed shape by means of a linear theory.

We end this section with the suggestion to the reader to check in a number of cases that the force action of the disk corresponds to the momentum added to the fluid per unit of time.

2.3. Thrust deduction

When a screw propeller is placed in the neighbourhood of the stern of a ship it is well known that the propulsive force in the propellershaft is larger than the force necessary to tow the ship without propeller, [10] page 390. One of the causes is that by the action of the propeller a low pressure region is created at the stern which induces a force in the wrong direction on the hull. We will show by using a linear theory, that such an effect follows qualitatively from the interaction of an actuator disk with a body.

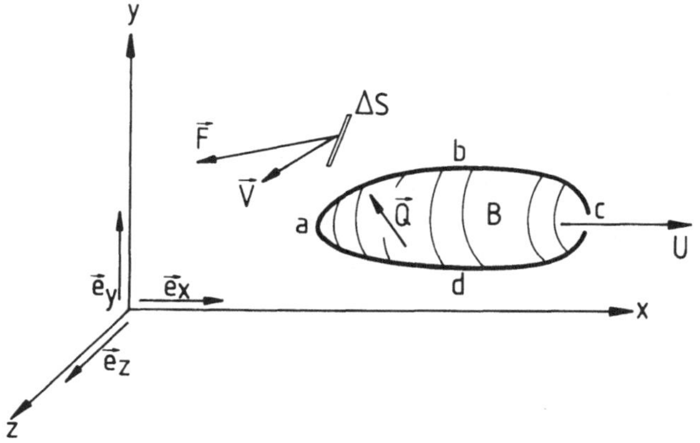

Fig. 2.3.1. Body B with nearby actuator surface ΔS.

Consider a body B translating with velocity U in the positive x direction of a Cartesian coordinate system (x, y, z), through an unbounded, inviscid and incompressible fluid. The fluid is assumed to be at rest at infinity with respect to the coordinate system. The body has finite dimensions hence its disturbance velocities $\boldsymbol{V}$ are $O(\varepsilon^0)$. In the neighbourhood of the body we have a small actuator surface of area ΔS by which a force $\boldsymbol{F}$ of $O(\varepsilon)$ per unit of area is exerted on the fluid (figure 2.3.1).

It is most easy to consider the actuator surface moving besides the body B at a fixed relative position without any mechanical contact with B. We discuss the force $\boldsymbol{Q}$ of $O(\varepsilon)$ exerted at B by the actuator surface ΔS, which is transferred by the fluid. It is this force that can cause thrust deduction. In "reality" the actuator surface is coupled to B and then also the reaction force $-\boldsymbol{F}\Delta S$ is acting at B as a force with a propulsive component.

We now use formula (1.8.9) which reads in this case

$$\frac{\mathrm{d}E}{\mathrm{d}t} = \int_{\Delta S} \boldsymbol{v}\cdot\boldsymbol{F}\,\mathrm{d}S - \int_{\partial B} p(\boldsymbol{v}\cdot\boldsymbol{n})\,\mathrm{d}S. \tag{2.3.1}$$

The total velocity $\boldsymbol{v}$ of the fluid can be split into two parts

$$\boldsymbol{v} = \boldsymbol{V} + \boldsymbol{v}_a, \tag{2.3.2}$$

where $\boldsymbol{v}_a$ of $O(\varepsilon)$ is the velocity induced by the presence of the actuator disk. Also the pressure can be divided into a part P of $O(\varepsilon^0)$ induced by the steady motion of the body B in an undisturbed fluid and a part p_a of $O(\varepsilon)$ induced by the presence of the actuator disk,

$$p = P + p_a. \tag{2.3.3}$$

The condition for the velocity field $\boldsymbol{v}$ at the boundary of B, reads

$$\boldsymbol{v}\cdot\boldsymbol{n} = U\boldsymbol{e}_x\cdot\boldsymbol{n}. \tag{2.3.4}$$

Hence (2.3.1) changes into

$$\frac{\mathrm{d}E}{\mathrm{d}t} = \int_{\Delta S} (\boldsymbol{V} + \boldsymbol{v}_a)\cdot\boldsymbol{F}\,\mathrm{d}S - U\boldsymbol{e}_x\int_{\partial B} (P + p_a)\boldsymbol{n}\,\mathrm{d}S. \tag{2.3.5}$$

Because the resistance of a rigid body translating with a constant velocity in an inviscid and incompressible fluid is zero we find

$$\frac{\mathrm{d}E}{\mathrm{d}t} = \int_{\Delta S} (\boldsymbol{V} + \boldsymbol{v}_a)\cdot\boldsymbol{F}\,\mathrm{d}S - U\boldsymbol{e}_x\cdot\boldsymbol{Q}. \tag{2.3.6}$$

The left hand side of (2.3.6) is the rate of change of the kinetic energy

E in the fluid. Far behind B we have disturbance velocities of $O(\varepsilon)$ caused by the free vorticity shed by the actuator surface ΔS. Hence this rate of change is $O(\varepsilon^2)$ because the finite velocity field induced by B tends to zero at large distances from B. Comparing in (2.3.6) quantities of $O(\varepsilon)$ we find by means of our linearized theory

$$\boldsymbol{Q} \cdot \boldsymbol{e}_x = \frac{\Delta S}{U} \boldsymbol{F} \cdot \boldsymbol{V}, \tag{2.3.7}$$

where we have approximated the integration over ΔS, which is allowed when ΔS is sufficiently small.

In case that $\boldsymbol{F}$ is a propulsive force, hence is predominantly in the negative x-direction, we can state that $\boldsymbol{Q} \cdot \boldsymbol{e}_x$ will be negative for positions of ΔS where the x-component of $\boldsymbol{V}$ is positive, then we have thrust deduction. For instance at the positions a and c of figure 2.3.1. Inversely we have thrust augmentation when ΔS is at the positions b or d. Remember that in case the actuator surface ΔS is connected to the body B the total force on B becomes $-\boldsymbol{F}\Delta S + \boldsymbol{Q}$. If we consider a larger actuator surface we have to carry out an integration over its area.

Loosely stated we can give the following slightly different version of the conclusion. If a lightly loaded propeller is placed in a region where the fluid velocities with respect to the propeller are mainly lower than the translational velocity U of the propelled body, linearized potential theory predicts thrust deduction. Inversely if the propeller is placed in a region where these velocities are mainly larger than U we have thrust augmentation. This result is in agreement with the expected influence of the pressure field of the propeller on the body.

For papers which, although in an approximate way, discuss the non linear interaction of a propeller and a hull we refer for instance to [66] and [17].

2.4. Non linear actuator disk theory, 1

In this section we discuss a non linear theory for actuator disks. The following is based on an article of Wu [72]. The theory developed in [72] has been used by Greenberg as a basis for carrying out numerical calculations, the results of which are given in [19]. The theory is restricted to flow fields induced by circular actuator disks with a large axisymmetric external force field concentrated on it. Insight in the induced flow is important for a better understanding of the global flow pattern caused by a heavily loaded screw propeller.

We use again a cylindrical coordinate system (x, r, φ) as in figure 2.2.3. At the disk we have an axisymmetric force field $\boldsymbol{F}(r)$ per unit of area. The velocity field is denoted by $\boldsymbol{v}=(v_x, v_r, v_\varphi)$ where the components are in the x, r and φ direction. The disk region is given by $x=0$, $0 \le r \le b$. In some parts of the development of the theory, however it is more easy to keep in mind not a concentrated force field at the disk, but a sufficiently smooth force field, per unit of volume for instance in the region B with $0 \le x \le \tilde{\varepsilon}$, $0 \le r \le b$, where $\tilde{\varepsilon}$ is a positive number, small with respect to b. Afterwards we can consider the limit $\tilde{\varepsilon} \to 0$. We do not use a special notation for this distributed force field. The external force field is placed in a homogeneous flow with velocity U in the positive x direction. Because the problem is axisymmetric its quantities depend only on the two coordinates x and r. Then we can introduce a stream function $\Psi = \Psi(x, r)$ ([3], page 78) with

$$v_x = \frac{1}{r}\frac{\partial \Psi}{\partial r}, \qquad v_r = -\frac{1}{r}\frac{\partial \Psi}{\partial x}, \tag{2.4.1}$$

hence the flow field is determined by Ψ and the tangential velocity component v_φ.

We write the equation of motion (1.3.8) for this stationary case as

$$\boldsymbol{v} * \boldsymbol{\omega} = \text{grad}\, H - \frac{1}{\mu}\boldsymbol{F}, \tag{2.4.2}$$

where the "head" H is given by

$$H = \frac{p}{\mu} + \tfrac{1}{2}(\boldsymbol{v})^2, \tag{2.4.3}$$

and by appendix B, formula (B.2.4) the physical components of grad H are

$$\text{grad}\, H = \left(\frac{\partial H}{\partial x}, \frac{\partial H}{\partial r}, 0\right). \tag{2.4.4}$$

The physical components of $\boldsymbol{\omega}=(\omega_x, \omega_r, \omega_\varphi)$ are in this coordinate system (B.2.7)

$$\begin{aligned} \omega_x &= \frac{1}{r}\frac{\partial}{\partial r}(r v_\varphi), \qquad \omega_r = -\frac{\partial}{\partial x} v_\varphi, \\ \omega_\varphi &= -\frac{\partial}{\partial r}\left(\frac{1}{r}\frac{\partial \Psi}{\partial r}\right) - \frac{1}{r}\frac{\partial^2 \Psi}{\partial x^2}. \end{aligned} \tag{2.4.5}$$

Outside the external force region, hence when $\boldsymbol{F}=0$, we find from (2.4.2) by taking the scalar product with $\boldsymbol{v}$ and also with $\boldsymbol{\omega}$

$$\boldsymbol{v} \cdot \text{grad}\, H = \boldsymbol{\omega} \cdot \text{grad}\, H = 0. \tag{2.4.6}$$

This means that the axisymmetric streamtubes and vortex tubes coincide outside the disk region B with surfaces $H=$ const.. Vortex tubes generally exist within the slipstream region S of the disk, also a vortex tube can occur with a concentrated vortex sheet at the boundary of S. The concept slipstream region apparently has an obvious meaning, as the region of fluid particles which have passed through B. It will be shown however in the next section that there are simple disks where, theoretically, fluid particles pass back and forth the disk an arbitrarily chosen large number of times before finally entering the region "downstream" of the disk. However in this section this does not need to bother us too much.

We now restrict our attention to a meridional plane in which we have the coordinates x and r. In this plane we introduce two directions of differentiation, one along the lines $\Psi=$ const. and one perpendicular to these lines. Along the lines $\Psi=$ const. we introduce a length parameter s. When $G(x, r)$ is any sufficiently smooth function of x and r we find

$$\frac{\Delta G}{\Delta s}=\left(\frac{\partial G}{\partial x}, \frac{\partial G}{\partial r}\right)\left(\frac{\Delta x}{\Delta s}, \frac{\Delta r}{\Delta s}\right)=\left(\frac{\partial G}{\partial x}, \frac{\partial G}{\partial r}\right)\left(\frac{v_x}{V_s}, \frac{v_r}{V_s}\right), \tag{2.4.7}$$

where $V_s=(v_x^2+v_r^2)^{1/2}$ and the latter equality is correct because s measures length. By this we find

$$\frac{\partial}{\partial s}=\frac{1}{V_s}\left(v_x\frac{\partial}{\partial x}+v_r\frac{\partial}{\partial r}\right). \tag{2.4.8}$$

The differentiation in the direction perpendicular to the lines $\Psi=$ const. follows from

$$\frac{\Delta G}{\Delta \Psi}=\left(\frac{\partial G}{\partial x}, \frac{\partial G}{\partial r}\right)\left(\frac{\Delta x}{\Delta \Psi}, \frac{\Delta r}{\Delta \Psi}\right), \tag{2.4.9}$$

and from

$$\Delta\Psi=V_s r\Delta l, \tag{2.4.10}$$

where Δl is the small distance used in the process of differentiation. Combining (2.4.9) and (2.4.10) we obtain

$$\frac{\Delta G}{\Delta \Psi}=\frac{1}{V_s r}\left(\frac{\partial G}{\partial x}, \frac{\partial G}{\partial r}\right)\left(\frac{\Delta x}{\Delta l}, \frac{\Delta r}{\Delta l}\right)=\frac{1}{V_s r}\left(\frac{\partial G}{\partial x}, \frac{\partial G}{\partial r}\right)\left(-\frac{v_r}{V_s}, \frac{v_x}{V_s}\right), \tag{2.4.11}$$

the latter equality follows from the fact that both $(\frac{\Delta x}{\Delta l}, \frac{\Delta r}{\Delta l})$ and $(-\frac{v_r}{V_s}, \frac{v_x}{V_s})$ are unit vectors perpendicular to $\Psi=$ const. Hence we find

$$\frac{\partial}{\partial \Psi}=\frac{1}{rV_s^2}\left(v_x\frac{\partial}{\partial r}-v\frac{\partial}{\partial x}\right). \tag{2.4.12}$$

We remark that it is not possible to introduce instead of the coordinates (x, r) in the meridional plane the quantities s and Ψ as new coordinates. This new coordinate system has to be locally orthogonal and s has to measure length. Then it can be proved that the lines $\Psi = \text{const.}$ have to be straight, which conflicts our problem.

The φ component of (2.4.2) becomes by (2.4.8)

$$\mu V_s \frac{\partial}{\partial s}(v_\varphi r) = rF_\varphi, \tag{2.4.13}$$

which describes the rate of change of moment of momentum about the x axis (compare (1.1.13)). Hence outside the external force region where $F_\varphi = 0$, the quantity $v_\varphi r$ can only depend on Ψ

$$rv_\varphi = f(\Psi), \qquad (x, r, \varphi) \in S, \tag{2.4.14}$$

$$v_\varphi = 0, \qquad (x, r, \varphi) \notin S + B. \tag{2.4.15}$$

When we take the scalar product of $\boldsymbol{v}$ with equation (2.4.2) we obtain

$$V_s \frac{\partial H}{\partial s} = \frac{\boldsymbol{v} \cdot \boldsymbol{F}}{\mu}, \tag{2.4.16}$$

because grad H has no component in the φ direction. This represents the fact that as long as a fluid particle is outside the force region, its value of H does not change because $\partial/\partial s$ is along a streamtube.

We next assume that the force field $\boldsymbol{F}$ arises from radially directed axisymmetric bound vorticity which rotates with an angular velocity Ω around the x axis in the region B. This means that the force field has to be perpendicular to the relative velocity of the fluid with respect to the bound vorticity, hence

$$(\boldsymbol{v} - \boldsymbol{e}_\varphi \Omega r) \cdot \boldsymbol{F} = 0, \tag{2.4.17}$$

where $\boldsymbol{e}_\varphi$ is the unit vector in the φ direction. By this it follows from (2.4.16) that

$$V_s \frac{\partial H}{\partial s} = \frac{\Omega r F_\varphi}{\mu}. \tag{2.4.18}$$

Combination of (2.4.13) and (2.4.18) yields

$$H = \Omega r v_\varphi + H_\infty, \qquad H_\infty = \frac{p_\infty}{\mu} + \tfrac{1}{2}U^2, \tag{2.4.19}$$

where H_∞ is the head of the undisturbed parallel flow.

We now determine the φ component of the vector product of $\boldsymbol{v}$ and equation (2.4.2), we find

$$-\omega_\varphi r = r^2 \frac{\partial H}{\partial \Psi} - (rv_\varphi)\frac{\partial}{\partial \Psi}(rv_\varphi) - \frac{rF_\Psi}{\mu V_s}, \tag{2.4.20}$$

where

$$F_\Psi = (v_x F_r - v_r F_x)/V_s, \tag{2.4.21}$$

is the Ψ component of $\boldsymbol{F}$.

Outside the force region B we find by combining (2.4.5) (ω_φ), (2.4.14), (2.4.19) and (2.4.20)

$$\frac{\partial^2 \Psi}{\partial r^2} - \frac{1}{r}\frac{\partial \Psi}{\partial r} + \frac{\partial^2 \Psi}{\partial x^2} = (\Omega r^2 - rv_\varphi)\frac{\mathrm{d}}{\mathrm{d}\Psi}(rv_\varphi). \tag{2.4.22}$$

The streamfunction Ψ has to satisfy the limit conditions

$$\frac{1}{r}\frac{\partial \Psi}{\partial r} \to U, \qquad \frac{\partial \Psi}{\partial x} \to 0; \qquad x \to -\infty \quad \text{or} \quad r \to \infty \tag{2.4.23}$$

and

$$\frac{\partial \Psi}{\partial x} \to 0, \qquad x \to +\infty. \tag{2.4.24}$$

In order to calculate the induced fluid flow, it is clear that we need information about the rotating bound vorticity in the force region B. We consider the limit case that the thickness of the external force region is zero ($\tilde{\varepsilon} = 0$). Then we have radial bound vorticity concentrated at the disk, say of strength $\Gamma(r)$ per unit of length in the φ direction. We assume $\Gamma(r) > 0$ to be coupled with a right hand screw to the negative r direction. In front of the disk, when the fluid particles have not yet crossed the disk, we know by (2.4.15) that $v_\varphi = 0$. By the definition of vorticity it is known that when the particles have crossed the disk, v_φ has the value $\Gamma(r)$. So in order to complete the formulation we prescribe v_φ just behind the disk.

It is however here that difficulties can arise, because as has been observed already with respect to the definition of the slipstream region, it can happen that particles cross the disk even from behind and then back again a large number of times before they cross for the last time the disk in the positive x direction. Again, we leave out of consideration this phenomenon and assume that it does not occur at all for the chosen $\Gamma(r)$ or at an narrow insignificant region near the edge of the disk.

The problem is now: determine the functions $\Psi = \Psi(x, r)$ and $v_\varphi = v_\varphi(x, r)$ and the unknown slipstream region S, such that the equations (2.4.14), (2.4.15), (2.4.22) and the limit conditions (2.4.23), (2.4.24) are satisfied and v_φ assumes its prescribed values just behind the disk.

In order to bring this problem in a form which is more amenable to numerical calculations by an iteration method, equation (2.4.22) is in [72] converted into an integral equation.

First the following dimensionless variables are introduced

$$\begin{aligned} r^* &= r/b, \qquad x^* = x/b, \qquad \mathbf{v}^* = \mathbf{v}/U, \\ \Psi^* &= \Psi/Ub^2, \quad \lambda = U/\Omega b. \end{aligned} \tag{2.4.25}$$

Substitution of (2.4.25) in (2.4.22) and (2.4.14) and neglecting the *, yields for the dimensionless form of the equations

$$\frac{\partial^2 \Psi}{\partial r^2} - \frac{1}{r}\frac{\partial \Psi}{\partial r} + \frac{\partial^2 \Psi}{\partial x^2} = \left(\frac{r^2}{\lambda} - r v_\varphi\right)\frac{d}{d\Psi}(r v_\varphi), \tag{2.4.26}$$

$$r v_\varphi = f(\Psi). \tag{2.4.27}$$

Introducing the perturbation stream function ψ by

$$\Psi(x, r) = \Psi_\infty(r) + \psi(x, r), \qquad \Psi_\infty(r) = \tfrac{1}{2} r^2, \tag{2.4.28}$$

we find using (2.4.26) and (2.4.27)

$$L\left(\frac{\psi}{r}\right) \overset{\text{def}}{=} r\left(\frac{\partial^2}{\partial r^2} + \frac{1}{r}\frac{\partial}{\partial r} - \frac{1}{r^2} + \frac{\partial^2}{\partial x^2}\right)\left(\frac{\psi}{r}\right) = \left(\frac{r^2}{\lambda} - f(\Psi)\right) f'(\Psi) \tag{2.4.29}$$

The limit conditions (2.4.23) and (2.4.24) for ψ become homogeneous and read

$$\frac{\partial \psi}{\partial r}, \frac{\partial \psi}{\partial x} \to 0; \qquad x \to -\infty \quad \text{or} \quad r \to \infty, \tag{2.4.30}$$

$$\frac{\partial \psi}{\partial x} \to 0; \qquad x \to +\infty. \tag{2.4.31}$$

We next introduce the Green function belonging to the differential operator L at the left hand side of (2.4.29), hence which is the solution of

$$L(K(x - \xi, r, \rho)) = -\delta(x - \xi)\delta(r - p), \tag{2.4.32}$$

where δ is the delta function of Dirac. In connection with (2.4.30) and

(2.4.31) K has to vanish at infinity. Using the method of Hankel transforms [52] the following solution can be found

$$K(x-\xi, r, \rho)=\tfrac{1}{2}\int_0^{\infty} e^{-|x-\xi|t} J_1(rt) J_1(\rho t)\, dt$$

$$=\frac{1}{2\pi\sqrt{r\rho}} Q_{1/2}\left(\frac{(x-\xi)^2+r^2+\rho^2}{2r\rho}\right), \qquad (2.4.33)$$

where $Q_{1/2}$ is a Legendre function of the second kind ([21] II page 203).
Using (2.4.33) we can write (2.4.29) as

$$\psi(x, r)= -r\iint_S K(x-\xi, r, \rho)\left\{\frac{\rho^2}{\lambda}-f(\Psi(\xi, \rho))\right\} f'(\Psi(\xi, \rho))\, d\xi\, d\rho, \qquad (2.4.34)$$

where the region of integration is S because outside S the tangential velocity $v_\varphi \equiv 0$ and hence by (2.4.14) $f(\Psi)=0$.

The boundary of the slipstream region S or of the streamtube which passes through the edge $x=0$, $r=1$ of the disk (dimensionless case) is defined by

$$\Psi(x, r)=\Psi(0, 1) \quad \text{or} \quad \psi(x, r)=\psi(0, 1)+\tfrac{1}{2}(1-r^2). \qquad (2.4.35)$$

A possible iteration method is as follows. First we prescribe just behind the disk some function $v_\varphi(+0, r)$ for the tangential velocity. Next we choose an approximation $\psi_0(x, r)$ of the perturbation stream function, hence we find by (2.4.35) an approximation of the slipstream region S which is the domain of integration in (2.4.34). Then also the function $f(\Psi)$ as a function of ξ and ρ is known, hence the integral in (2.3.34) can be calculated which yields another approximation $\psi_1(x, r)$ of $\psi(x, r)$. Repeating this procedure by replacing ψ_0 by ψ_1, yields again an approximation $\psi_2(x, r)$, etc.. Then it is hoped that this iteration will converge and ultimately give the desired solution of the problem.

We conclude this section by showing some of the numerical results obtained in [19] by using a method based on the preceding considerations. The calculations have been carried out, among others, for the case

$$rv_\varphi(+0, r)=C=\text{const.}, \qquad 0\le r\le 1. \qquad (2.4.36)$$

This means that the rotating bound vorticity at the disk $\Gamma(r)$ per unit of length in the φ direction, has the value C/r. This two dimensional vorticity field is for $r\neq 0$ and $r\neq 1$ without divergence, hence no free vorticity will be shed in the slipstream with the exception of the positive x axis where a concentrated hub vortex of strength $2\pi C$ is present and at the boundary of the slip stream. When we approximate the disk by a finite number of N concentrated radial bound vortices which rotate with angular velocity Ω in the plane of the disk, the strength of each of these vortices has to be independent of r namely $2\pi C/N$ $(0 \leq r \leq 1)$. Hence the disk under consideration is to some extent a representation of a screw propeller of which the circulation around its blades is independent of r. Because $v_\varphi = 0$ outside the slipstream, the function $rv_\varphi = f(\Psi)$, has a jump across the slipstream boundary, hence $\mathrm{d}f/\mathrm{d}\Psi$ has a delta function of Dirac character at this boundary. This makes it possible to carry out one of the integrations in (2.4.34) and a one dimensional integral equation is obtained. For more details of the calculations we refer to [19] where also the vortex representation of the disk is discussed.

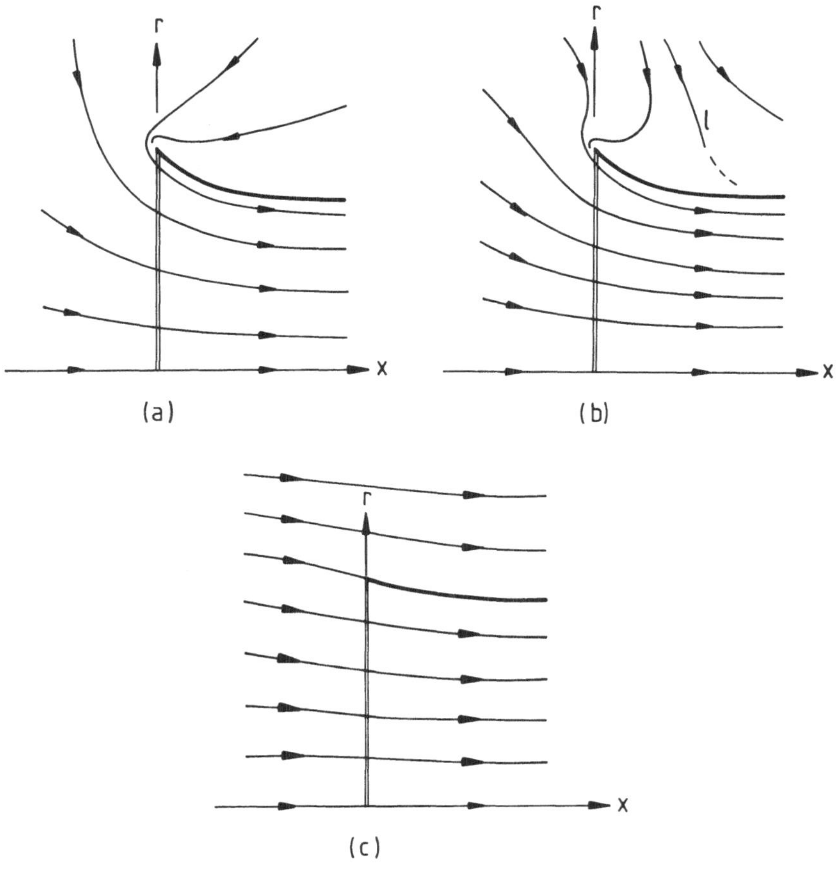

Fig. 2.4.1. Streamlines induced by actuator disks; $rv_\varphi(+0, r) = C$; (a) $\lambda = 0$; (b) $\lambda = 0.01$; (c) $\lambda = 0.1$.

In figure 2.4.1 are given the results for $rv_\varphi(+0, r) = C$ and for three values of $\lambda = U/\Omega b$ (2.4.25), $\lambda = 0$, $\lambda = 0.01$ and $\lambda = 0.1$. It is stated in [19] that the shape of the streamlines is nearly independent of the constant C.

It is seen that in the case of $\lambda = 0.01$ (figure 2.4.1b) a dividing stream line l seems to exist. Fluid particles to the left of l will pass through the disk while fluid particles to the right of l do not. From the calculations it was not clear if l ended at the boundary of the slipstream or that l turned downstream to $x = +\infty$. In section 2.6 we make it plausible that l ends at the boundary.

For small values of λ, which correspond to small values of the incoming velocity or to heavy loading of the disk, the boundary of the slipstream is very steep in the neighbourhood of the edge of the disk. It was assumed in the calculations that the vorticity at the boundary of the slipstream had a singular behaviour at the edge, which as a function of x had the character $0(x^{-1/2})$ for $x \to 0$. In the next section we return to the behaviour of the flow at the rim of the disk. In [19] still other radial loads or $rv_\varphi(+0, r)$ distributions have been considered so that a better approximation to the flow induced by a real propeller is obtained.

2.5. Non linear actuator disk theory, 2

We will now try to obtain insight in the singular behaviour of the flow at the edge of an actuator disk. We consider the special case of a constant normal load and no incoming velocity, hence $U = 0$, or in comparison with (2.4.25) $\lambda = 0$.

In section 1.3 we discussed that two different external force fields $\boldsymbol{F}_1$ and $\boldsymbol{F}_2$ will induce the same velocity field when (1.3.6)

$$\boldsymbol{F}_1 - \boldsymbol{F}_2 = \operatorname{grad} \psi, \tag{2.5.1}$$

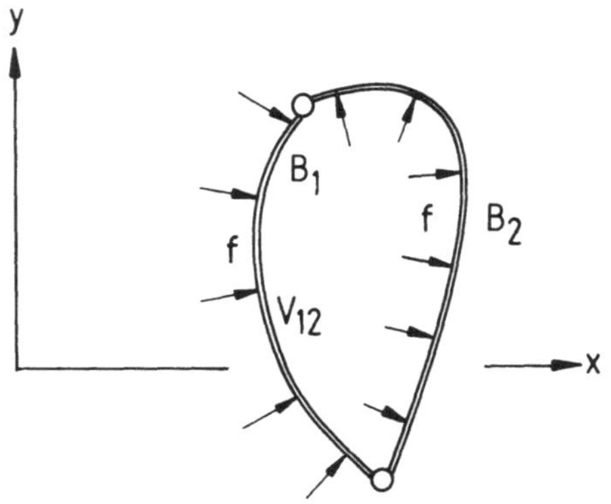

Fig. 2.5.1. Two different actuator surfaces with the same edge.

where ψ is some scalar function. Now consider two actuator surfaces B_1 and B_2 of arbitrary shape however with the same edge. In figure 2.5.1 is given a cross section of this situation. B_1 and B_2 are loaded by an equal normal load of constant strength f per unit of area. Then the assertion is that these actuator surfaces, when each of them is placed separately in an unbounded fluid, induce the same velocity field.

The proof is simple. The force field acting at B_i is called $\boldsymbol{F}_i$ $(i = 1, 2)$, its strength in space is a delta function of Dirac, concentrated at B_i. Hence their difference $\boldsymbol{F}_1 - \boldsymbol{F}_2$ is the difference of these delta functions. Now consider the scalar function ψ which is zero everywhere except in the region V_{12} between B_1 and B_2 where it has the value f. Then it follows that (2.5.1) is satisfied and hence B_1 and B_2 induce the same velocity fields only their pressure fields differ in the region V_{12} by the constant value f (1.3.7).

Note that this independency of the velocity field of the shape of the actuator surface holds also when the normal load is a function of time $f = f(t)$ and when B is time dependent, provided that the edge of B does not change.

Because the incoming velocity is assumed to be zero it seems by the foregoing considerations, that there is not a local direction at the edge of such an actuator surface which is specific for the problem. The equivalent actuator surfaces can meet the edge under any direction while their flow field is the same. Then the idea arises that also there cannot be a specific direction of the vortex sheet by which it leaves the edge. This happens when the vortex sheet has a spiral behaviour around it.

To check if this is true we consider the more simple case of the actuator half plane, then the problem is two dimensional. Schmidt and Sparenberg [50] have given an exact solution of this problem from which

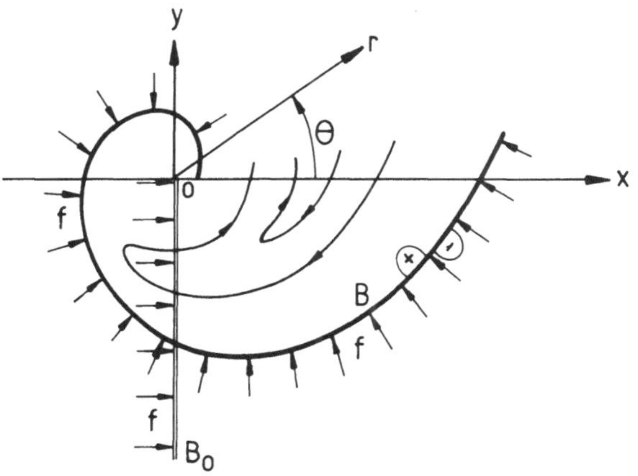

Fig. 2.5.2. The deformed actuator surface B with stream lines.

follows for this case, the singular behaviour of the flow at the edge. In order to fix the attention, we consider (figure 2.5.2), the half plane B_0 $(x=0, y \leq 0, -\infty < z < +\infty)$, on which the force field f per unit of area acts in the positive x direction. We now deform B_0 such that it lies along its own, still unknown free vortex sheet, which is assumed to spiral from the origin O (or in fact from the z axis) towards infinity. This spiral is the new actuator surface B, which replaces the original half plane B_0 and on which we assume the same normal load f. By definition of a free vortex sheet the normal component of the velocity at this sheet is zero, or $v_n = 0$ and because of the applied normal load the pressure at this sheet exhibits a jump $p^+ - p^- = [p]_-^+ = f$.

The problem being two dimensional, we use a complex representation in the (x, y) plane. In the remaining part of this section z will be

$$z = (x + iy) = r\,\mathrm{e}^{i\theta}. \tag{2.5.2}$$

We assume the vortex sheet B to have the representation

$$z = \mathrm{e}^{\lambda\theta}\,\mathrm{e}^{i\theta}, \qquad \lambda > 0, \qquad -\infty < \theta < +\infty. \tag{2.5.3}$$

In order to formulate the conditions for v_n and p at B, we introduce the complex velocity potential

$$h(z) = \varphi(x, y) + i\psi(x, y), \tag{2.5.4}$$

where φ is the real velocity potential and ψ the stream function. Then the velocity components v_x, v_y follow from

$$v_x - iv_y = \frac{\mathrm{d}h}{\mathrm{d}z}. \tag{2.5.5}$$

Because B is a streamline, $v_n = 0$, the stream function has to be constant along B. This constant we take zero, hence

$$\lim_{z \to B} \operatorname{Im} h(z) = 0. \tag{2.5.6}$$

In the whole field of flow we have $H = p/\mu + \frac{1}{2}v^2 = \text{const.}$ because the fluid particles do not cross anymore the actuator surface. Then the condition at B for the pressure becomes

$$[p]_-^+ = -\frac{\mu}{2}\left[\left|\frac{\mathrm{d}h}{\mathrm{d}z}\right|^2\right]_-^+ = f. \tag{2.5.7}$$

Consider the conformal mapping from the $z = x + iy$ plane to the $\zeta = \xi + i\eta$ plane (figure 2.5.3) by

$$z = e^{(\lambda + i)\zeta} = e^{(\lambda\xi - \eta)}\, e^{i(\xi + \lambda\eta)}. \tag{2.5.8}$$

This transformation maps the strip

$$0 < \operatorname{Im} \zeta < \frac{2\lambda\pi}{\lambda^2 + 1} \tag{2.5.9}$$

of the ζ plane uniquely into the region (z plane $- B$). On the region (2.5.9) we consider the transformed complex potential

$$g(\zeta) = h(z) = h(e^{(\lambda + i)\zeta}). \tag{2.5.10}$$

Condition (2.5.6) then becomes

$$\operatorname{Im} g(\zeta) = 0; \qquad \eta = 0, \qquad \eta = \frac{2\lambda\pi}{\lambda^2 + 1}. \tag{2.5.11}$$

In order to be able to apply condition (2.5.7) we first have to determine which points ζ_1 and ζ_2 on the two lines $\eta = 0$ and $\eta = 2\lambda\pi/(\lambda^2 + 1)$ in the ζ plane correspond to one point at B in the z plane. A substitution shows that this happens to be the case for the points

$$\zeta_1 = \xi, \quad \zeta_2 = \xi + \frac{2\pi}{(\lambda^2 + 1)} + i\frac{2\lambda\pi}{(\lambda^2 + 1)}, \qquad -\infty < \xi < +\infty. \tag{2.5.12}$$

Using the relation

$$\frac{dh}{dz} = \frac{e^{-(\lambda + i)\zeta}}{(\lambda + i)} \frac{dg}{d\zeta}, \tag{2.5.13}$$

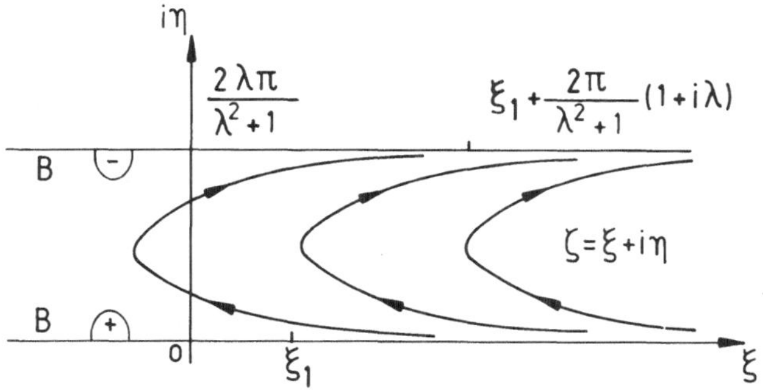

Fig. 2.5.3. Flow region in ζ plane with streamlines.

we can write (2.5.7) as

$$\frac{e^{-2\lambda\xi}}{(\lambda^2+1)}\left\{\left|\frac{dg}{d\zeta}\right|^2\Bigg|_{\zeta=\zeta_1}-\left|\frac{dg}{d\zeta}\right|^2\Bigg|_{\zeta=\zeta_2}\right\}=-\frac{2}{\mu}f. \tag{2.5.14}$$

As a function $g(\zeta)$ which satisfies (2.5.11) and (2.5.14) we try

$$g(\zeta)=\alpha\,e^{\beta\zeta}, \tag{2.5.15}$$

with α and β real. Then condition (2.5.11) becomes

$$\alpha\,e^{\beta\xi}\sin\beta\eta=0;\ -\infty<\xi<+\infty,\quad \eta=0 \text{ or } \eta=\frac{2\lambda\pi}{(\lambda^2+1)}. \tag{2.5.16}$$

Hence

$$\beta=\frac{(\lambda^2+1)}{2\lambda}k,\qquad k=\pm1,\qquad \pm2,\ \ldots\ . \tag{2.5.17}$$

Condition (2.5.14) yields

$$\alpha^2\beta^2\frac{e^{2(\beta-\lambda)\xi}}{(\lambda^2+1)}\left(1-e^{4\pi\beta/(\lambda^2+1)}\right)=-\frac{2}{\mu}f,\qquad -\infty<\xi<+\infty, \tag{2.5.18}$$

from which it follows

$$\lambda=\beta \tag{2.5.19}$$

and

$$\frac{\alpha^2\beta^2}{(\lambda^2+1)}\left(1-e^{4\pi\beta/(\lambda^2+1)}\right)=-\frac{2}{\mu}f. \tag{2.5.20}$$

Equations (2.5.17), (2.5.19) and (2.5.20) are three equations for the unknowns α, β and λ. Substituting (2.5.19) into (2.5.17), we find

$$(2-k)\lambda^2=k. \tag{2.5.21}$$

Because k assumes the values $\pm1,\ \pm2,\ldots,$ the only possibility is

$$\lambda=k=\beta=1. \tag{2.5.22}$$

Then from (2.5.20) we obtain by taking the positive root

$$\alpha=2\left\{\frac{f}{\mu}\left(e^{2\pi}-1\right)^{-1}\right\}^{1/2}. \tag{2.5.23}$$

After some elementary calculations we find

$$\left[\frac{dh}{dz}\right]_-^+ = \frac{\alpha}{(1+i)}\, e^{-i\xi}(1+e^{\pi}). \tag{2.5.24}$$

Because the velocities at both sides of B (figure 2.5.2) are tangential to B, the vorticity at B follows from (2.5.24). We find the constant value

$$\left(\frac{2}{\mu} f \frac{e^{\pi}+1}{e^{\pi}-1}\right)^{1/2} \tag{2.5.25}$$

reckoned positive with a right hand screw in anticlockwise direction.

The absolute value $|dh/dz|$ of the velocity is uniformly bounded in the whole plane

$$\frac{\alpha}{\sqrt{2}} \le \left|\frac{dh}{dz}\right| \le \frac{\alpha}{\sqrt{2}}\, e^{\pi}. \tag{2.5.26}$$

Since H is constant in the whole flowfield belonging to the deformed actuator surface B, the pressure p is also uniformly bounded. This is not true for the original half infinite flat actuator plane B_0 (figure 2.5.2) as we will discuss now.

The streamlines in the ζ plane are drawn in figure 2.5.3, which are the lines

$$\operatorname{Im} g(\zeta) = \alpha\, e^{\xi} \sin \eta = \text{const.} \tag{2.5.27}$$

These lines transformed back to the original (x, y) plane, are drawn in figure 2.5.2. They enter the spiral deeper and deeper and come out again. This means that for each integer $N > 0$ we can find streamlines which cut B_0, N consecutive times in one direction and then N times in the other direction. However when a fluid particle crosses B_0 in a direction opposite to f its pressure suddenly drops by the amount f. This means that very close to O there are regions with arbitrarily low pressures. It is not difficult to show that the pressure in case of B_0, tends logarithmically to minus infinity when we approach the edge of the disk.

By (2.5.3) and (2.5.22) we find for the spiral on which the vorticity is located

$$z = e^{\theta}(\cos\theta + i \sin\theta), \qquad -\infty < \theta < +\infty. \tag{2.5.28}$$

Hence for half a turn around the origin O, the radius is multiplied by

$$e^{\pi} \approx 23.2. \tag{2.5.29}$$

The above mentioned results have been used in [50] for numerical calculation of the flow pattern of a circular actuator disk of radius 1 with a constant normal load, placed in a fluid without incoming velocity. It has been assumed that sufficiently close to the edge the aforementioned spiralling behaviour (2.5.28) of the sheet dominates, while its vorticity is given by (2.5.25). The original disk is replaced by an actuator surface which is axi-symmetric and situated along the unknown boundary of the slipstream (figure 2.5.4a). This means in essence that we have calculated the shape of a rigid tube over which we have a constant pressure jump and inside of which we have at $x = +\infty$ a parallel flow in the positive x direction. Approaching the orifice from inside, the tube widens, curves backwards and tends spiralling to the edge of the original disk.

In figure 2.5.4 we have also drawn the computed streamlines of the flow pattern induced by the disk. At large distances from the disk and sufficiently away from the x axis the pattern resembles the flow induced by a sink. It is interesting to compare this with the remark in the last paragraph of section 1.7. The number at a streamline denotes the relative value of the streamfunction, it is the value of the streamfunction at that line divided by the value of the streamfunction at the boundary of the slipstream formed by the vortex sheet. In the case of zero incoming velocity, the shape of the vortex sheet and the streamlines depend neither on the intensity of the force field nor on the density of the fluid, as follows from dimensional analysis.

It is noted that in the neighbourhood of the edge we have a region where the fluid particles cross the disk more than once. This region occupies about 23 percent of the total area of the disk. The amount of fluid passing this region however is relatively very small, as follows from the number on each streamline. In other words the fluid is nearly stagnant at that region of the disk. This result gives perhaps some

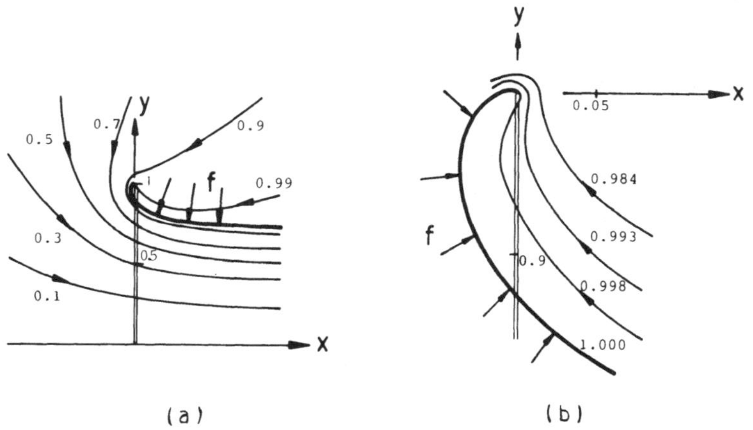

Fig. 2.5.4. Streamlines for constant normal load f and no incoming flow, (a) survey, (b) flow in neighbourhood of edge.

information on the flow of a propeller working in the static condition at bollard.

2.6. Discussion of results and methods

We now reconsider the general method of section 2.4 from the point of view of the more limited theory of section 2.5, which however could be carried out in greater detail. We start with the remark that the circular actuator disk of finite dimensions of section 2.5 is a limit case of the actuator disk considered in section 2.4. The value of $\lambda = U/\Omega b$ (2.4.20) has to be zero, while we have to let increase Ω unboundedly. Then the load of the disk, for $\Gamma(r) = v_\varphi(+0, r) = C/r$ with $C \to 0$ in a suitable way, becomes normal and constant over the disk. This means that it is not improbable that flow characteristics found in section 2.5 will also occur to some extent in the more realistic theory of section 2.4.

The difference in the calculated flow fields in the two sections is most striking at the edge of the disk. In [19] the flow is determined by solving a one dimensional integral equation (not discussed in section 2.4, formula (33) in [19]) which has as its domain $0 \leq x < \infty$. It is assumed there that the vorticity does not come in front of the disk. However it is seen that the sheet "tries to linger in the disk plane". Hence it seems not contradictory with these results that when more freedom is given to the shed vortex sheet it will come in front of the disk as is drawn in figure 2.5.4.

In section 2.4 the concept slipstream has been used. From the results of section 2.5 we find that this concept can be very complicated. By means of (2.4.36) a condition is put on the tangential velocity at the place where the slipstream starts, hence just behind the disk at $x = +0$, $0 \leq r \leq 1$. Now consider (figure 2.5.4b) the annular region at $x = +0$, in between the smallest radius where the sheet intersects the disk for the first time and the smallest radius but one, where the sheet intersects the disk for the second time. In that region the fluid particles have not yet crossed the disk or they crossed the disk the same number of times from $x = +0$ towards $x = -0$ as in the opposite direction. Now it is reasonable from continuity considerations, that such a region will also occur in the exact theory with a very slight swirl. Then however at the corresponding region the tangential velocity $(v_\varphi(+0, r) = 0$ and not C/r as in (2.4.36). At the opposite side of the disk hence for $x = -0$, for this annular region we find $v_\varphi(-0, r) = -C/r$ and not zero as is demanded in (2.4.15). Such phenomena will complicate very much an exact theory as described in section 2.4.

In figure 2.6.1 we have drawn the measured mean values of the velocity field induced by a four bladed aircraft propeller working in the static condition (incoming flow is zero) as given by Adams in [1]. The axial direction is drawn vertically and has an expanded scale. The velocity vectors however have been drawn at their true angle and show the relative magnitude of the velocity. The curvature of the flow for small values of r/R, where R is the radius of the propeller, is caused by the hub. It is seen that in the neighbourhood of the edge of the propeller disk there exists a reverse flow which is not unlikely to cross the disk from behind. Flow visualization by means of narrow jets of smoke confirm this phenomenon.

From the calculations carried out in [19], as shown in figure 2.4.1b for $\lambda = 0.01$, it seems that a dividing streamline l can occur. Also this phenomenon is compatible with the theory of section 2.5. There the disk was replaced by the actuator surface along the unknown free vortex sheet, across which now a pressure jump of strength f has to occur. This means that a cross section of this sheet with a plane through the x axis has become a "profile" with constant pressure jump. For profiles however dividing streamlines are common and they end at the stagnation point of the flow at the profile, hence in this case at the free vortex sheet.

It is suggested by these results that an infinitely thin profile with a constant pressure jump, as sometimes is used in hydrodynamics, has to have in the non linear theory a spiralling behaviour in the neighbourhood of the leading and of the trailing edge.

Another application of the change of the shape of an actuator disk without changing its induced velocity field, is given in [37] for the unsteady case of a pulsating disk placed in a nozzle (see section 3.6).

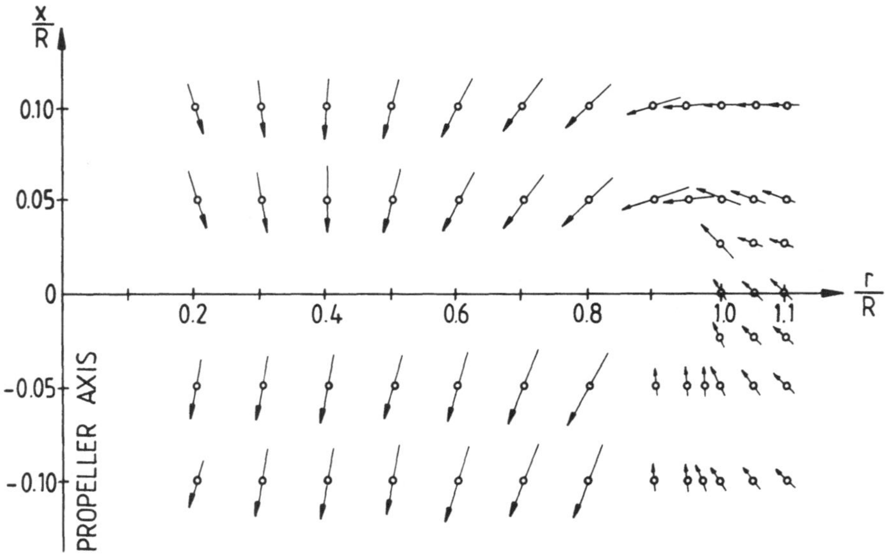

Fig. 2.6.1. Experimental investigation of velocity field.

3. The ship screw

Our next subject will be the ship screw which is upto now the most important device for hydrodynamic propulsion. It consists of a number of helicoidally shaped blades connected to a hub. The number of blades can vary from two upto about six. The hub is mounted on a shaft (figure 3.1) which is rotated by the engine. The blades have to be designed so that at a given rotational velocity, a prescribed thrust is produced which moves the ship with the desired speed. This yields a difficult problem because the screw propeller behind a ship is working in a field of flow which differs from a homogeneous flow for several reasons. We indicate some of these.

First, the water has to follow the ship's form, hence it has to converge at the stern. Second, the water flowing closely along the ship is dragged with it by viscosity and becomes turbulent. Third, the wave pattern induced by the hull at the free surface causes a velocity field which varies with depth. When at the stern a crest is formed this velocity field is in the direction of the motion of the ship. When a trough is formed the inverse

Fig. 3.1. A three bladed screw propeller.

happens. The resulting disturbances, in the absence of the propeller, are called the nominal wake of the hull ([10] page 388). It has to be observed that when the propeller is present and developing thrust, this wake is changed.

The hull of the ship has influence still in another way. As a rigid boundary of the flow domain it will hamper the water to be set into motion by the propeller. The same holds with respect to the rudder. The free surface also acts as a boundary of the region in which the propeller operates.

As a result of the mentioned inhomogeneities of the inflow and because of the motion of the propeller with respect to the boundaries of its region of operation unsteady loading of the propeller blades will occur and possibly also unsteady cavitation. The latter phenomenon is undesirable from the point of view of erosion caused by the collapsing cavitation bubbles.

Another essential difficulty in propeller theory is that for realistic ship screws non-linear effects are important. For instance the contraction of the flow behind the propeller (figure 2.4.1b), which deforms the free vortex sheets shed by the blades. So, in design procedures, often a linear lifting surface theory for propeller blades is only used to obtain corrections of the blade profiles which are calculated by means of a "more or less" non linear lifting line theory.

Considering such difficulties it seems sensible in a first approach to the problem to make simplifications. Our theory will be linear, we neglect viscosity and assume no influence of the ship's hull and the free surface. The incoming flow will be homogeneous. We neglect the influence of the hub because the thrust is mostly delivered by parts of the blades which are not too close to the axis of rotation. Hence the blades are moving freely through the fluid, however along prescribed helicoidal paths. We also assume that the fluid keeps contact with the blades, so that no cavitation as a result of low pressures will occur. Finally we consider only one blade, this already shows the mathematical difficulties which can be encountered.

In the last section of this chapter we refer to some articles which take into consideration one or more of the aspects of real ship screws, which have been neglected in our simple model. However a huge amount of literature is devoted to the screw propeller itself and to its interaction with the environment in which it is situated, so only a very small part of the existing literature is mentioned. Also in that section some more complicated propeller arrangements are touched upon.

3.1. The geometry of the ship screw

We will give a description of the geometry of a ship screw, adapted to our intention to derive a linear hydrodynamical theory. For a more technical description we refer to [10] page 397. We use a cylindrical coordinate system (x, r, φ) of which the x axis is along the axis of rotation of the screw (figure 3.1.1). With respect to the coordinate system we have an incoming homogeneous parallel flow with velocity U in the positive x direction. Our first aim is to find impermeable surfaces which can rotate about the x axis without disturbing the parallel flow.

Suppose we have a surface

$$G(x, r, \varphi\ t) = 0 \tag{3.1.1}$$

moving through a fluid with velocity field $(U + v_x, v_r, v_\varphi)$ (components in x, r and φ direction). We ask for the conditions which the velocity field has to satisfy in order that the fluid flows along this surface. Consider $G = G(x, r, \varphi, t)$ as a function defined in space. Then each particle of the fluid perceives at the place where it is at a certain moment, some value of G. When this particle moves on, the rate of change of this value is calculated as

$$\frac{dG}{dt} = \frac{\partial G}{\partial t} + (U + v_x)\frac{\partial G}{\partial x} + v_r\frac{\partial G}{\partial r} + v_\varphi\frac{\partial G}{r\partial\varphi}. \tag{3.1.2}$$

However a particle moving along the surface (3.1.1) has to observe the

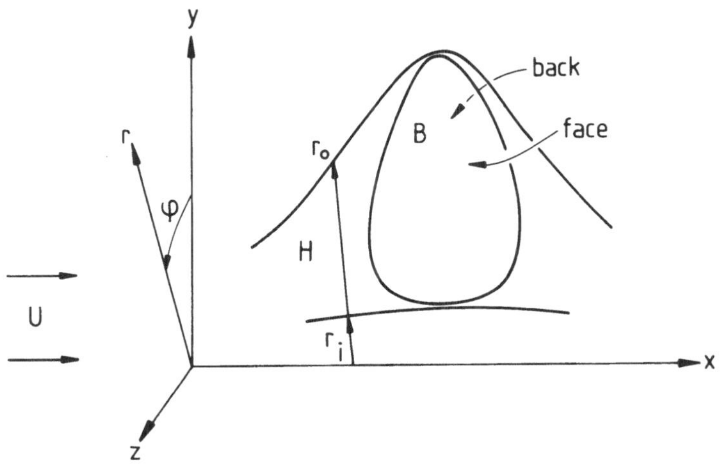

Fig. 3.1.1. The planform B of a screwpropeller blade.

constant value $G=0$ during its motion. This means that the velocity $(U+v_x, v_r, v_\varphi)$ of it has to satisfy

$$\frac{\partial G}{\partial t}+(U+v_x)\frac{\partial G}{\partial x}+v_r\frac{\partial G}{\partial r}+v_\varphi\frac{\partial G}{r\partial\varphi}=0. \tag{3.1.3}$$

Now consider a rigid surface rotating around the x axis

$$G(x, r, \varphi-\omega t)=0, \tag{3.1.4}$$

where G is an arbitrary function of the three arguments x, r and $\varphi-\omega t$. In order that this surface does not disturb the homogeneous incoming parallel flow, it has to satisfy (3.1.3) with $v_x=v_r=v_\varphi=0$, hence

$$-\omega\frac{\partial G}{\partial\varphi}(x, r, \varphi-\omega t)+U\frac{\partial G}{\partial x}(x, r, \varphi-\omega t)=0. \tag{3.1.5}$$

The general solution of this equation is

$$G=G(\varphi+ax-\omega t, r), \qquad a=\frac{\omega}{U}, \tag{3.1.6}$$

where now G is an arbitrary function of the two arguments $(\varphi+ax-\omega t)$ and r. Hence by (3.1.1) $G(\varphi+ax-\omega t, r)=0$ is a surface rotating around the x axis which does not disturb the fluid. We restrict ourselves in this chapter to the special case of the simple helicoidal surface

$$H=\varphi+ax-\omega t=0, \qquad \omega>0. \tag{3.1.7}$$

This surface will be called the helicoidal reference surface.

On H we choose as a two dimensional coordinate system x and r. This means that a point (x, r) at H is the point $(x, r, -ax+\omega t)$ in three dimensional space. Then we define on H the planform **B** of the propeller blade by

$$B\colon x_l(r)\le x\le x_t(r), \qquad r_i\le r\le r_0, \tag{3.1.8}$$

where $x_l(r)$ and $x_t(r)$ are given functions.

The planform (3.1.7), (3.1.8) as an impermeable rigid surface of finite extent does not disturb the incoming flow. A screw blade however has to produce thrust and being a body of finite extent it has to shed vorticity (section 1.2) and hence disturbs the parallel flow. Because we want to develop a linear theory, the assumption is made that the disturbance velocities (v_x, v_r, v_φ) are small. This happens when the blade is in the neighbourhood of the planform B. We represent the propeller blade by

$$\varphi - \omega t + ax + \varepsilon f_j(x, r) = 0, \qquad (x, r) \in B, \qquad j = 1, 2, \tag{3.1.9}$$

$$f_1(x, r) \geq f_2(x, r), \tag{3.1.10}$$

$$f_1(x_l(r), r) = f_2(x_l(r), r), \qquad f_1(x_t(r), r) = f_2(x_t(r), r), \tag{3.1.11}$$

where $\varepsilon f_1(x, r)$ and $\varepsilon f_2(x, r)$ describe the small deviations of the "back" and the "face" respectively of the blade, from the planform. The number ε is a small parameter, used to linearize the theory, we will neglect quantities of $O(\varepsilon^{n+1})$ with respect to quantities of $O(\varepsilon^n)$, $n = 0, 1, 2, \ldots$ The leading edge of the blade consists of points $(x_l(r), r, \varphi)$ which satisfy

$$\varphi - \omega t + ax_l(r) + \varepsilon f_1(x_l(r), r) = 0. \tag{3.1.12}$$

The trailing edge is defined analogously with $x_l(r)$ replaced by $x_t(r)$.

We introduced below (3.1.11) the back and the face of a blade. A more physical definition reads, the back of a blade is its low pressure side, the face of a blade is its high pressure side. This characterization is independent of $\omega > 0$ in (3.1.7).

The thickness D of the blade for a point P of the planform B will now be defined. Determine the points of intersection of the normal at B through P, with the back and the face of the blade. The thickness D will be the distance between these two points of intersection. Introducing the unit normal $\boldsymbol{n}$ on the planform, pointing in the direction of decreasing values of φ,

$$\boldsymbol{n} = (n_x, n_r, n_\varphi) = -\frac{(ar, 0, 1)}{(1 + a^2 r^2)^{1/2}}, \tag{3.1.13}$$

D is found to be equal to

$$D(x, r) = \varepsilon r \frac{\{f_1(x, r) - f_2(x, r)\}}{(1 + a^2 r^2)^{1/2}}, \qquad (x, r) \in B. \tag{3.1.14}$$

At the leading edge and the trailing edge $D = 0$ (3.1.11).

The mean plane H_m, lying in the middle of the back and the face is given by

$$H_m\colon \varphi - \omega t + ax + \tfrac{1}{2}\varepsilon(f_1(x, r) + f_2(x, r))$$

$$\stackrel{\text{def}}{=} \varphi - \omega t + ax + \varepsilon f_3(x, r) = 0, \tag{3.1.15}$$

where $\varepsilon f_3(x, r)$ is called the camber of the blade. Next we introduce the geometrical local angle of incidence α_i of H_m at some point $(\tilde{x}, \tilde{r}, \tilde{\varphi}) \in H_m$. This is the angle formed by the helicoidal line at H for $r = \tilde{r}$ at the point $(\tilde{x}, \tilde{r}) \in B$ with the tangent plane of H_m at $(\tilde{x}, \tilde{r}, \tilde{\varphi})$. This angle follows from the scalar product of the unit vector tangent to the helicoidal line at $(\tilde{x}, \tilde{r}) \in$ B with the unit vector normal at H_m at $(\tilde{x}, \tilde{r}, \tilde{\varphi})$,

$$\alpha_i(x, r) = -\frac{(1, 0, -ar)}{(1 + a^2r^2)^{1/2}} \frac{\left\{r\left(a + \varepsilon\frac{\partial}{\partial x}f_3\right), r\varepsilon\frac{\partial}{\partial r}f_3, 1\right\}}{\left\{r^2\left(a + \varepsilon\frac{\partial}{\partial x}f_3\right)^2 + \left(r\varepsilon\frac{\partial}{\partial r}f_3\right)^2 + 1\right\}^{1/2}}$$

$$= \frac{-\varepsilon r}{(1 + a^2r^2)} \frac{\partial}{\partial x} f_3(x, r) + \mathrm{O}(\varepsilon^2). \tag{3.1.16}$$

A profile of the blade is defined as the cross section of the blade with a cylinder $r =$ const.. The skeleton line of the profile is the intersection of the middle plane H_m (3.1.15) with this cylinder.

At the back and the face of the blade we have to satisfy the boundary conditions for the flow, stating that the fluid velocities are tangent to these surfaces. Because our theory is linear, these conditions will not be demanded at points $(\tilde{x}, \tilde{r}, \tilde{\varphi})$ of the back or the face but at points $(\tilde{x}, \tilde{r})$ of the planform B.

In the following we generally identify the blade of the propeller and its planform B. A chord of the blade is by definition a line $r =$ const. at B. Chord lines can be chosen as reference lines for the profiles because they lie in their neighbourhood and are in the direction of the undisturbed relative fluid velocity.

We now state the problem which we want to solve by lifting surface theory. Assume that the load Q, which is defined as the difference in pressure between the back and the face of the blade, is given as a function of position

$$Q = \tilde{Q}(x, r), \qquad (x, r) \in B. \tag{3.1.17}$$

Besides this we assume that, for instance by demands on strength and stiffness of the blade, its thickness D is known

$$D = \tilde{D}(x, r), \qquad (x, r) \in B. \tag{3.1.18}$$

Then the question is, how do we have to choose the middle surface H_m of the blade so that we obtain the desired load (3.1.17). When H_m is found we can construct the back and the face of the blade, because the thickness is known.

This problem can be split into two separate parts. First, what is the camber $\varepsilon f_{3D}(x, r)$ (3.1.15) which yields a load $Q(x, r)=0$, while the blade has the thickness $\tilde{D}(x, r)$. Second, what is the camber $\varepsilon f_{3Q}(x, r)$ which yields the prescribed load $\tilde{Q}(x, r)$ while $D(x, r)=0$. Then the total camber needed to satisfy (3.1.17) under the demand (3.1.18) follows from

$$f_3(x, r)=f_{3D}(x, r)+f_{3Q}(x, r). \tag{3.1.19}$$

This is allowed because our theory is linear.

The splitting of the problem into two independent parts is interesting from the following point of view. By changing either the thickness D or the load Q separately, we need to take into account the changing quantity only. A multiplication of D or Q by a constant results in a simple multiplication of f_{3D} or f_{3Q} by the same constant. When instead of a screw propeller we consider a flat wing we have as is well known, $f_{3D}=0$.

3.2. The screw blade with thickness and without load

We discuss now the first part of the problem as described in the last paragraph but one of the previous section; the construction of a screw blade with a prescribed thickness distribution $D(x, r)\neq 0$, without pressure difference between back and face or $Q(x, r)=0$. Hence we have to calculate for $Q=0$ the mean surface f_{3D}.

Consider a layer of sources placed at the planform B. We will show that such a layer is sufficient to represent the induced flow in this case. The layer induces a disturbance potential Φ (1.7.3)

$$\Phi=-\frac{1}{4\pi}\iint_B \frac{\sigma}{R}\,\mathrm{d}S, \tag{3.2.1}$$

where σ is the local strength per unit of area of the source layer and R is the distance from the point where we calculate the potential towards the element of area $\mathrm{d}S$. The limiting values of the normal derivative of Φ at the back (1) and at the face (2) respectively become ([32], page 164)

$$\frac{\partial \Phi_1}{\partial n}=\frac{\sigma}{2}-\frac{1}{4\pi}\iint_B \sigma\frac{\partial}{\partial n}\frac{1}{R}\,\mathrm{d}S, \tag{3.2.2}$$

$$\frac{\partial \Phi_2}{\partial n}=\frac{-\sigma}{2}-\frac{1}{4\pi}\iint_B \sigma\frac{\partial}{\partial n}\frac{1}{R}\,\mathrm{d}S, \tag{3.2.3}$$

where $\partial/\partial n$ means differentiation in the direction of the normal $\boldsymbol{n}$ given in (3.1.13). The double integrals exist because the singular behaviour of the integrand is only $O(R^{-1})$. For the difference of the normal components of the disturbance velocity at both sides of the blade B we find from (3.2.2) and (3.2.3) the well known relation

$$\frac{\partial \Phi_1}{\partial n} - \frac{\partial \Phi_2}{\partial n} = \sigma. \tag{3.2.4}$$

The condition that the fluid flows along the back and along the face of the blade, follows from substitution of (3.1.9) in (3.1.3). This yields when we neglect terms of $O(\varepsilon^2)$,

$$\varepsilon U \frac{\partial f_j}{\partial x} = -\left(a v_{xj} + \frac{1}{r} v_{\varphi j} \right), \qquad j = 1, 2, \tag{3.2.5}$$

where v_{xj} and $v_{\varphi j}$ denote the disturbance velocity at the back $j = 1$ or at the face $j = 2$. The difference of the normal components of the disturbance velocity at both sides has the value

$$\frac{\partial \Phi_1}{\partial n} - \frac{\partial \Phi_2}{\partial n} = \boldsymbol{n} \cdot (v_{x1} - v_{x2}, v_{r1} - v_{r2}, v_{\varphi 1} - v_{\varphi 2}). \tag{3.2.6}$$

Using (3.2.4), (3.1.13) and (3.2.5) we find from (3.2.6)

$$\sigma = \frac{\varepsilon U r}{(1 + a^2 r^2)^{1/2}} \cdot \left(\frac{\partial f_1}{\partial x} - \frac{\partial f_2}{\partial x} \right). \tag{3.2.7}$$

By the definition of the thickness D (3.1.14) of the blade, this can be written as

$$\sigma(x, r) = U \frac{\partial}{\partial x} D(x, r) = U(1 + a^2 r^2)^{1/2} \frac{\partial D}{\partial s}(x, r), \tag{3.2.8}$$

where s is a length parameter along a helicoidal line in the planform B, hence for $r = \text{const.}$.

This formula could have been derived more directly. The quantity

$$U(1 + a^2 r^2)^{1/2} = (U^2 + \omega^2 r^2)^{1/2}, \tag{3.2.9}$$

is the relative velocity of the fluid with respect to B. The right hand side of (3.2.8) can be interpreted as the difference in normal velocity of the fluid at both sides of the blade, because $\partial D / \partial s$ is the difference in slope of the sides of the blade. Then by (3.2.4) we obtain (3.2.8).

Now we derive a relation between the middle surface H_{m} (3.1.15) and the disturbance velocities. Consider the sum of the normal components of the disturbance velocities,

$$\frac{\partial \Phi_1}{\partial n} + \frac{\partial \Phi_2}{\partial n} = \boldsymbol{n} \cdot \left(v_{x1} + v_{x2},\, v_{r1} + v_{r2},\, v_{\varphi 1} + v_{\varphi 2} \right). \tag{3.2.10}$$

Using (3.2.2), (3.2.3), (3.1.13), (3.2.5) and (3.2.15) we find

$$\frac{\varepsilon r U}{\left(1 + a^2 r^2\right)^{1/2}} \frac{\partial}{\partial x} f_{3D} = \varepsilon r U \frac{\partial f_{3D}}{\partial s} = -\frac{1}{4\pi} \iint_B \sigma \frac{\partial}{\partial n} \frac{1}{R}\, \mathrm{d}S. \tag{3.2.11}$$

The physical meaning of this result can be understood as follows. We calculate the normal component of the relative flow with velocity $U(1 + a^2 r^2)^{1/2}$ with respect to the middle plane H_m, from (3.1.16) we find the value

$$-\frac{\varepsilon r U}{\left(1 + a^2 r^2\right)^{1/2}} \frac{\partial}{\partial x} f_{3D} = -\varepsilon r U \frac{\partial}{\partial s} f_{3D}. \tag{3.2.12}$$

This however equals minus the left hand side of (3.2.11). Hence when we disturb the parallel flow by the second term of the right hand side of (3.2.2) or of (3.2.3), which is the influence of sources at a distance from the point under consideration, there results a flow tangent to the middle surface. The first term of the right hand sides of (3.2.2) and (3.2.3), which is the influence of the source distribution at the place of the point under consideration, takes care of the difference in slope at both sides of the blade.

The middle surface H_m is not defined uniquely by (3.2.11) and (3.2.8). We have the freedom to choose at the planform B some suitable line along which H_m is supposed to intersect B, for instance a line $x = \text{const.}$ Then by integration with respect to the length parameter s along helicoidal lines $r = \text{const.}$, we determine f_{3D} from (3.2.11), hence H_m is known.

Summing up we arrive at the following result. From the prescribed thickness distribution $D = D(x, r)$ of the blade, we find by (3.2.8) the source distribution $\sigma(x, r)$ (sinks when $\sigma < 0$) on the planform B. Then by (3.2.11) we construct a middle surface H_m, around which we have to build symmetrically the blade with the prescribed thickness.

One question is left; are the pressures at both sides of the blade constructed in this way, equal to each other so that $Q = 0$? To answer this question we consider the linearized version of Bernoulli's law (1.1.6) for the instationary case,

$$p = -\mu\left(U \frac{\partial \Phi}{\partial x} + \frac{\partial \Phi}{\partial t} \right), \tag{3.2.13}$$

which we apply in the neighbourhood of the blade. Because the blade rotates, hence $\Phi = \Phi(x, r, \varphi - \omega t)$, we can replace the derivative with respect to t by a derivative with respect to φ,

$$-\frac{p}{\mu} = \left(U\frac{\partial \Phi}{\partial x} - \frac{\partial \Phi}{r \partial \varphi} r\omega \right). \tag{3.2.14}$$

This is the inner product of grad Φ and $(U, 0, -\omega r)$. The latter vector is the relative velocity of a particle of the undisturbed flow with respect to the planform B. This means that (3.2.14) is the rate of change of Φ observed by a fluid particle moving with respect to B along a helicoidal line ($r = \text{const.}$) with a velocity $(U^2 + \omega^2 r^2)^{1/2}$. Hence we can write

$$-\frac{p}{\mu} = (U^2 + \omega^2 r^2)^{1/2} \frac{\partial \Phi}{\partial s}. \tag{3.2.15}$$

It is well known however that a tangential derivative of the potential of a layer of sources is continuous across the layer. Hence p has at both sides of the blade the same value.

The specific influence of the thickness of a screwblade in lifting surface theory is discussed for instance by Jacobs and Tsakonas in [30] and by Kerwin and Leopold in [36]. It turns out that for conventional screw propellers this influence on the total camber of the blade is small.

3.3. The velocity field induced by rotating force

After having discussed the screw blade with thickness and zero load we consider the blade with load and zero thickness. First we determine in this section explicitly the Green function used in the integral representation of the geometry of such a blade. This Green function is the velocity field induced by a rotating unit force acting at the fluid, perpendicular to and at a fixed point of the rotating planform B (3.1.7) and (3.1.8).

The position of the point of application of the force, will be

$$x = \xi, \qquad r = \rho, \qquad \varphi = \omega t - a\xi. \tag{3.3.1}$$

The components of the force in the local directions of the cylindrical coordinate system are

$$\boldsymbol{h} = (h_x, h_r, h_\varphi) = -h\boldsymbol{n} = \frac{h \cdot (a\rho, 0, 1)}{(1 + a^2\rho^2)^{1/2}}, \tag{3.3.2}$$

where h is the magnitude of the force and $\boldsymbol{n}$ is defined in (3.1.13). For $h > 0$ the reaction at the screw blade is a propulsive force.

We now want to use (1.4.21). There is one slight complication, in section 1.4 the fluid is assumed to be at rest at infinity with respect to the coordinate system while here we have an incoming parallel flow with velocity U. When however we refer our considerations to a translating coordinate system $(\tilde{x}, \tilde{r}, \tilde{\varphi})$ with

$$\tilde{x} = x - Ut, \qquad \tilde{r} = r, \qquad \tilde{\varphi} = \varphi, \tag{3.3.3}$$

the incoming flow has disappeared and the force moves with a constant velocity $V = U(1 + a^2\rho^2)^{1/2}$, along the helicoidal line

$$\tilde{r} = \rho, \qquad \tilde{\varphi} + a\tilde{x} = 0, \tag{3.3.4}$$

along which it sheds its vorticity.

The force is assumed to rotate already an infinitely long time, hence it follows from (3.3.3) and (3.3.4) that in the original (x, r, φ) system its free vorticity stretches along the helicoidal line $L^*(t)$

$$L^*(t): r = \rho, \qquad \varphi - \omega t + ax = 0, \qquad x \geq \xi. \tag{3.3.5}$$

It is now easily concluded from (1.4.21) that for $t = 0$, hence when the force is at the point $(\xi, \rho, -a\xi)$ the induced velocity field with respect to the (x, r, φ) system, reads

$$\begin{aligned} \boldsymbol{v}(x, r, \varphi, t = 0) &= (v_x, v_r, v_\varphi) \\ &= -\frac{1}{4\pi\mu}\int_L \left\{ \frac{\boldsymbol{h}(s)}{VR^3} - \frac{3\boldsymbol{R}\cdot(\boldsymbol{h}(s)\cdot\boldsymbol{R})}{VR^5} \right\} \mathrm{d}s, \end{aligned} \tag{3.3.6}$$

where $L = L^*(0)$ and s is a length parameter along L.

In order to carry out the integrations in (3.3.6), we will take the components of the vectors in the integrand with respect to the coordinate directions at the point (x, r, φ) at which we calculate the velocity. We denote the components of $\boldsymbol{h}$ at the point $(\tilde{\xi}, \rho, \tilde{\theta})$ of L by $(h_{\tilde{\xi}}, h_\rho, h_{\tilde{\theta}})$ and the components of the same vector $\boldsymbol{h}$ with respect to the coordinate directions at (x, r, φ) by (h_x, h_r, h_φ), it is easily found that

$$\begin{aligned} &h_x = h_{\tilde{\xi}}, \qquad h_r = h_\rho \cos(\varphi - \tilde{\theta}) + h_{\tilde{\theta}} \sin(\varphi - \tilde{\theta}), \\ &h_\varphi = -h_\rho \sin(\varphi - \tilde{\theta}) + h_{\tilde{\theta}} \cos(\varphi - \tilde{\theta}). \end{aligned} \tag{3.3.7}$$

For the components of the vector $\boldsymbol{R}$ from $(\tilde{\xi}, \rho, \tilde{\theta})$ towards (x, r, φ) we find

$$\boldsymbol{R} = (R_x, R_r, R_\varphi) = ((x - \tilde{\xi}), r - \rho\cos(\varphi - \tilde{\theta}), \rho\sin(\varphi - \tilde{\theta})). \tag{3.3.8}$$

Substitution of (3.3.7) and (3.3.8) into (3.3.6), with $\tilde{\theta} = -a\tilde{\xi}$ and choosing $\tilde{\xi}$ as the variable of integration, we obtain

$$\boldsymbol{v}(x, r, \varphi, t=0) = (v_x, v_r, v_\varphi)$$

$$= -\frac{1}{4\pi\mu}\frac{h}{U(1+a^2\rho^2)^{1/2}}\int_{\xi}^{\infty}\left[\frac{\{a\rho, \sin(\varphi + a\tilde{\xi}), \cos(\varphi + a\tilde{\xi})\}}{R^3}\right.$$

$$-3\{x - \tilde{\xi}, r - \rho\cos(\varphi + a\tilde{\xi}), \rho\sin(\varphi + a\tilde{\xi})\}$$

$$\left.\times\frac{\{a\rho(x-\tilde{\xi}) + r\sin(\varphi + a\tilde{\xi})\}}{R^5}\right]\mathrm{d}\tilde{\xi}, \tag{3.3.9}$$

where $R = |\boldsymbol{R}|$ and we have used $V(s) = U(1 + a^2\rho^2)^{1/2} = \text{const.}$ and $\mathrm{d}s = (1 + a^2\rho^2)^{1/2}\,\mathrm{d}\tilde{\xi}$.

In order to bring this integral into a form in which it often occurs in literature we put $\tilde{\xi} = -\tau + x$, then

$$\boldsymbol{v}(x, r, \varphi, t=0) = (v_x, v_r, v_\varphi)$$

$$= \frac{1}{4\pi\mu}\frac{h}{U(1+a^2\rho^2)^{1/2}}$$

$$\times\int_{-\infty}^{(x-\xi)}\left[\frac{-\{a\rho, \sin(\varphi + a(x-\tau)), \cos(\varphi + a(x-\tau))\}}{R^3}\right.$$

$$+3\{\tau, r - \rho\cos(\varphi + a(x-\tau)), \rho\sin(\varphi + a(x-\tau))\}$$

$$\left.\times\frac{\{a\rho\tau + r\sin(\varphi + a(x-\tau))\}}{R^5}\right]\mathrm{d}\tau, \tag{3.3.10}$$

where

$$R = \{\tau^2 + r^2 + \rho^2 - 2r\rho\cos(\varphi + a(x-\tau))\}^{1/2}. \tag{3.3.11}$$

We will give another representation of this velocity field by using the vortex model of section (1.6). In this case the velocity field consists of two parts. First we have a contribution by the short bound vortex at the place of the force, given by the upper bound in (1.6.10). Relevant quantities in that formula are $V(s) = U(1 + a^2\rho^2)^{1/2}$ and the vector $\boldsymbol{k}$

pointing in the $+\rho$ direction, hence analogous to (3.3.7)

$$\begin{aligned} \boldsymbol{k} &= \left(k_\xi, k_\rho, k_\theta\right) = (0, 1, 0), \\ \boldsymbol{k} &= \left(k_x, k_r, k_\varphi\right) = (0, \cos(\varphi-\theta), \; -\sin(\varphi-\theta)), \end{aligned} \tag{3.3.12}$$

where $\theta = -a\xi$. The vector $\boldsymbol{R}$ is given by (3.3.8), omitting the tilde because here we consider the point $(\xi, \rho, -a\xi)$ where at $t=0$ the force is acting. In this way we find for the contribution $\boldsymbol{v}_1$, of the upper bound of (1.6.10)

$$\boldsymbol{v}_1(x, r, \varphi, t=0) = \left(v_{1x}, v_{1r}, v_{1\varphi}\right) = \frac{-h}{4\pi\mu U\left(1+a^2\rho^2\right)^{1/2}}$$
$$\times \frac{\{r\sin(\varphi+a\xi), \; -(x-\xi)\sin(\varphi+a\xi), \; -(x-\xi)\cos(\varphi+a\xi)\}}{\left\{(x-\xi)^2 + r^2 + \rho^2 - 2r\rho\cos(\varphi+a\xi)\right\}^{3/2}}. \tag{3.3.13}$$

Second, the contribution $\boldsymbol{v}_2$ from the vorticity along L. From the left hand side of (1.6.13) it follows

$$\boldsymbol{v}_2(x, r, \varphi, t=0) = \frac{h}{4\pi\mu U\left(1+a^2\rho^2\right)^{1/2}}$$
$$\times \int_\xi^\infty \frac{\mathrm{d}}{\mathrm{d}\lambda}\left\{\left(\boldsymbol{i}+\lambda\frac{\mathrm{d}\boldsymbol{k}}{\mathrm{d}s}\right) * \frac{(\boldsymbol{R}-\lambda\boldsymbol{k})}{|\boldsymbol{R}-\lambda\boldsymbol{k}|^3}\right\}\Bigg|_{\lambda=0} \cdot \left(1+a^2\rho^2\right)^{1/2} \mathrm{d}\tilde{\xi}, \tag{3.3.14}$$

where we changed from $\mathrm{d}s$ to $\mathrm{d}\tilde{\xi} = (1+a^2\rho^2)^{-1/2}\,\mathrm{d}s$. First we determine the quantities in the integrand, $\boldsymbol{R}$ is given in (3.3.8),

$$\boldsymbol{i} = \left(i_{\tilde{\xi}}, i_\rho, i_{\tilde{\theta}}\right) = \frac{(-1, 0, a\rho)}{\left(1+a^2\rho^2\right)^{1/2}},$$

$$\boldsymbol{i} = \left(i_x, i_r, i_\varphi\right) = \frac{\left(-1, a\rho\sin(\varphi+a\tilde{\xi}), a\rho\cos(\varphi+a\tilde{\xi})\right)}{\left(1+a^2\rho^2\right)^{1/2}}, \tag{3.3.15}$$

from (3.3.12) with $\theta \to \tilde{\theta}$ and $\tilde{\theta} = -a\tilde{\xi}$

$$\frac{\mathrm{d}\boldsymbol{k}}{\mathrm{d}s} = \frac{\mathrm{d}\boldsymbol{k}}{\mathrm{d}\tilde{\xi}}\frac{\mathrm{d}\tilde{\xi}}{\mathrm{d}s} = \frac{1}{\left(1+a^2\rho^2\right)^{1/2}}\left(0, \; -a\sin(\varphi+a\tilde{\xi}), \; -a\cos(\varphi+a\tilde{\xi})\right), \tag{3.3.16}$$

which are the components in the local coordinate directions at (x, r, φ).

We now carry out the vector product in (3.3.14) and find

$$\boldsymbol{v}_2(x, r, \varphi, t=0) = (v_{2x}, v_{2r}, v_{2\varphi}) = \frac{h}{4\pi\mu U(1+a^2\rho^2)^{1/2}} \frac{\mathrm{d}}{\mathrm{d}\lambda}$$

$$\int_\xi^\infty \{a(\rho+\lambda)^2 - ar(\rho+\lambda)\cos(\varphi + a\tilde{\xi}), (x-\xi)a(\rho+\lambda)\cos(\varphi + a\tilde{\xi})$$

$$+(\rho+\lambda)\sin(\varphi + a\tilde{\xi}), -r + (\rho+\lambda)\cos(\varphi + a\tilde{\xi})$$

$$-(x-\tilde{\xi})a(\rho+\lambda)\sin(\varphi + a\tilde{\xi})\} |\boldsymbol{R} - \lambda \boldsymbol{k}|^{-3} \, \mathrm{d}\xi|_{\lambda=0}. \tag{3.3.17}$$

In order to bring this formula in the form used in literature we change again the integration variable into $\tau = -\tilde{\xi} + \mathrm{x}$ and remark that in the integrand only the combination $(\rho+\lambda)$ occurs, hence we can put $\lambda = 0$ and change the differentiation with respect to λ into a differentiation with respect to ρ. We find

$$\boldsymbol{v}_2(x, r, \varphi, t=0) = \frac{h}{4\pi\mu U(1+a^2\rho^2)^{1/2}} \frac{\partial}{\partial\rho}$$

$$\int_{-\infty}^{x-\xi} \{a\rho^2 - ar\rho\cos(\varphi + a(x-\tau)), \tau a\rho\cos(\varphi + a(x-\tau))$$

$$+\rho\sin(\varphi + a(x-\tau)), -r + \rho\cos(\varphi + a(x-\tau))$$

$$-\tau a\rho\sin(\varphi + a(x-\tau))\} R^{-3} \, \mathrm{d}\tau, \tag{3.3.18}$$

where R is given in (3.3.11).

Then we have found that the force $\boldsymbol{h}$ at $(\xi, \rho, -a\xi)$ induces at (x, r, φ) the velocity

$$\boldsymbol{v}(x, r, \varphi, t=0) = (v_x, v_r, v_\varphi) = \boldsymbol{v}_1 + \boldsymbol{v}_2. \tag{3.3.19}$$

Both expressions (3.3.10) and (3.3.19) will be used in the next sections. First by means of the pressure dipole representation (3.3.10) a two dimensional integral is obtained which yields the component normal to the helicoidal surface, of the disturbance velocities induced by the load. This integral has to be defined by means of the Hadamard principal value. Then the vortex representation (3.3.19) is used to obtain an equivalent two dimensional integral of which the integrand has a less singular behaviour than the former one.

3.4. Screw blade of zero thickness, prescribed load, 1

We first discuss the screw blade of zero thickness and prescribed load $Q(x, r)$ by using (3.3.10). It is only necessary to consider the blade in its position at $t = 0$ because during its rotation in a homogeneous incoming flow the pressures on it are independent of time. All positions are equivalent in the sense that the whole disturbance field of the propeller rotates with it.

Suppose the screw blade (zero thickness) is defined by

$$\varphi - \omega t + ax + \varepsilon f_{3Q}(x, r) = 0, \tag{3.4.1}$$

$$x_l(r) \leq x \leq x_t(r), \qquad r_i \leq r \leq r_0. \tag{3.4.2}$$

In the following we drop the indices $3Q$ from the part of the camber f_{3Q} (3.1.19), which yields the prescribed load. The condition (3.1.3) for the fluid flow to be tangent to the blade, becomes

$$\varepsilon U \frac{\partial f}{\partial x} = -\left(av_x + \frac{v_\varphi}{r}\right). \tag{3.4.3}$$

We assumed that the blade experiences a load $Q = Q(x, r)$ perpendicular to the blade as a result of the fluid action. This load is positive $Q > 0$, when it has a component in the negative x direction, hence when it contributes to the thrust of the propeller. Then an elementary area $\mathrm{d}S$ of the blade exerts a force $\boldsymbol{h}$ (3.3.2) of magnitude $Q\,\mathrm{d}S$ on the fluid with a component in the positive x direction. By this it is allowed to replace h by $Q\,\mathrm{d}S$ in formula (3.3.10) in order to find the velocity induced by the elementary force.

We want to determine the right hand side of (3.4.3) at the blade, then $\partial f/\partial x$ is known. Again we remark that only after the choice of a suitable line at the planform B, for which we assume $f(x, r)$ to be zero, this function can be determined from $\partial f/\partial x$ by integration. This procedure is the same as the one described for H_m in the new paragraph below formula (3.2.12).

First we calculate the right hand side of (3.4.3) for a point (x, r, φ) at a non zero distance away from the blade. By (3.3.10) we find

$$-4\pi\mu U r\left(av_x + \frac{v_\varphi}{r}\right) = \iint_B Q(\xi, \rho) \int_{-\infty}^{(x-\xi)} \frac{\{a^2 r\rho + \cos(\varphi + a(x-\tau))\}}{R^3}$$

$$- \frac{3\{ar\tau + \rho\sin(\varphi + a(x-\tau))\}\{a\rho\tau + r\sin(\varphi + a(x-\tau))\}}{R^5}\,\mathrm{d}\tau\,\mathrm{d}\xi\,\mathrm{d}\rho$$

$$\overset{\text{def}}{=} \iint_B Q(\xi, \rho) \int_{-\infty}^{(x-\xi)} G(x, r, \varphi, \rho, \tau)\,\mathrm{d}\tau\,\mathrm{d}\xi\,\mathrm{d}\rho$$

$$\overset{\text{def}}{=} \iint_B Q(\xi, \rho) K^*(x, r, \varphi, \xi, \rho)\,\mathrm{d}\xi\,\mathrm{d}\rho, \tag{3.4.4}$$

where we replaced the element of area dS by $(1+a^2\rho^2)^{1/2}\,d\xi\,d\rho$, introduced the functions G and K^* and where R is given in (3.3.11).

In order to obtain the value of (3.4.4) at the planform B we have to consider the limit $(x, r, \varphi)\to(x, r, -ax)$, where x and r satisfy (3.4.2). Then the point (x, r, φ) tends to B. From (3.4.3) we find

$$4\pi\mu\varepsilon U^2 r\frac{\partial f}{\partial x}(x, r)=\lim_{\varphi\to -ax}\iint_B Q(\xi,\rho)K^*(x, r, \varphi, \xi, \rho)\,d\xi\,d\rho. \quad (3.4.5)$$

Changing the order of the limit procedure and the integration we have to consider the kernel function

$$K(x, r, \xi, \rho)\overset{\text{def}}{=}K^*(x, r, -ax, \xi, \rho)$$

$$=\int_{-\infty}^{(x-\xi)}\left[\frac{\{a^2r\rho+\cos a\tau\}}{\{\tau^2+r^2+\rho^2-2r\rho\cos a\tau\}^{3/2}}\right.$$

$$\left.-\frac{3\{ar\tau-\rho\sin a\tau\}\{a\rho\tau-r\sin a\tau\}}{\{\tau^2+r^2+\rho^2-2r\rho\cos a\tau\}^{5/2}}\right]d\tau. \quad (3.4.6)$$

When $(x-\xi)>0$ and $\rho\to r$ this function becomes infinite. The singularity arises from a small part of the range of integration in the neighbourhood of $\tau=0$, because there the denominator tends to zero for $\tau\to 0$ and $\rho\to r$. This singular behaviour arises from the fact that we meet the highly singular free vorticity shed by the loading of the blade upstream of the point $(x, r, -ax)$. Hence for the study of the singular behaviour, the range of integration in (3.4.6) can be changed into $-\alpha<\tau<\alpha$, where α is a sufficiently small but positive number. Putting $\rho=r+\nu$, where $|\nu|$ is supposed to be small, we expand the numerators in (3.4.6) with respect to τ and ν, then we obtain integrals of the form

$$\nu^l\int_{-\alpha}^{\alpha}\frac{\tau^m\,d\tau}{\{\tau^2+r^2+(r+\nu)^2-2r(r+\nu)\cos a\tau\}^{q/2}}$$

$$\le|\nu|^l\int_{-\alpha}^{\alpha}\frac{\tau^m\,d\tau}{\{\tau^2+\nu^2\}^{q/2}}\le 2|\nu|^{l+m+1-q}\int_0^{\alpha/|\nu|}\frac{\xi^m}{\{\xi^2+1\}^{q/2}}\,d\xi$$

$$\le 2|\nu|^{l+m+1-q}\left\{1+\int_1^{\alpha/|\nu|}\xi^{m-q}\,d\xi\right\}, \quad (3.4.7)$$

where l and m assume the values 0, 1, 2,... and $q=3$ or 5. It is seen that for $l+m\ge q-1$ this expression remains finite or increases only logarithmically when $\nu \to 0$. From this it follows that we have to expand, with respect to ν and τ, the numerators of the first and the second term under the integral sign in (3.4.6), only up to and including terms of the second and fourth order, respectively. By doing this we find after an estimation of the resulting integrals

$$K(x, r, \xi, \rho) = \frac{2(1+a^2r^2)^{1/2}}{(r-\rho)^2} - \frac{ra^2}{(1+a^2r^2)^{1/2}(r-\rho)} + \mathrm{O}(\ln|r-\rho|),$$

$$\xi < x, \qquad \rho \to r. \tag{3.4.8}$$

This singularity cannot be integrated in the ρ direction. Hence the interchange of the limit $\varphi \to -ax$ and the integration in (3.4.5) is not allowed. A more careful method for giving a meaning to this limit will be discussed in the following.

We divide the planform B of the blade into three regions (figure 3.4.1). The strip B_1 defined by $x_l(r) \le \xi \le x - \gamma$, $|\rho - r| \le \beta$, where $x_l(r)$ is the leading edge (3.2.8), γ and β are sufficiently small positive quantities. Next the rectangle B_2 defined by $|x - \xi| \le \gamma$, $|\rho - r| \le \beta$ and finally the remaining part B_3.

The domain B_3 does not yield any difficulty in the limiting process $\varphi \to -ax$, because its points (ξ, ρ) cannot come close to the fixed point (x, r).

The integration in (3.4.4) over B_1 when first $\varphi \to -ax$ and then $\beta \to 0$, can be written as

$$\lim_{\varphi \to -ax} \int_{x_l(r)}^{x-\gamma} \int_{r-\beta}^{r+\beta} Q(\xi, \rho) \left\{ \int_{-\alpha}^{\alpha} G(x, r, \varphi, \rho, \tau)\, \mathrm{d}\tau \right\} \mathrm{d}\rho\, \mathrm{d}\xi, \tag{3.4.9}$$

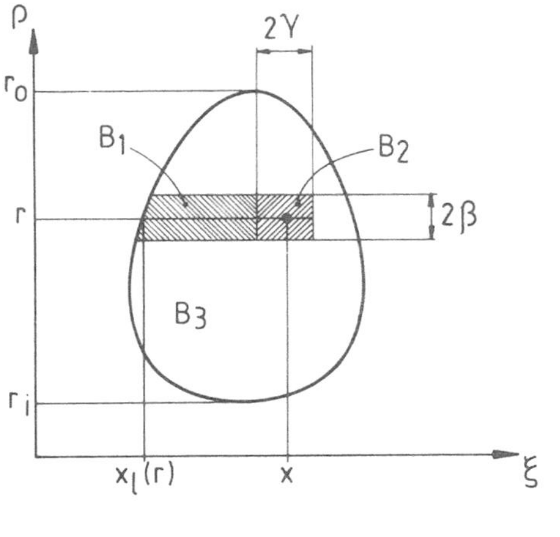

Fig. 3.4.1. The domains of integration B_1, B_2 and B_3.

where $\alpha > 0$ is a small, however fixed quantity. The restriction of the integration over τ to the interval $(-\alpha, \alpha)$ is valid because only the singular part of the kernel can possibly yield a contribution to the integral over the strip of vanishing width, $\beta \to 0$. We introduce new variables δ and ν by

$$\delta = \varphi + ax, \qquad \nu = \rho - r. \tag{3.4.10}$$

Substitution of (3.4.10) into (3.4.9) yields

$$\lim_{\delta \to 0} \int_{x_l(r)}^{x-\gamma} \int_{-\beta}^{\beta} Q(\xi, r+\nu) \left\{ \int_{-\alpha}^{\alpha} G(x, r, -ax+\delta, r+\nu, \tau)\, \mathrm{d}\tau \right\} \mathrm{d}\nu\, \mathrm{d}\xi. \tag{3.4.11}$$

First we consider the integrations with respect to τ and ν. When α, β and δ are sufficiently small, we assume that we can expand $Q(\xi, r+\nu)$ and the two numerators in the function $G(x, r, -ax+\delta, r+\nu, \tau)$, with respect to τ, ν and δ. Then we obtain integrals of the form

$$I(l, m, n, q) = \delta^l \int_{-\beta}^{\beta} \int_{-\alpha}^{\alpha} \frac{\tau^m \nu^n \, \mathrm{d}\tau\, \mathrm{d}\nu}{\left\{ \tau^2 + r^2 + (r+\nu)^2 - 2r(r+\nu)\cos(\delta - a\tau) \right\}^{q/2}}, \tag{3.4.12}$$

where the integers l, m, n and q satisfy

$$l \geq 0, \qquad m \geq 0, \qquad n \geq 0, \qquad q = 3 \text{ or } 5. \tag{3.4.13}$$

We have to keep in mind that we consider the limiting procedure, first $\delta \to 0$ for fixed β, then $\beta \to 0$. There exists a small constant $k > 0$ independent of β, such that

$$|I(l, m, n, q)| \leq |\delta|^l \int_{-\beta}^{\beta} \int_{-\alpha}^{\alpha} \frac{|\tau|^m |\nu|^n \, \mathrm{d}\tau\, \mathrm{d}\nu}{\left\{ \tau^2 + \nu^2 + kr^2(a\tau - \delta)^2 \right\}^{q/2}}. \tag{3.4.14}$$

Next we introduce new variables of integration τ^* and ν^* by

$$\tau = |\delta| \left\{ \tau^* (1 + ka^2 r^2)^{-1/2} \pm akr^2 (1 + ka^2 r^2)^{-1} \right\} \overset{\text{def}}{=} |\delta| (a_1 \tau^* \pm a_2),$$

$$\delta \gtrless 0, \tag{3.4.15}$$

$$\nu = |\delta| \nu^*, \tag{3.4.16}$$

where the upper signs refer to $\delta > 0$ and the lower signs to $\delta < 0$, this convention is also used in the following. Substitution of (3.4.15) and (3.4.16) into (3.4.14) and omitting the asterisks, yields

$$|I(l,m,n,q)| \leq a_1 |\delta|^{(l+m+n-q+2)} \int_{-\beta/|\delta|}^{\beta/|\delta|} |\nu|^n \, \mathrm{d}\nu$$

$$\times \int_{-[(\alpha/|\delta|) \mp a_2] a_1^{-1}}^{[(\alpha/|\delta|) \mp a_2] a_1^{-1}} \frac{|a_1 \tau \pm a_2|^m}{\left(\tau^2 + \nu^2 + a_3\right)^{q/2}} \, \mathrm{d}\tau, \tag{3.4.17}$$

where

$$a_3 = \frac{kr^2}{(1 + ka^2 r^2)}, \tag{3.4.18}$$

$a_3 \neq 0$ for $r \geq r_i > 0$. It is clear that in (3.4.17) the singular behaviour of the two dimensional integral is now determined by the behaviour of the integrand for large values of τ and ν. This means that we can replace (3.4.17) by

$$|I(l,m,n,q)| \leq C_1 |\delta|^{(l+m+n-q+2)} \int_0^{\beta/|\delta|} \nu^n \, \mathrm{d}\nu$$

$$\times \int_0^{[(\alpha/|\delta|) \mp a_2] a_1^{-1}} \frac{\tau^m}{\left(\tau^2 + \nu^2 + a_3\right)^{q/2}} \, \mathrm{d}\tau, \tag{3.4.19}$$

where C_1 is a suitable constant independent of β, again $|\delta|$ is small.

Introducing the integer j we will consider the case

$$j \overset{\text{def}}{=} l + m + n - q + 2 \geq 1. \tag{3.4.20}$$

Then it is allowed to neglect in (3.4.19) the area of integration $0 \leq \tau \leq 1$, $0 \leq \nu \leq 1$ (figure 3.4.2) of which the contribution tends to zero with δ because $a_3 > 0$. The region of integration $0 \leq \tau \leq 1$, $1 \leq \nu \leq \beta/|\delta|$, yields

$$C_1 |\delta|^j \int_1^{\beta/|\delta|} \nu^n \, \mathrm{d}\nu \int_0^1 \frac{\tau^m \, \mathrm{d}\tau}{\left(\tau^2 + \nu^2 + a_3\right)^{q/2}} \leq C_2 |\delta|^j \int_1^{\beta/|\delta|} \nu^{n-q+\tilde{\varepsilon}} \, \mathrm{d}\nu$$

$$= C_2 |\delta|^j \left. \frac{\nu^{n-q+1+\tilde{\varepsilon}}}{(n-q+1+\tilde{\varepsilon})} \right|_1^{\beta/|\delta|}, \tag{3.4.21}$$

where C_2 is a suitable constant independent of β and the constant $\tilde{\varepsilon}$ a

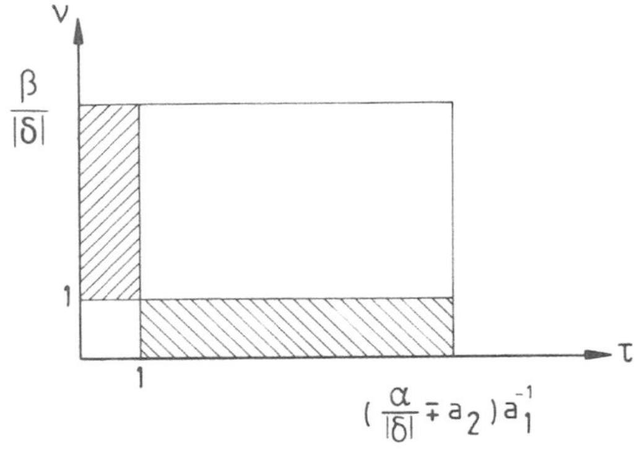

Fig. 3.4.2. Several parts of the domain of integration in (3.4.19), $|\delta| \ll \beta$.

sufficiently small positive fixed number, introduced to keep the denominator of the last expression away from zero. Then we find (3.4.21) $\rightarrow 0$ when $\delta \rightarrow 0$. Analogously we estimate the contribution of the area of integration $0 \leq \nu \leq 1$, $1 \leq \tau \leq (\frac{\alpha}{|\delta|} + a_2)a_1^{-1}$, which also tends to zero for $\delta \rightarrow 0$.

In this way we can replace (3.4.19) by

$$|I(l, m, n, q)| \leq C_1 |\delta|^l \int_1^{\beta/|\delta|} \nu^n \, d\nu \int_1^{[(\alpha/\delta) \mp a_2]a_1^{-1}} \frac{\tau^m}{\left(\tau^2 + \nu^2 + a_3\right)^{q/2}} \, d\tau$$
$$+ O(\delta). \tag{3.4.22}$$

We estimate the denominator of the integral as follows

$$\left(\tau^2 + \nu^2 + a_3\right)^{q/2} = \tau^{\frac{1}{2}q(2-\tilde{\varepsilon})} \nu^{\frac{1}{2}q\tilde{\varepsilon}} \left\{ \left(\frac{\tau}{\nu}\right)^{\tilde{\varepsilon}} + \left(\frac{\nu}{\tau}\right)^{2-\tilde{\varepsilon}} + \frac{a_3}{\tau^{2-\tilde{\varepsilon}} \nu^{\tilde{\varepsilon}}} \right\}^{q/2}$$
$$\geq \tau^{\frac{1}{2}q(2-\tilde{\varepsilon})} \nu^{\frac{1}{2}q\tilde{\varepsilon}}, \tag{3.4.23}$$

where $\tilde{\varepsilon}$ is again some sufficiently small positive fixed number. This holds because $\tau/\nu \geq 1$ or $\nu/\tau \geq 1$ in (3.4.22). Hence we find from (3.4.22)

$$|I(l, m, n, q)| \leq C_1 |\delta|^l \int_1^{\beta/|\delta|} \nu^{(n - \frac{1}{2}\tilde{\varepsilon}q)} \, d\nu \int_1^{[(\alpha/\delta) \mp a_2]a_1^{-1}} \tau^{(m - q + \frac{1}{2}\tilde{\varepsilon}q)} \, d\tau$$
$$+ O(\delta) = O\left(|\delta|^l \beta^{(n - \frac{1}{2}\tilde{\varepsilon}q + 1)}\right) + O(\delta). \tag{3.4.24}$$

For $l \geq 1$ this tends to zero with $\delta \to 0$, for $l = 0$ it tends to zero with $\beta \to 0$.

We have found the result that under condition (3.4.20) the integrals (3.4.12) tend to zero under the limits first $\delta \to 0$, or what is the same $\varphi \to -ax$ and then $\beta \to 0$.

Next we consider more closely the expansions of the load of the blade represented by $Q(\xi, r+\nu)$, and of the two numerators in $G(x, r, -ax + \delta, r+\nu, \tau)$, occurring in (3.4.11). From the foregoing result it follows that we have to develop these functions to such an extent that (3.4.20)

$$l + m + n \leq 1, \qquad q = 3; \, l + m + n \leq 3, \qquad q = 5, \tag{3.4.25}$$

because higher order terms will not give a contribution in the limit procedure under consideration.

First we discuss the numerators of $G(x, r, -ax + \delta, r+\nu, t)$. In connection with (3.4.25) and with the definitions of l, m and n (3.4.12) we take

$$\{a^2 r(r+\nu) + \cos(\delta - a\tau)\} \approx \{(1 + a^2 r^2) + a^2 r\nu\}, \qquad q = 3, \tag{3.4.26}$$

$$-3\{ar\tau + (r+\nu)\sin(\delta - a\tau)\}\{a(r+\nu)\tau + r\sin(\delta - a\tau)\}$$
$$\approx -3(r^2\delta^2 + r\nu\delta^2), \qquad q = 5, \tag{3.4.27}$$

where from now on ν and τ have again their original meaning. From (3.4.25) it follows that we have to expand $Q(\xi, r+\nu)$ only up to and including terms of the first order

$$Q(\xi, r+\nu) \approx Q(\xi, r) + \nu \frac{\partial Q}{\partial r}(\xi, r). \tag{3.4.28}$$

Then we write (3.4.11), when we still disregard for a while the integration with respect to ξ and replace the variable of integration τ by $-\tau$,

$$\lim_{\delta \to 0} \int_{-\beta}^{\beta} \left\{ Q(\xi, r) + \nu \frac{\partial Q}{\partial r}(\xi, r) \right\}$$
$$\times \int_{-\alpha}^{\alpha} \left[\frac{\{(1 + a^2 r^2) + a^2 r\nu\}}{\{\tau^2 + r^2 + (r+\nu)^2 - 2r(r+\nu)\cos(\delta + a\tau)\}^{3/2}} \right.$$
$$\left. - \frac{3\{r^2\delta^2 + r\nu\delta^2\}}{\{\tau^2 + r^2 + (r+\nu)^2 - 2r(r+\nu)\cos(\delta + a\tau)\}^{5/2}} \right] \mathrm{d}\tau \, \mathrm{d}\nu. \tag{3.4.29}$$

Next we expand the cosine in the denominators of (3.4.29)

$$\{\tau^2+r^2+(r+\nu)^2-2r(r+\nu)\cos(\delta+a\tau)\}$$

$$=\{\tau^2+\nu^2+r^2(\delta+a\tau)^2\}\{1+\mathrm{O}((\delta+a\tau)^2(\nu+(\delta+a\tau)^2))\}. \tag{3.4.30}$$

From this it follows in connection with (3.4.25) that we can replace (3.4.29) by

$$\lim_{\delta\to 0} Q(\xi,r)\int_{-\beta}^{\beta}\int_{-\alpha}^{\alpha}\left[\frac{(1+a^2r^2)}{\{\tau^2+\nu^2+r^2(\delta+a\tau)^2\}^{3/2}}\right.$$

$$\left.-\frac{3r^2\delta^2}{\{\tau^2+\nu^2+r^2(\delta+a\tau)^2\}^{5/2}}\right]\mathrm{d}\tau\,\mathrm{d}\nu. \tag{3.4.31}$$

We now choose new variables of integration λ and σ as follows

$$\nu=\frac{r\delta}{(1+a^2r^2)^{1/2}}\lambda,\qquad \tau=\frac{r\delta}{(1+a^2r^2)}(\sigma-ar). \tag{3.4.32}$$

Then (3.4.31) changes into

$$\frac{(1+a^2r^2)}{r}Q(\xi,r)\lim_{\delta\to 0}\frac{1}{\delta}\int_{-\beta_1/\delta}^{\beta_1/\delta}\int_{-(\alpha_1/\delta)+\alpha_2}^{(\alpha_1/\delta)+\alpha_2}\left\{\frac{1}{(\sigma^2+\lambda^2+1)^{3/2}}\right.$$

$$\left.-\frac{3}{(\sigma^2+\lambda^2+1)^{5/2}}\right\}\mathrm{d}\lambda\,\mathrm{d}\sigma, \tag{3.4.33}$$

where

$$\beta_1=\frac{(1+a^2r^2)^{1/2}}{r}\beta,\qquad \alpha_1=\frac{(1+a^2r^2)}{r}\alpha,\qquad \alpha_2=ar \tag{3.4.34}$$

and where we assume $\delta\to 0$ through positive values.

First we remark that

$$\int_{-\infty}^{\infty}\int_{-\infty}^{\infty}\left\{\frac{1}{(\sigma^2+\lambda^2+1)^{3/2}}-\frac{3}{(\sigma^2+\lambda^2+1)^{5/2}}\right\}\mathrm{d}\lambda\,\mathrm{d}\sigma$$

$$=2\pi\int_0^{\infty}\left\{\frac{1}{(\rho^2+1)^{3/2}}-\frac{3}{(\rho^2+1)^{5/2}}\right\}\rho\,\mathrm{d}\rho=0, \tag{3.4.35}$$

hence the limit in (3.4.33) for $\delta\to 0$ can have a finite value. The integrals in (3.4.33) can be calculated in closed form ([21], I, pages 48, 49). We do not enter into details but state that the limit $\delta\to 0$ of (3.4.33) assumes the value

$$\frac{-4(1+a^2r^2)^{1/2}Q(\xi,r)}{\beta}+\mathrm{O}(\beta). \tag{3.4.36}$$

This result substituted into (3.4.11) yields for the contribution of the strip B_1 (figure 3.4.1) to the integral (3.4.4)

$$\frac{-4(1+a^2r^2)^{1/2}}{\beta}\int_{x_l(r)}^{x-\gamma}Q(\xi,r)\,\mathrm{d}\xi+\mathrm{O}(\beta). \tag{3.4.37}$$

The remaining part of the integration over the blade which still has to be discussed, is over the region B_2 (figure 3.4.1) in which the point (x,r) is situated. The contribution of B_2, when first $\varphi\to -ax$ and then $\beta\to 0$, has the form

$$\lim_{\varphi\to -ax}\int_{r-\beta}^{r+\beta}\mathrm{d}\rho\int_{x-\gamma}^{x+\gamma}\mathrm{d}\xi Q(\xi,\rho)\int_{-\infty}^{x-\xi}G(x,r,\varphi,\rho,\tau)\,\mathrm{d}\tau$$

$$=\lim_{\varphi\to -ax}\sum_{n,m=0}^{\infty}Q_{mn}\int_{r-\beta}^{r+\beta}(\rho-r)^n\,\mathrm{d}\rho\int_{x-\gamma}^{x+\gamma}(\xi-x)^m\,\mathrm{d}\xi$$

$$\times\int_{-\infty}^{x-\xi}G(x,r,\varphi,\rho,\tau)\,\mathrm{d}\tau, \tag{3.4.38}$$

where

$$Q_{mn}=\frac{1}{m!n!}\frac{\partial^{m+n}}{\partial x^m\partial r^n}Q(x,r), \tag{3.4.39}$$

we did not mention $\beta\to 0$ and assumed the expansion of Q to be valid, as well as the change of the order of summation and integration.

Introducing again the variables $\delta = \varphi + ax$ and $\nu = \rho - r$ and the new variable

$$\eta = \xi - x, \tag{3.4.40}$$

we write (3.4.38) as

$$\lim_{\delta\to 0} \sum_{n,m=0}^{\infty} Q_{mn} \int_{-\beta}^{\beta} \nu^n \,\mathrm{d}\nu \int_{-\gamma}^{\gamma} \eta^m \,\mathrm{d}\eta \int_{-\infty}^{-\eta} G(x, r, -ax+\delta, r+\nu, \tau)\,\mathrm{d}\tau. \tag{3.4.41}$$

First we discuss the integrations with respect to ν and τ. Partial integration with respect to η yields

$$\frac{1}{(m+1)}\left\{\eta^{m+1}\int_{-\infty}^{-\eta} G(x, r, -ax+\delta, r+\nu, \tau)\,\mathrm{d}\tau \,\Big|_{-\gamma}^{\gamma} + \int_{-\gamma}^{\gamma} \eta^{m+1} G(x, r, -ax+\delta, r+\nu, -\eta)\,\mathrm{d}\eta\right\}. \tag{3.4.42}$$

The value of the first term for the upper bound γ is finite because γ is fixed and positive, hence it yields no contribution to (3.4.41) in the limit $\beta \to 0$. Then the relevant part of (3.4.42) becomes

$$\frac{1}{(m+1)}\left\{-(-\gamma)^{m+1}\int_{-\infty}^{\gamma} G(x, r, -ax+\delta, r+\nu, \tau)\,\mathrm{d}\tau + \int_{-\gamma}^{\gamma} (-\tau)^{m+1} G(x, r, -ax+\delta, r+\nu, \tau)\,\mathrm{d}\tau\right\}. \tag{3.4.43}$$

Consider the first term of this expression. For its contribution to (3.4.41) we find

$$\lim_{\delta\to 0} - \sum_{n,m=0}^{\infty} \frac{(-\gamma)^{m+1}}{(m+1)} Q_{mn} \int_{-\beta}^{\beta}\int_{-\infty}^{\gamma} \nu^n G(x, r, -ax+\delta, r+\nu, \tau)\,\mathrm{d}\tau\,\mathrm{d}\nu. \tag{3.4.44}$$

Analogous to our previous reasoning we can show that only for an interval of τ enclosing $\tau = 0$ and for $n = 0$, we have a contribution in the case $\beta \to 0$. Hence instead of (3.4.44) we consider

$$\lim_{\delta \to 0} -\sum_{m=0}^{\infty} \frac{(-\gamma)^{m+1}}{(m+1)!} \frac{\partial^m Q(x,r)}{\partial x^m}$$

$$\times \int_{-\beta}^{\beta} \int_{-\gamma}^{\gamma} G(x, r, -ax+\delta, r+\nu, \tau)\, \mathrm{d}\tau\, \mathrm{d}\nu. \tag{3.4.45}$$

We compare this expression with (3.4.11) in which we replace $Q(\xi, \mathrm{r}+\nu)$ by the constant

$$-\sum_{m=0}^{\infty} \frac{(-\gamma)^{m+1}}{(m+1)!} \frac{\partial^m Q(x,r)}{\partial x^m}, \tag{3.4.46}$$

and the fixed constant α by γ, then we can use the result (3.4.36). By this we find for (3.4.45)

$$\frac{4(1+a^2r^2)^{1/2}}{\beta} \sum_{m=0}^{\infty} \frac{(-\gamma)^{m+1}}{(m+1)!} \frac{\partial^m Q(x,r)}{\partial x^m}$$

$$= -\frac{4(1+a^2r^2)^{1/2}}{\beta} \int_{x-\gamma}^{x} Q(\xi, r)\, \mathrm{d}\xi, \tag{3.4.47}$$

which is the extension upto x of the integral in (3.4.37).

It can be proved that the second term in (3.4.43) gives no contribution in the limit procedure, first $\delta \to 0$ and then $\beta \to 0$.

In this way we give the following meaning to (3.4.5)

$$4\pi\mu\varepsilon U^2 r \frac{\partial f}{\partial x}(x,r)$$

$$= \lim_{\beta \to 0} \left\{ \left(\int_{r_t}^{r-\beta} + \int_{r+\beta}^{r_0} \right) \int_{x_l(\rho)}^{x_t(\rho)} Q(\xi, \rho) K(x, r, \xi, \rho)\, \mathrm{d}\xi\, \mathrm{d}\rho \right.$$

$$\left. - \frac{4(1+a^2r^2)^{1/2}}{\beta} \int_{x_l(r)}^{x} Q(\xi, r)\, \mathrm{d}\xi \right\}, \tag{3.4.48}$$

where $K(x, r, \xi. \rho)$ is defined in (3.4.6) and the first two terms of its singular behaviour are given in (3.4.8).

The limit procedure of (3.4.48) is called the Hadamard principal value ([14] page 60) of the integration with respect to ρ. In the one dimensional case the Hadamard principal value of the "non existing" integral

$$\int_a^b \frac{f(\rho)}{(\rho - r)^2}\, \mathrm{d}\rho, \qquad a < r < b, \tag{3.4.49}$$

is defined by

$$\lim_{\tilde{\varepsilon}\to 0}\left\{\left(\int_a^{r-\tilde{\varepsilon}}+\int_{r+\tilde{\varepsilon}}^{b}\right)\frac{f(\rho)}{(\rho-r)^2}\,\mathrm{d}\rho-\frac{2f(r)}{\tilde{\varepsilon}}\right\}, \tag{3.4.50}$$

which corresponds to (3.4.48) for a fixed value of $\xi < x$.

That a finite limit value occurs for the "physical" procedure $\varphi \to -ax$ in (3.4.5) is caused by the fact that the singular behaviour of

$$K(x, r, \xi, \rho) = \lim_{\varphi \to -ax} K^*(x, r, \varphi, \xi, \rho), \tag{3.4.51}$$

is not correctly given by (3.4.8). When we draw K^* for fixed values of x, r and ξ and for small values of $|\varphi - ax|$ we find the picture of figure 3.4.3, it is seen that K^* assumes very "large" negative values at a small interval around $\rho = r$. For $\varphi \to -ax$, the width of this interval tends to zero while the values of K^* at that interval tend to $-\infty$. These large negative values balance the large positive values of K^* outside the small interval so that a finite limit value of the integral occurs.

As a simple example how a function of the type as drawn in figure 3.4.3 arises we consider two infinitely long parallel straight vortices at a distance $\tilde{\varepsilon}$ of each other and of opposite strength $\Gamma/\tilde{\varepsilon}$ and $-\Gamma/\tilde{\varepsilon}$. When we calculate the induced velocities which are normal to the plane through these vortices, we obtain a velocity distribution which is only a function of the distance to the vortices. For $\tilde{\varepsilon} \to 0$ this velocity distribution shows the behaviour of figure 3.4.3. Of course in case of our screw propeller these infinitely long vortices are curved and lying along helicoidal lines, however in essence the phenomenon is the same.

That our limiting result coincides with the Hadamard principle value

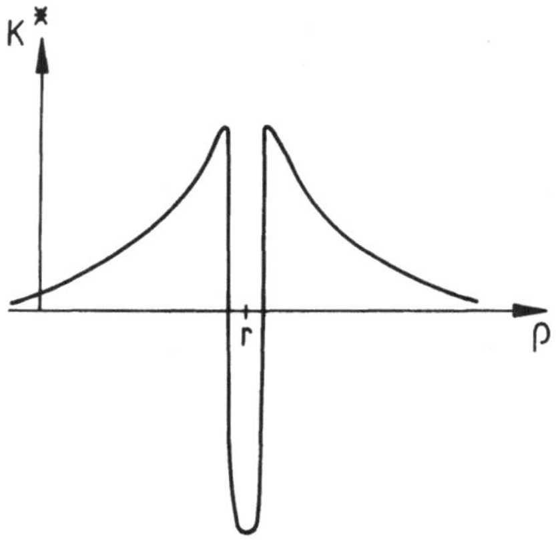

Fig. 3.4.3. Limiting behaviour of K^*.

could not be predicted from the singular behaviour given in (3.4.8). For instance when the function $K^*(x, r, \varphi, \xi, \rho)$ would in addition develop a $\delta(r-\rho)$ function of Dirac for $\varphi \to -ax$, this would not change (3.4.8), however it would alter the interpretation of the limit of the double integral. Hence it is necessary that in physical problems a careful limiting process is carried out, which starts from a position where everything is well defined, in this case from a point with $\varphi \neq -ax$.

3.5. Screw blade of zero thickness, prescribed load, 2

In this section we discuss once more the screw blade of zero thickness with prescribed load, however now we start from the representation (3.3.19), (3.3.13) and (3.3.18) of the velocity induced by a rotating force. Again the basic expression at the right hand side of (3.4.3) has to be calculated,

$$-4\pi\mu U r\left(av_x+\frac{v_\varphi}{r}\right)$$

$$=\iint_B Q(\xi,\rho)\left[\frac{\{ar^2\sin(\varphi+a\xi)-(x-\xi)\cos(\varphi+a\xi)\}}{\{(x-\xi)^2+r^2+\rho^2-2r\rho\cos(\varphi+a\xi)\}^{3/2}}\right.$$

$$\left.+\frac{\partial}{\partial\rho}M(x,r,\varphi,\xi,\rho)\right]\mathrm{d}\xi\,\mathrm{d}\rho, \tag{3.5.1}$$

where

$$M(x,r,\varphi,\xi,\rho)$$

$$=\int_{-\infty}^{x-\xi}\{\rho(a^2r^2-1)\cos(\varphi+a(x-\tau))+r(1-a^2\rho^2)$$

$$+\tau a\rho\sin(\varphi+a(x-\tau))\}$$

$$\times\{\tau^2+r^2+\rho^2-2r\rho\cos(\varphi+a(x-\tau))\}^{-3/2}\,\mathrm{d}\tau \tag{3.5.2}$$

and (x, r, φ) with $\varphi \neq -ax$, is still a point in space away from the planform B. By partial integration we can write (3.5.1) in the form

$$-4\pi\mu U r\left(a v_x + \frac{v_\varphi}{r}\right)$$

$$= \iint_B Q(\xi, \rho) \frac{\{ar^2 \sin(\varphi + a\xi) - (x-\xi)\cos(\varphi + a\xi)\}}{\{(x-\xi)^2 + r^2 + \rho^2 - 2r\rho\cos(\varphi + a\xi)\}^{3/2}}\, \mathrm{d}\xi\, \mathrm{d}\rho$$

$$- \iint_B \frac{\partial Q(\xi, \rho)}{\partial \rho} M(x, r, \varphi, \xi, \rho)\, \mathrm{d}\xi\, \mathrm{d}\rho$$

$$+ \int_{x_f}^{x_b} Q(\xi, \rho_u(\xi)) M(x, r, \xi, \rho_u(\xi))\, \mathrm{d}\xi$$

$$- \int_{x_f}^{x_b} Q(\xi, \rho_d(\xi)) M(x, r, \varphi, \xi, \rho_d(\xi))\, \mathrm{d}\xi, \qquad (3.5.3)$$

where (figure 3.5.1) x_f and x_b denote the smallest and the largest value of the ξ coordinate on the screw blade and $\rho_u(\xi)$ and $\rho_d(\xi)$ describe the upper and lower edge of the blade as a function of ξ. The above is correct in case the blade in the (ξ, ρ) plane is convex, which we assume to be true. Otherwise the boundaries of the integrals have to be specified more carefully.

In order to obtain information about the local angle of incidence of the screw blade we consider again the limit $\varphi \to -ax$. Substituting $\varphi = -ax$ into (3.5.3) we find for the singular behaviour of the kernel of the first integral, when $\xi \to x$ and $\rho \to r$

$$-\frac{(1 + a^2r^2)(x - \xi)}{\{(1 + a^2r^2)(x-\xi)^2 + (r-\rho)^2\}^{3/2}}. \qquad (3.5.4)$$

For the singularity of the kernel of the second integral of (3.5.3) we find

$$\lim_{\rho \to r} M(x, r, -ax, \xi, \rho) \approx 2\frac{(1 + a^2r^2)}{(r-\rho)}, \qquad x > \xi. \qquad (3.5.5)$$

By excluding a strip $|r - \rho| \leq \beta$ (figure 3.5.1) from the domain of integration, we exclude the singularity and it is possible to carry out the integrations in (3.5.3). From (3.5.4) and (3.5.5) it follows that the limit $\beta \to 0$ exists, for the latter one it yields the Cauchy principle value. However when the integrals can be interpreted in this way, we are not sure that the result is correct as we discussed in the last paragraph of the previous section. We need a more careful passing to the limit for the small strip, first $\varphi \to -ax$ and then $\beta \to 0$. This will not be carried out here, we content ourselves with the remark that this has been done and

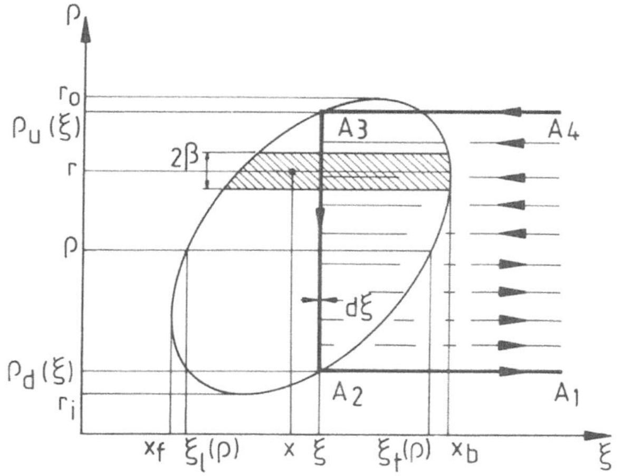

Fig. 3.5.1. A typical vortex system on the blade.

the result we obtain by excluding a strip and taking the just mentioned two limits is correct.

The last two integrals in (3.5.3) are one dimensional with ξ as the variable of integration. It is more natural to introduce ρ instead of ξ, because then the whole leading edge and the whole trailing edge are covered separately. By (3.4.3) we obtain as the ultimate form of our equation

$$4\pi\varepsilon\mu U^2 r\frac{\partial f}{\partial x} = \lim_{\beta\to 0}\left[\int_{x_f}^{x_b}\left(\int_{\rho_d(\xi)}^{r-\beta} + \int_{r+\beta}^{\rho_u(\xi)}\right)\right.$$

$$\frac{Q(\xi,\rho)\{-ar^2\sin a(x-\xi)-(x-\xi)\cos a(x-\xi)\}}{\{(x-\xi)^2+r^2+\rho^2-2r\rho\cos a(x-\xi)\}^{3/2}}\,\mathrm{d}\rho\,\mathrm{d}\xi$$

$$-\int_{x_f}^{x_b}\left(\int_{\rho_d(\xi)}^{r-\beta} + \int_{r+\beta}^{\rho_u(\xi)}\right)\frac{\partial Q}{\partial \rho}(\xi,\rho)M(x,r,-ax,\xi,\rho)\,\mathrm{d}\rho\,\mathrm{d}\xi$$

$$+\left(\int_{r_i}^{r-\beta} + \int_{r+\beta}^{r_0}\right)Q(\xi_l(\rho),\rho)M(x,r,-ax,\xi_l(\rho),\rho)\frac{\mathrm{d}\xi_l(\rho)}{\mathrm{d}\rho}\,\mathrm{d}\rho$$

$$-\left(\int_{r_i}^{r-\beta}+\int_{r+\beta}^{r_0}\right)Q(\xi_t(\rho),\rho)$$

$$\times M(x, r, -ax, \xi_t(\rho), \rho)\frac{\mathrm{d}\xi_t(\rho)}{\mathrm{d}\rho}\,\mathrm{d}\rho\Bigg]. \tag{3.5.6}$$

This equation and equation (3.4.48) were derived in [54].

We now discuss the physical meaning of the different parts of the right hand side of this equation. In the first integral consider the integration with respect to ρ for fixed ξ. This integral represents the velocity induced by a bound vortex of strength $Q(\xi, \rho)\,\mathrm{d}\xi$, denoted in figure 3.5.1 by A_2A_3. By varying ξ we obtain the contribution of all these vortices. The second integral gives the velocity induced by the free vorticity starting at the inner parts of A_2A_3, caused by the variation of $Q(\xi, \rho)$ as a function of ρ. This vorticity is denoted by the thin horizontal lines starting at A_2A_3 and stretching to the right. The last two integrals give the velocities induced by free vortices A_2A_1 and A_3A_4 which arise by the ending of the bound vortices A_2A_3 at the circumference of the blade. Possible directions of rotation of the vortices are denoted by arrows (right hand screw). In space the horizontally drawn vortices are lying along helicoidal lines. Although we used the word vortex in this description it is clear that in fact we have vortex densities.

We have the identity

$$\operatorname{div}\left(\operatorname{rot}\left(U+v_x, v_r, v_\varphi\right)\right)\equiv 0. \tag{3.5.7}$$

This means that when we consider lines which are everywhere tangent to the direction of the vorticity (figure 3.5.2) we get a picture which represents the flow of a two dimensional incompressible fluid on the curved surface consisting of the blade and the helicoidal surface behind it. However then the vorticity in the neighbourhood of the leading edge has to be tangent to it. That this happens follows from (3.5.6). The vorticity component (analogous to (1.5.16))

$$\Gamma(\xi, \rho)=\frac{Q(\xi, \rho)}{\mu U\left(1+a^2\rho^2\right)^{1/2}}, \tag{3.5.8}$$

is the density of bound vorticity in the ρ direction per unit of length along a line $\rho=$ const. at the helicoidal surface. This density per unit of ξ is larger and becomes $Q(\xi, \rho)/\mu\mathrm{U}$. The density of the free vorticity in the ξ direction (in space along the helicoidal lines) per unit of length in the ρ

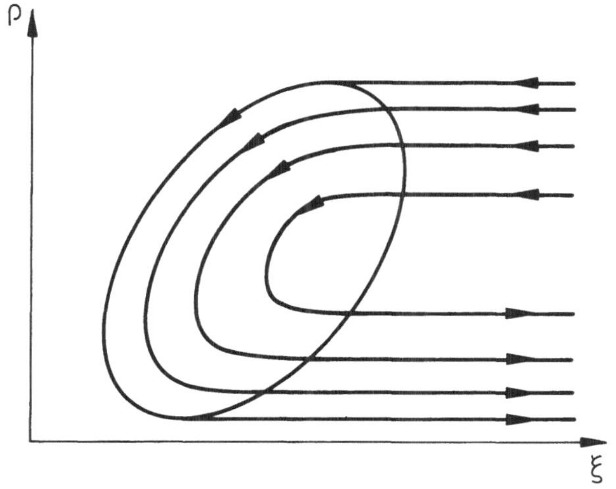

Fig. 3.5.2. Another picture of the vorticity on the blade.

direction, when $(\xi, \rho) = (\xi_l(\rho), \rho)$ is a point of the leading edge, has the value

$$\frac{1}{\mu U}\left\{Q(\xi_l(\rho), \rho)\frac{\mathrm{d}\xi_l(\rho)}{\mathrm{d}\rho}\right\}, \tag{3.5.9}$$

this follows from the third integral in (3.5.6) and the definition of M. Hence the tangent of the angle which the resultant vorticity forms with the ρ axis at the point $(\xi_l(\rho), \rho)$ has the value

$$\frac{1}{\mu U}\left\{Q(\xi_l, \rho)\frac{\mathrm{d}\xi_l(\rho)}{\mathrm{d}\rho}\right\}\left\{\frac{1}{\mu U}Q(\xi_l, \rho)\right\}^{-1} = \frac{\mathrm{d}\xi_l}{\mathrm{d}\rho}, \tag{3.5.10}$$

which had to be proved.

3.6. Some additional remarks

We now indicate very shortly the extension of the theory of the previous sections to make it applicable to a system of N equally spaced, identical blades. In the latter case each blade experiences the disturbance velocities of the other ones. Because the blades are identical, their pressure distributions are equal and we can replace for instance (3.4.5) by

$$4\pi\mu\varepsilon U^2 r\frac{\partial f}{\partial x}=\lim_{\varphi\to -ax}\iint_B Q(\xi,\rho)\sum_{n=0}^{N-1}K^*\left(x,r,\varphi+\frac{2\pi n}{N},\xi,\rho\right)\mathrm{d}\xi\,\mathrm{d}\rho. \tag{3.6.1}$$

The parts added to the original kernel have no singularities because the points of one blade are at a finite distance from the points of the other ones. From this it follows that all our previous limit considerations and statements about integrability remain valid. The same procedure can be followed with respect to the theory for the influence of the thickness of the added blades.

In lifting surface theory there are two main types of problems. First, the design problem, here the profiles of the blades have to be determined to meet a prescribed thickness distribution D and load distribution Q. This means mathematically that in (3.2.11) σ is given by (3.2.8) and in (3.4.48) or (3.5.6) Q is prescribed. Then the integrations have to be performed by which we can find the ultimate camber distribution $f_3(x, r)$ (3.1.19) of the blades.

Second, the calculation of the off-design behaviour of the propeller. Then the thickness D and total camber $f_3 = f_{3D} + f_{3Q}$ of the blades is given and it is asked to calculate under a given external situation (U and ω) the loading of the blades. This means that now we have to solve equation (3.4.48) or equation (3.5.6) which both are integral equations with respect to Q.

In practice the ship screw has to perform its task behind a ship. This means as we discussed in the beginning of this chapter, that the inflow is not homogeneous. The presence of the hull induces perturbation velocities at the screw disk. These perturbations depend besides on x and r also on the angular coordinate φ of our cylindrical coordinate system. Then the loading of the screw blades will possess a periodic character and possibly unsteady cavitation will occur (for instance van Gent and van Oossanen [16]). In order that the phenomena can be described by a linear theory the deviations at the place of the screw must remain "sufficiently" small. Hence the assumption has to be made that the total velocity $\boldsymbol{U}_0$ behind the ship, when the screw is absent, can be written as

$$\boldsymbol{U}_0 = (U + v_{0x}, v_{0r}, v_{0\varphi}), \tag{3.6.2}$$

where the components of the wake v_{0x}, v_{0r} and $v_{0\varphi}$ have to satisfy the relation

$$\frac{v_{0x}}{U}, \frac{v_{0r}}{U}, \frac{v_{0\varphi}}{U} \ll 1. \tag{3.6.3}$$

Possibly this condition can be relieved to some extent by comparing the disturbance velocities at a radius r not with the incoming velocity U but with the velocity $U(1 + a^2r^2)^{1/2}$ of the blade relative to the water. When these disturbances are not small, essential difficulties arise. Then the velocity field in the sense of Biot and Savart of a free vortex does not satisfy anymore the equations of motion.

In order to calculate the fluctuations of the load during each revolution of the screw, we can consider the N planforms of the blades to be impermeable and to have no camber hence to be parts of exactly helicoidal surfaces

$$\varphi + \frac{2\pi n}{N} - \omega t + ax = 0, \qquad n = 0,\ldots,N-1. \tag{3.6.4}$$

Having found in one way or another, the fluctuating pressures in this case, we can add them to the pressures of the screw with thickness and camber, working in the undisturbed parallel flow. This is allowed because our theory is linear. We refer to van Gent [15] and to [27] and [28], where a large number of references can be found.

A consequence of the disturbance velocity field is that unsteady cavitation is likely to occur. This is undesirable for reasons of erosion and deformation of the blades, vibration and noise. We refer to [29], to Szantyr [59] and again to [16]. The first reference also gives information about super cavitating propellers, for this subject we also mention Kruppa [40].

A more direct approach to lifting surface theory is possible by replacing the continuous vorticity at the blades and the trailing vorticity by concentrated vortex lines. These vortex lines consist of two different types, bound vortex lines perpendicular to the local relative velocity $\boldsymbol{V}$ of the fluid and free vortices aligned with $\boldsymbol{V}$. Suppose that the loading Q of the blade is prescribed as a function of position. Then we can assign to each elementary bound vortex $(A–D)$ (figure 3.6.1) a small region ΔS of

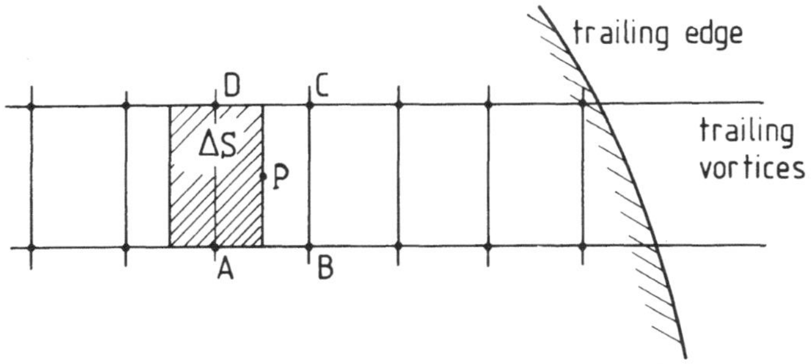

Fig. 3.6.1. Vortex lattice.

the blade and demand that the vortex strength Γ of $(A-D)$ is such that

$$\mu\Gamma|V|\,|(A-D)|=Q\Delta S. \tag{3.6.5}$$

From the elementary bound vortices follow the elementary free vortices, for instance $(A-B)$ and $(D-C)$ by the property that a vortex field is free of divergence. Then at the midpoints of the vortex rectangles such as the point P, the component of the velocity normal to the blade is calculated. From these normal component follows again the camber of the profiles of the blade. Care has to be taken in the neighbourhood of the boundary of the blade. A lot of experience has been obtained with this method, which has the advantage of being mathematically simple and of which the results seem to be reliable. An analogous treatment can be given to the influence of the thickness of the blades, by replacing the continuous layer of sources and sinks of section 3.2 by concentrated source (sink) lines coinciding with the bound vortex lines.

Non linear effects can be treated also by this theory. For instance the vortex grid need not to be placed at the helicoidal reference surface but can be placed at the profile itself. Also in the neighbourhood of the relatively thick roots of the blades a grid can be placed at both sides of the blade. For more information about the vortex lattice method we mention [4] and Kerwin [34] and [35].

General considerations about the non linear effect of the contraction of the free vortex sheets behind a screw propeller are given by Cummings in [13].

We conclude this chapter by mentioning some more complicated propulsion devices in which screw propellers are used. First the very important combination of a screw propeller with an annular "hydrofoil" [4], called duct, shroud or nozzle. We distinguish between the accelerating duct in which the fluid is accelerated and the decelerating duct in which the fluid is decelerated. The first one can be used in the case of a heavily loaded screw, in order to improve the efficiency of the propeller. The second one is used to increase the pressure in the neighbourhood of the screw, hence it can be used to delay cavitation. We do not enter here into the complicated hydrodynamical problems connected with this configuration but refer to Weissinger and Maass [70] and to Ordway and Sluyter [48]. In connection with the optimization theory of propulsion systems we will discuss in chapter 6 some of the properties of a duct, particularly its ability of spreading the vorticity which is shed by the tips of the screw blades.

We also mention the combination of two screw propellers one working behind the other. Tandem propellers consist of two propellers mounted on the same shaft hence rotating in the same direction with the same angular speed. In order to regain the kinetic energy connected to the rotation of the fluid behind a propeller contra rotating propellers have been constructed which are mounted on coaxial, contrary-turning shafts. The theory of this latter configuration is very complicated because the after propeller blades cut the free vortex sheets of the forward one. These types of propellers are for instance discussed in [10] page 433 and by Hadler, Morgan and Meyers in [23] and by Hadler in [22].

Finally we mention the screw propeller with vane wheel. The vane wheel consists of a freely rotating hub installed on the propeller shaft behind the propeller. On its hub are mounted a number of blades or vanes. The diameter of the vane wheel is larger than the diameter of the screw propeller. The inner parts of the vanes are driven by the slipstream of the screw propeller, while their tips contribute to the propulsion. For an elaborate discussion of this interesting device we refer to Grimm [20].

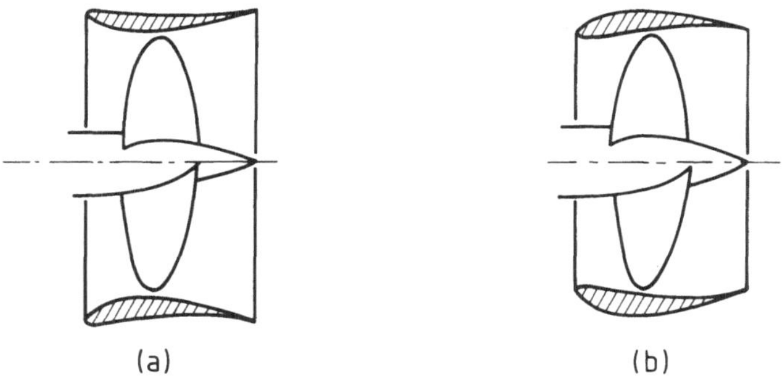

Fig. 3.6.2. Ducted propeller, (a) accelerating type, (b) decelerating type.

4. Unsteady propulsion

In this chapter we discuss some aspects of unsteady propulsion. First we have to give a meaning to the expression "unsteady propeller". The most simple one seems to be: a propeller is unsteady when no inertial reference frame exists with respect to which the induced flow is time independent. This however is not appropriate because then nearly all conceivable propulsion systems are unsteady with the only exception probably of the sails of a yacht in steady motion. Even such a reference frame does not exist for the screw propeller working in a homogeneous flow. A better description is: a propeller is unsteady when the fluid flow relative to its lifting surfaces is time dependent while this time dependency is essential for its functioning. The second part of this definition is vague to some degree, it is intended to exclude for instance the screw propeller in the wake behind a ship. Essential unsteady propulsion occurs in the case of the Voith-Schneider propeller (section 4.7), contra rotating propellers, the propulsion wheels of a paddle boat and the sculling by means of an oar.

In nature almost all animal propulsion in water or in air is unsteady, the flapping motion of the tail of a fish or of the wings of a bird. Perhaps an exception is the "rotating" motion of the helicoidal tail of some species of bacteria which screw themselves through a fluid. In that case however the inertia forces of the fluid can be neglected with respect to the viscous forces (Stokes flow, Reynolds number of $O(10^{-6})$).

The unsteady propulsion we will consider here is assumed to happen in an incompressible and inviscid fluid while the propulsion occurs by the lift of profiles and the suction forces at their leading edge. This excludes the paddle wheels which produce thrust by resistance. What is left are propulsion systems consisting of possibly flexible lifting surfaces making flapping motions. These motions are assumed to be periodic and are allowed to have a small $O(\varepsilon)$ or a large $O(\varepsilon^0)$ amplitude.

An interesting subject for investigation is the possibility of designing unsteady propulsion by means of two or more oscillating wings behind a boat or ship, which can compete with the screw propeller. We discuss this somewhat further in the last section of this chapter.

4.1. Some types of unsteady propulsion

Up to now only unsteady propulsion systems which shed a small amount O(ε) of free vorticity per unit of time can be approached by analytic means. This is the reason that here we will restrict ourselves to this type of systems working, as we mentioned already, in an incompressible and inviscid fluid. In the following we discuss a number of different regimes of working of these propellers.

Regime i a: Small amplitude, one wing.

Consider a lifting surface W without thickness moving along a flat reference strip H (figure 4.1.1) which is part of the plane $y=0$ and stretches along the x axis. This possibly flexible lifting surface moves in an ε neighbourhood of H, which means that its amplitude, its slopes and its curvatures are O(ε). It has a velocity U of O(ε^0) in the positive x direction.

If W moves exactly along the strip H, it does not cause any fluid flow and hence does not shed free vorticity. This motion will be called the "base motion". Small periodic deviations of O(ε) of the base motion, by which thrust can be generated will be called the "added motion". Because our theory will be linear the boundary conditions related to this lifting surface are demanded to be satisfied at the planform of W, which is its projection at H. The bound and free vorticity of W are assumed to be situated at H, the same holds for the free vorticity $\gamma=\gamma(x, z)$ behind W.

The splitting of the motion of lifting surfaces into two parts, namely the base motion and the added motion, is important for the way in which we will treat the linearized theory of propulsion systems. The base motion is always assumed to be such that it sheds no vorticity into the

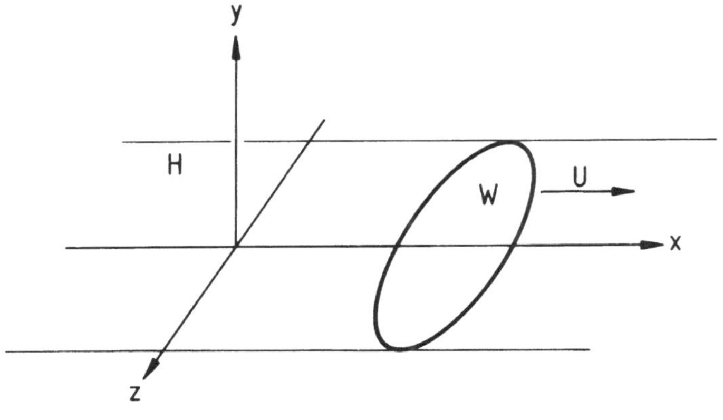

Fig. 4.1.1. Small amplitude flexible wing.

fluid, while the added motion sheds vorticity of $O(\varepsilon)$ by which thrust is generated. These two parts of the motion of a lifting surface are essential in connection with the optimization theory discussed in the next chapter.

The pressure differences between the two sides of W are $O(\varepsilon)$ and also the slopes of W in the x direction are $O(\varepsilon)$, hence the component in the x direction of the force exerted at the lifting surface is $O(\varepsilon^2)$. The leading edge suction force is also $O(\varepsilon^2)$, because Γ_l (1.11.14) is $O(\varepsilon)$. Hence the time dependent total thrust $T = T(t)$ is $O(\varepsilon^2)$.

We assume in this section and whenever it simplifies the reasoning in the following sections, that the propelled body has a large mass. Then its velocity U is nearly constant and the useful power of the propeller equals $U\bar{T}$, where $\bar{T}$ is the mean value with respect to time of the thrust $T(t)$.

The shed free vorticity is $O(\varepsilon)$, hence the mean value of the kinetic energy E added to the fluid per unit of time is $O(\varepsilon^2)$. This means that for the efficiency η_T of such a propulsion system holds

$$\eta_T = \frac{U\bar{T}}{U\bar{T} + E} = 1 - O(\varepsilon^0). \tag{4.1.1}$$

It is seen that this efficiency is independent of $\bar{T}$ because both $\bar{T}$ and E are proportional to ε^2.

Regime i b: Small amplitude, two wings, one behind the other.

Suppose we have another wing W_2 of the same type moving behind W. Then W_2 can deform periodically in such a way that its shed free vorticity $\gamma_2 = \gamma_2(x, z)$ is opposite to the free vorticity $\gamma = \gamma(x, z)$ shed by W. Hence in the linearized theory we do not have shed vorticity at H behind the two wings. By section 1.2 it seems that then the mean value of the thrust has to be zero. This however is not true, the reason is that $\gamma(x, z)$ and $\gamma_2(x, z)$ do not coincide exactly but are a distance of $O(\varepsilon)$ apart, which is caused by the fact that W_2 passes periodically above and below γ. The linearized theory however does not take into account separately these two close free vortex layers.

In figure 4.1.2 we have made a cross section of the two layers. When we look far behind the two wings, we find that in between the two layers the velocity is $O(\varepsilon)$, while outside the layers the velocity is only $O(\varepsilon^2)$. The enclosed area by γ and γ_2 per period of length in the x direction is of $O(\varepsilon)$. Hence the mean value of the lost kinetic energy E per unit of time is of $O(\varepsilon^3)$. This means that we have to replace (4.1.1) by

$$\eta_T = \frac{U(\bar{T} + \bar{T}_2)}{U(\bar{T} + \bar{T}_2) + E} = 1 - O(\varepsilon), \tag{4.1.2}$$

where $\bar{T}_2$ of $O(\varepsilon^2)$ is the mean value of the thrust of W_2. It follows that

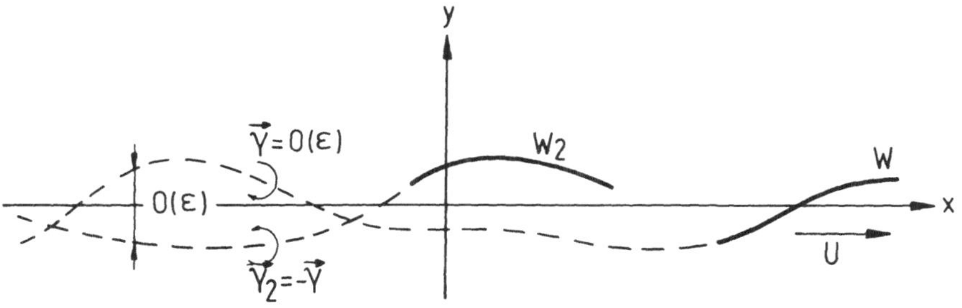

Fig. 4.1.2. Two free vortex layers, a distance of O(ε) apart.

the deviation of the efficiency η_T from 1 is now only O(ε). For sufficiently small values of ε, when the linearized theory is accurate, the efficiency (4.1.2) will be better than the efficiency (4.1.1). In this case η_T decreases with increasing values of $\bar{T}+\bar{T}_2$.

Regime ii: Large amplitude propulsion, linearized theory.

We consider a periodically curved reference strip H (figure 4.1.3) which is at rest with respect to the undisturbed fluid and of which the deviations from the (x, z) plane are O(ε^0). In an ε neighbourhood of H moves a flexible wing W without thickness. This means, as in the previous cases, that the deviations of W from H and deviations of its slopes and its curvatures from those of H, are of O(ε). If W moves exactly along H, it does not disturb the fluid at all and hence does not shed free vorticity. This motion is the base motion of H. Small deviations of O(ε) from the base motion, which induce velocities and vorticity and by which a mean value of thrust can be generated, is the added motion. The boundary conditions related to the lifting surface will be satisfied at the projection of W on H. The bound and free vorticity of W is assumed to be situated at H.

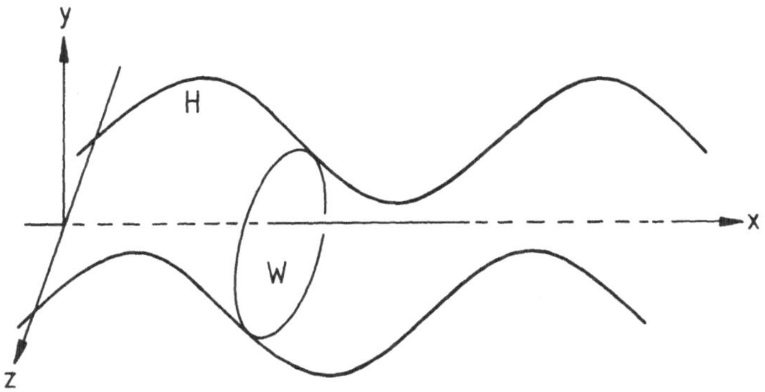

Fig. 4.1.3. Large amplitude flexible wing.

The pressure differences between the two sides of W are again $O(\varepsilon)$. With respect to the calculation of the thrust the angles of W with the x axis can be replaced, within the accuracy of the theory by the angles of H with the x axis and are $O(\varepsilon^0)$. From this it follows that the mean value $\bar{T}$ of the thrust (in the x direction) is $O(\varepsilon)$. Because the shed free vorticity is $O(\varepsilon)$ it follows that the mean value of the energy losses per unit of time are $O(\varepsilon^2)$, hence for the efficiency holds

$$\eta_T = \frac{U\bar{T}}{U\bar{T}+E} = 1 - O(\varepsilon), \tag{4.1.3}$$

where U is the constant velocity with which W moves in the x direction. From a comparison of (4.1.3) and (4.1.1) it follows that in the linearized case the large amplitude propulsion will have a higher efficiency for sufficiently small values of ε than the small amplitude propulsion of regime i a. It is clear that the proportion between the thrust and the lateral forces in case of large amplitude propulsion is $O(\varepsilon^0)$ while for regime i a this proportion is $O(\varepsilon)$. This means that also from this point of view the large amplitude propulsion is more favourable.

It makes no sense to consider a second wing W_2 as we did in regime i b, moving behind W, which annihilates the free vorticity of W. Then we would obtain an analogous picture as is drawn in figure 4.1.2, but now in the neighbourhood of the curved strip H. It is seen that then the momentum of the fluid in the x direction becomes $O(\varepsilon^2)$, hence also the thrust of both wings together becomes $O(\varepsilon^2)$. However for one wing the thrust can be $O(\varepsilon)$, which is "much larger". So in order to destroy the vorticity of the first wing, the second wing has to have a thrust in the wrong direction and of nearly equal magnitude as the thrust of the first one.

Regime iii: Large amplitude propulsion, semi linear theory.

We restrict ourselves to the two dimensional case in which the phenomena are independent of the z coordinate (figure 4.1.4). The main difference with regime ii is that now the profile which carries out the large amplitude motion is rigid, hence causes finite disturbance velocities in the fluid which are of $O(\varepsilon^0)$. We first describe a method to obtain a base motion for a rigid profile. This is as before a motion whereby the profile does not shed free vorticity, hence its circulation has to be constant. We take this constant equal to zero. In the following we assume that the Kutta condition is satisfied at the sharp trailing edge of the profile.

We prove the existence of a point $\tilde{Q}$ such that when rotating the profile around $\tilde{Q}$ its circulation is zero. First, consider, in case of the situation of figure 4.1.4 a point $\tilde{Q}_1$ at the y axis, with a sufficiently large negative value of y. A clockwise rotational velocity of the profile around $\tilde{Q}_1$ will cause a positive or anticlockwise circulation round the profile. Next we consider a point $\tilde{Q}_2$, which we take for instance at a line through O making an angle of $-45°$ with the x axis at a sufficiently large distance from O. Then a clockwise rotational velocity of the profile around $\tilde{Q}_2$ will cause a negative circulation. Hence when we connect $\tilde{Q}_1$ and $\tilde{Q}_2$ by a line, there will be by continuity a point $\tilde{Q}$ on this line such that the circulation is zero when the profile rotates clockwise around it.

For each profile we can also find a direction of translation so that the circulation is zero, this direction is denoted by the line m. When we have found the point $\tilde{Q}$ and a line m we can construct a straight line r of points with the same property as $\tilde{Q}$, namely the line through $\tilde{Q}$ and perpendicular to m. This is correct because a rotational velocity of the profile around any point Q of r can be represented by a rotational velocity around $\tilde{Q}$ and a translational velocity in the direction m, hence the circulation of the profile is zero. Therefore we can assign to some chosen profile a line r which has a fixed position with respect to it and a direction m with the above mentioned properties.

We now take at r some point Q, construct through Q the line m which is perpendicular to r and force the profile to move in the following way. Choose a line L and because we are interested in periodic motions we assume L to be periodic in the x direction. The point Q moves along L and we keep the line m tangent to L. Then irrespective of the shape of L the circulation round the profile remains zero. This follows from the fact that the motion of the profile at each moment consists of a rotation around Q and a translation in the direction of m. The motion defined in this way is a base motion of the profile with respect to L. Making

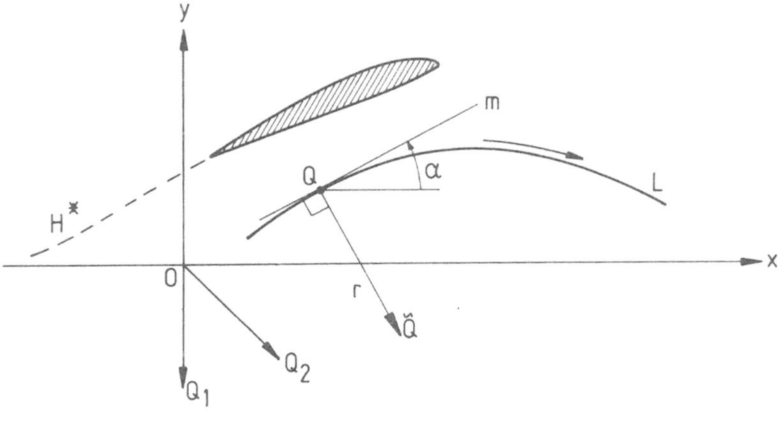

Fig. 4.1.4. Rigid profile moving along L.

another choice of Q yields another base motion. The previous result is obtained in a different way in [11] (page 112).

Next we consider the fluid particles which have passed along the boundary of the profile and have left it at the trailing edge. These particles are lying at a line which will be called the wake H^* (figure 4.1.4). In the neighbourhood of the moving profile the shape of H^* is still changing. When the profile however is a number of periods away, H^* obtains a fixed form which becomes periodic with the same period as L. The wake with its ultimate periodic shape is denoted by H. It is not very difficult to calculate H^* and H because no vorticity is in the fluid and the velocity field depends only on the momentary motion of the profile.

When the profile has to deliver a non-zero mean value of the thrust however, an added motion has to be considered. This motion sheds free vorticity of $O(\varepsilon)$ which is assumed to be situated at H^* and in process of time at H. This is the linearization assumption which facilitates calculations because now we know where to place the shed free vorticity.

In this case the mean value of the thrust $\bar{T} = O(\epsilon)$ and the kinetic energy losses are $O(\varepsilon^2)$, hence the efficiency η_T of this type of propulsion satisfies (4.1.3).

The foregoing regimes i a, i b, ii and iii have in common that in some way a linearization procedure is applied by which the position of the free vorticity in the fluid is determined. This simplification makes these systems more accessible for analysis than is possible for propellers with large amounts of shed free vorticity which drifts away also by its own induced velocities. Of course in practice these propellers can be more effective and certainly deserve to be investigated.

We next make a remark with respect to the formulas for the efficiency and direct our attention to (4.1.3), the other two can be treated analogously. The objection can be made that the mean value of the thrust $\bar{T}(\varepsilon)$ of $O(\varepsilon)$ is accurate only up to errors $\tilde{T}(\varepsilon^2)$ of $O(\varepsilon^2)$. These errors are of the same order of magnitude as the kinetic energy $E(\varepsilon^2)$ of $O(\varepsilon^2)$ left behind per unit of time. So it seems that the efficiency η_T cannot be calculated at all. However we find

$$\eta_T = \frac{U\{\bar{T}(\varepsilon) + \tilde{T}(\varepsilon^2)\}}{U\{\bar{T}(\varepsilon) + \tilde{T}(\varepsilon^2)\} + E(\varepsilon^2)} = \frac{U\bar{T}(\varepsilon)}{U\bar{T}(\varepsilon) + E(\varepsilon^2)} + O(\varepsilon^2). \qquad (4.1.4)$$

The first term at the right hand side of (4.1.4) however is $(1 - O(\varepsilon))$ as we discussed in relation with (4.1.3), hence we can state that the efficiency can be calculated correctly up to and including $O(\varepsilon)$. The faults $\tilde{T}(\varepsilon^2)$ do not disturb the $O(\varepsilon)$ deviation of η_T from 1.

In the remaining part of this chapter we will consider the four mentioned linearized approaches to unsteady propulsion more or less in detail.

4.2. Small amplitude propulsion, two dimensional

We discuss first the two dimensional case of regime i a of the previous section which is a classical problem. We mention for instance *Wu* [73] where also accelerated swimming motions are considered. The fluid is inviscid and incompressible. In figure 4.2.1 we have drawn the swimming profile, without thickness, stretching from $x = -l$ towards $x = l$. The motion of the profile is given by

$$y = h(x, t), \qquad |x| \leq l, \tag{4.2.1}$$

where $h(x, t)$ as well as its first and second derivatives with respect to x and t are assumed to be $O(\varepsilon)$. The profile is placed in a parallel flow of velocity U which is $O(\varepsilon^0)$.

The thickness of the profile is neglected because in linearized theory for a thin profile, it does not change the thrust or lift production. Its influence on the flow field can be described by a source and sink distribution at $|x| \leq l$, $y = 0$, of which the velocities can be added simply to the flow described here. We remark that this does not hold for propellers of which the lifting surfaces have a finite curvature (section 3.2) or can influence each other sideways. Also it is not true for single profiles with a large thickness of $O(\varepsilon^0)$, for instance Uldrick and Siekmann [62].

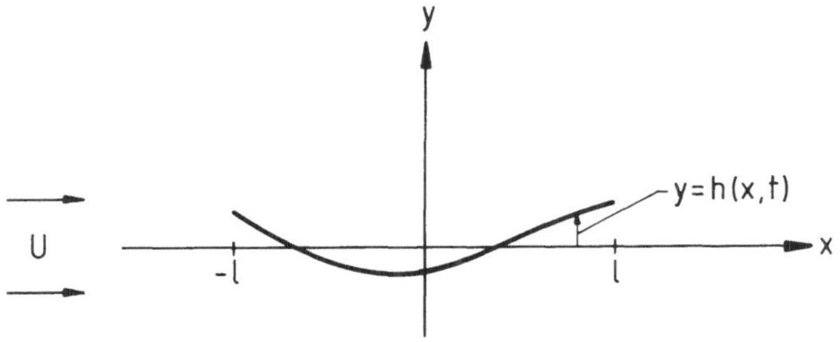

Fig. 4.2.1. Two dimensional small amplitude swimming motion.

The disturbance velocity components v_x, v_y satisfy the linearized equations of motion and the equation of conservation of mass,

$$\frac{\partial v_x}{\partial t} + U\frac{\partial v_x}{\partial x} = -\frac{1}{\mu}\frac{\partial p}{\partial x}, \tag{4.2.2}$$

$$\frac{\partial v_y}{\partial t} + U\frac{\partial v_y}{\partial x} = -\frac{1}{\mu}\frac{\partial p}{\partial y}, \tag{4.2.3}$$

$$\operatorname{div}(v_x, v_y) = \frac{\partial v_x}{\partial x} + \frac{\partial v_y}{\partial y} = 0. \tag{4.2.4}$$

For convenience we introduce the function

$$\varphi(x, y, t) = -\frac{1}{\mu} p(x, y, t), \tag{4.2.5}$$

called acceleration potential because its gradient yields the components of the acceleration of a fluid particle, given at the left hand sides of (4.2.2) and (4.2.3). Differentiation of (4.2.2) with respect to x and of (4.2.3) with respect to y and using (4.2.4) yields

$$\frac{\partial^2 \varphi}{\partial x^2} + \frac{\partial^2 \varphi}{\partial y^2} = 0. \tag{4.2.6}$$

Next we introduce the complex variable $z = x + iy$ and the analytic function

$$f(z, t) = \varphi(x, y, t) + i\psi(x, y, t), \tag{4.2.7}$$

where ψ is the complex conjugate of φ, hence φ and ψ are connected by the Cauchy-Riemann relations. The complex velocity vector is denoted by

$$w(z, t) = v_x(x, y, t) - iv_y(x, y, t). \tag{4.2.8}$$

Using (4.2.2) and (4.2.3) we find

$$\frac{\partial f}{\partial z} = \frac{\partial f}{\partial x} = \frac{\partial \varphi}{\partial x} + i\frac{\partial \psi}{\partial x} = \frac{\partial \varphi}{\partial x} - i\frac{\partial \varphi}{\partial y} = \frac{\partial w}{\partial t} + U\frac{\partial w}{\partial z}. \tag{4.2.9}$$

The boundary condition for our two dimensional problem has the form

$$\left(\frac{\partial}{\partial t} + (U + v_x)\frac{\partial}{\partial x} + v_y\frac{\partial}{\partial y}\right)(y - h(x, t)) = 0. \tag{4.2.10}$$

This formula corresponds to (3.1.3) but is written down here with respect to the coordinates x and y. Neglecting terms of $O(\varepsilon^2)$ we find

$$v_y(x, +0, t) = v_y(x, -0, t)$$

$$= \left(\frac{\partial}{\partial t} + U\frac{\partial}{\partial x}\right)h(x, t) \overset{\text{def}}{=} V(x, t), \qquad |x| < l, \tag{4.2.11}$$

where we introduced $V(x, t)$, which is a known function when $h(x, t)$ is prescribed. From the equation of motion (4.2.3) it follows by (4.2.10)

$$-\frac{\partial\psi}{\partial x}(x, 0, t) = \frac{\partial\varphi}{\partial y}(x, 0, t) = -\frac{1}{\mu}\frac{\partial p}{\partial y}(x, 0, t)$$

$$= \left(\frac{\partial}{\partial t} + U\frac{\partial}{\partial x}\right)V(x, t), \qquad |x| < l, \tag{4.2.12}$$

hence also this expression is known for $-l < x < l$ when $h(x, t)$ is given. We remark that $-\partial\psi/\partial x = \partial\varphi/\partial y$ is continuous at $y = 0$ for all values of x. This follows from (4.2.4) because by conservation of fluid, v_y has to be continuous at $y = 0$ for all values of x. We can represent the profile by a distribution of external forces parallel to the y axis. These forces are, because our problem is two dimensional in fact forces per unit of length in the x direction, concentrated at a line perpendicular to the (x, y) plane. Then it follows from (1.4.15)

$$\varphi(x, 0, t) = 0, \qquad l < |x|, \tag{4.2.13}$$

and

$$\varphi(x, -0, t) = -\varphi(x, +0, t), \qquad |x| < l. \tag{4.2.14}$$

Because $\partial\psi/\partial x$ is continuous at the x axis (4.2.12) we obtain that also

$$\psi(x, -0, t) = \psi(x, +0, t), \qquad |x| < \infty. \tag{4.2.15}$$

For Re $z \to -\infty$ we know that $w(z, t) \to 0$, hence by integration of (4.2.9)

$$f(z, t) = Uw(z, t) + \int_{-\infty}^{z} \frac{\partial w}{\partial t}(\zeta, t)\, d\zeta. \tag{4.2.16}$$

Comparing in this equation the imaginary parts we find

$$\psi(x, y, t) = -Uv_y(x, y, t) - \int_{-\infty}^{x} \frac{\partial}{\partial t} v_y(\xi, y, t)\, d\xi, \tag{4.2.17}$$

where we have chosen a simple path of integration, namely a straight line parallel to the x axis.

Solving (4.2.9) with respect to w we find

$$w(z, t) = \frac{1}{U} f(z, t) - \frac{1}{U^2}\int_{-\infty}^{z} \frac{\partial f}{\partial t}\left(\zeta, t + \frac{\zeta - z}{U}\right) d\zeta. \tag{4.2.18}$$

Comparing imaginary parts in (4.2.18)

$$v_y(x, y, t) = -\frac{1}{U}\psi(x, y, t) + \frac{1}{U^2}\int_{-\infty}^{x} \frac{\partial\psi}{\partial t}\left(\xi, y, t + \frac{\xi - x}{U}\right) \mathrm{d}\xi. \quad (4.2.19)$$

Substitution of (4.2.11) into (4.2.17) for $y = 0$ and $|x| < l$, hence on the profile, yields

$$\psi(x, 0, t) = -\left(U\frac{\partial}{\partial x} + \frac{\partial}{\partial t}\right)\int_{-l}^{x} V(\xi, t)\,\mathrm{d}\xi - \int_{-\infty}^{-l} \frac{\partial}{\partial t} v_y(\xi, 0, t)\,\mathrm{d}\xi$$

$$= \psi_1(x, t) + A(t), \qquad |x| < l, \quad (4.2.20)$$

where

$$\psi_1(x, t) = -\left(U\frac{\partial}{\partial x} + \frac{\partial}{\partial t}\right)\int_{-l}^{x} V(\xi, t)\,\mathrm{d}\xi, \quad (4.2.21)$$

is a known function for $|x| < l$ and the remaining part of (4.2.20) denoted by $A(t)$ is an unknown real function of time.

We derive still another expression for $A(t)$ which will be used in the following sections. From (4.2.17) we have

$$\psi(-l, 0, t) = -Uv_y(-l, 0, t) - \int_{-\infty}^{-l} \frac{\partial}{\partial t} v_y(\xi, 0, t)\,\mathrm{d}\xi, \quad (4.2.22)$$

and from (4.2.19)

$$\psi(-l, 0, t) = -Uv_y(-l, 0, t) + \frac{1}{U}\int_{-\infty}^{-l} \frac{\partial\psi}{\partial t}\left(\xi, 0, t + \frac{\xi + l}{U}\right) \mathrm{d}\xi. \quad (4.2.23)$$

Combination of (4.2.22) and (4.2.23) yields with the definition of $A(t)$

$$A(t) = \frac{1}{U}\int_{-\infty}^{-l} \frac{\partial\psi}{\partial t}\left(x, 0, t + \frac{x + l}{U}\right) \mathrm{d}x. \quad (4.2.24)$$

With respect to the unknown function $f(z, t)$ (4.2.7) we have the following data. Its real part φ satisfies (4.2.13) and (4.2.14) and for its imaginary part ψ holds (4.2.20). From this it follows

$$f^+(z, t) + f^-(z, t) = 2i(\psi_1(x, t) + A(t)), \qquad |x| < l, \quad (4.2.25)$$

$$f^+(z, t) - f^-(z, t) = 0, \qquad |x| > l, \quad (4.2.26)$$

where "$+$" and "$-$" denote the limit of $f(z, t)$ for $y \to 0$ through positive and negative values, respectively. The type of problem stated in (4.2.25) and (4.2.26) is called a Hilbert problem for the function $f(z, t)$. We remark that $f(z, t)$ will be analytic in the whole complex plane with the exception of the line $|x| < l$, $y = 0$, where it exhibits a jump discontinuity. Such a function is called "sectionally holomorfic". In the next section we will discuss this Hilbert problem.

4.3. Solution of the Hilbert Problem

The theory of the Hilbert problem is thoroughly discussed in [47]. For direct reference we mention appendix A, where some results are derived without entering into details.

First we consider the homogeneous part of (4.2.25),

$$X^+(z) + X^-(z) = 0, \qquad |x| < l. \tag{4.3.1}$$

The general solution of this equation which has an algebraic behaviour at infinity, a square root singularity at $x = -l$ and which is zero at $x = l$, has the form (A.2.5)

$$X(z) = P(z)\left(\frac{z-l}{z+l}\right)^{1/2}, \tag{4.3.2}$$

where $P(z)$ is some polynomial. This choice of $X(z)$ already anticipates the demand that the Kutta condition has to be satisfied at the trailing edge $x = l$ of the profile. The definition of the square root in (4.3.2) is as follows. For real values of z hence $z = x$, we assume for $x > l$

$$\left(\frac{z-l}{z+l}\right)^{1/2} = \left(\frac{x-l}{x+l}\right)^{1/2} > 0. \tag{4.3.3}$$

Then $X(z)$ is uniquely defined in the z plane by analytic continuation when the segment $|x| < l$, $y = 0$ is a cut in the plane. In the following a square root of a positive number is assumed to be positive.

By (A.2.7) it is seen that a solution of (4.2.25) can be written as

$$f(z, t) = \frac{X_1(z)}{2\pi i}\int_{-l}^{l} \frac{2i\{\psi_1(\xi, t) + A(t)\}\, d\xi}{X_1^+(\xi)(\xi - z)} + X_2(z), \tag{4.3.4}$$

where $X_1(z)$ and $X_2(z)$ are solutions of (4.3.1) of the form (4.3.2) with different choices of $P(z)$ namely $P_1(z)$ and $P_2(z)$ respectively. It follows from (4.3.3) that

$$X_1^+(\xi) = iP_1(\xi)\left(\frac{l-\xi}{l+\xi}\right)^{1/2}, \qquad |\xi| < l. \tag{4.3.5}$$

For $X_1(z)$ we take $P_1(z) = (z + l)$ and for $X_2(z)$ we take $P_2(z) = C(t)$, which is a time dependent constant. We find for $f(z, t)$

$$f(z,t)=\frac{(z^2-l^2)^{1/2}}{\pi i}\int_{-l}^{l}\frac{\{\psi_1(\xi,t)+A(t)\}}{(l^2-\xi^2)^{1/2}(\xi-z)}\,\mathrm{d}\xi+C(t)\left(\frac{z-l}{z+l}\right)^{1/2}. \tag{4.3.6}$$

These choices of $X_1(z)$ and $X_2(z)$ are not uniquely determined by the mathematical problem, however besides that the Kutta condition has to be satisfied they have to be such that the disturbance pressures tend to zero at infinity, or

$$\lim_{|z|\to\infty} f(z,t)=0. \tag{4.3.7}$$

Using the integrals

$$\begin{aligned}&\int_{-l}^{l}\frac{\mathrm{d}\xi}{(l^2-\xi^2)^{1/2}}=\pi,\\&\int_{-l}^{l}\frac{\mathrm{d}\xi}{(l^2-\xi^2)^{1/2}(\xi-z)}=-\frac{\pi}{(z^2-l^2)^{1/2}},\end{aligned} \tag{4.3.8}$$

we find from (4.3.6) and (4.3.7)

$$C(t)=-\frac{i}{\pi}\int_{-l}^{l}\frac{\psi_1(\xi,t)}{(l^2-\xi^2)^{1/2}}\,\mathrm{d}\xi-iA(t). \tag{4.3.9}$$

Substitution of (4.3.9) into (4.3.6) yields

$$f(z,t)=iA(t)\left\{1-\left(\frac{z-l}{z+l}\right)^{1/2}\right\}$$
$$-\frac{i}{\pi}\int_{-l}^{l}\frac{\psi_1(\xi,t)}{(l^2-\xi^2)^{1/2}}\left\{\frac{(z^2-l^2)^{1/2}}{(\xi-z)}+\left(\frac{z-l}{z+l}\right)^{1/2}\right\}\mathrm{d}\xi. \tag{4.3.10}$$

We now have to determine the still unknown function $A(t)$. Taking the imaginary part of (4.3.10), we find for $x<-l$ and $y=0$,

$$\psi(x,0,t)=A(t)\left\{1-\left(\frac{x-l}{x+l}\right)^{1/2}\right\}$$
$$-\frac{1}{\pi}\int_{-l}^{l}\frac{\psi_1(\xi,t)}{(l^2-\xi^2)^{1/2}}\left\{-\frac{(x^2-l^2)^{1/2}}{(\xi-x)}+\left(\frac{x-l}{x+l}\right)^{1/2}\right\}\mathrm{d}\xi. \tag{4.3.11}$$

Substitution of (4.3.11) into (4.2.24) yields

$$A(t)=\frac{1}{U}\int_{-\infty}^{-l}\left[A'\left(t+\frac{x+l}{U}\right)\left\{1-\left(\frac{x-l}{x+l}\right)^{1/2}\right\}-\frac{1}{\pi}\int_{-l}^{l}\frac{\frac{\partial}{\partial t}\psi_1\left(\xi,t+\frac{x+l}{U}\right)}{(l^2-\xi^2)^{1/2}}\right.$$

$$\left.\times\left\{-\frac{(x^2-l^2)^{1/2}}{(\xi-x)}+\left(\frac{x-l}{x+l}\right)^{1/2}\right\}\mathrm{d}\xi\right]\mathrm{d}x. \tag{4.3.12}$$

We assume that for $t \leqslant t_0$ the profile and the fluid are at rest, hence by (4.2.11), (4.2.21) and (4.2.24)

$$\psi_1(x,t)=0, \qquad A(t)=0, \qquad t<t_0. \tag{4.3.13}$$

Furthermore we assume that from $t=t_0$ up to $t=t_1>t_0$ the motion of the profile starts sufficiently smooth and for $t>t_1$ we have

$$h(x,t)=h(x,t+\tau), \qquad t>t_1>t_0 \tag{4.3.14}$$

hence the motion has become periodic.

When we replace x in (4.3.12) by $U(\xi-t)-l$ and ξ by x, we obtain

$$\int_{t_0}^{t}A'(\xi)K(t-\xi)\,\mathrm{d}\xi \overset{\text{def}}{=} \int_{t_0}^{t}A'(\xi)\left\{\frac{U(\xi-t)-2l}{U(\xi-t)}\right\}^{1/2}\mathrm{d}\xi$$

$$=\frac{1}{\pi}\int_{t_0}^{t}\int_{-l}^{l}\frac{\frac{\partial}{\partial\xi}\psi_1(x,\xi)}{(l^2-x^2)^{1/2}}$$

$$\times\left[\frac{\left\{(U(\xi-t)-l)^2-l^2\right\}^{1/2}}{(x-U(\xi-t)+l)}-\left\{\frac{U(\xi-t)-2l}{U(\xi-t)}\right\}^{1/2}\right]\mathrm{d}x\,\mathrm{d}\xi \overset{\text{def}}{=} \Psi(t),$$

(4.3.15)

where we introduced the kernel $K(t)$ and the known function $\Psi(t)$. This equation is an integral equation of the Volterra type for $A'(\xi)$ and can be solved by the method of the Laplace transform ([52], page 207).

We do not pursue this general problem but change to a more simple version by assuming

$$h(x,t)=h(x)e^{j\omega t}, \qquad h(x)=h_1(x)+jh_2(x), \tag{4.3.16}$$

where j is the imaginary unit in the "time domain". Hence we consider a simple time harmonic motion. In order to have convergence of a number of integrals which follow and to be able to carry out partial integrations when necessary, we assume that ω has a small negative imaginary part. We tacitly take the limit in the results such that these belong to a simple periodic motion (4.3.16) with a real value of ω. When t_0 and t_1 tend to $-\infty$ we assume

$$A(t) = Ae^{j\omega t}, \tag{4.3.17}$$

A being an unknown constant.

We first calculate the left hand side of (4.3.15). Substitution of (4.3.17) and $t_0 = -\infty$, into this left hand side yields

$$\int_{-\infty}^{t} j\omega A \, \mathrm{e}^{j\omega\xi} K(t-\xi) \, \mathrm{d}\xi = j\omega A \, \mathrm{e}^{j\omega t} \int_0^\infty \mathrm{e}^{-j\omega\eta} K(\eta) \, \mathrm{d}\eta$$

$$= j\alpha A \, \mathrm{e}^{j(\alpha+\omega t)} \int_1^\infty \mathrm{e}^{-j\alpha\xi} \left\{ \frac{1}{(\xi^2-1)^{1/2}} + \frac{\xi}{(\xi^2-1)^{1/2}} \right\} \mathrm{d}\xi, \tag{4.3.18}$$

$$\alpha = \frac{\omega l}{U},$$

where α is the dimensionless reduced frequency. Making use of the following relations for Hankel functions ([67] pages 45, 170)

$$H_0^{(2)}(x) = \frac{2j}{\pi} \int_1^\infty \frac{\mathrm{e}^{-jx\xi}}{(\xi^2-1)^{1/2}} \mathrm{d}\xi, \qquad (\mathrm{Re}\, x > 0), \tag{4.3.19}$$

$$\frac{\mathrm{d}}{\mathrm{d}x} H_0^{(2)}(x) = -H_1^{(2)}(x), \tag{4.3.20}$$

we write (4.3.18) as

$$\tfrac{1}{2}\pi\alpha A \, \mathrm{e}^{j(\alpha+\omega t)} \left\{ H_0^{(2)}(\alpha) - jH_1^{(2)}(\alpha) \right\}. \tag{4.3.21}$$

Next we consider the right hand side of (4.3.15). Introducing

$$V(x,t) = V(x) \, \mathrm{e}^{j\omega t}, \qquad \psi_1(x,t) = \psi_1(x) \, \mathrm{e}^{j\omega t}, \tag{4.3.22}$$

we find by substituting $\xi = l(-\zeta+1)/U + t$ into (4.3.15), after replacing x by ξ

$$\Psi(t) = -\frac{j\alpha}{\pi} \mathrm{e}^{j(\alpha+\omega t)} \int_{-l}^{l} \Bigg[\int_1^\infty \frac{\psi_1(\xi)\, \mathrm{e}^{-j\alpha\zeta}}{(l^2-\xi^2)^{1/2}}$$

$$\times \left\{ \left(\frac{\zeta+1}{\zeta-1} \right)^{1/2} - \frac{l(\zeta^2-1)^{1/2}}{(\xi+l\zeta)} \right\} \mathrm{d}\zeta \Bigg] \mathrm{d}\xi. \tag{4.3.23}$$

Using (4.3.19) and (4.3.20) we write the first term at the right hand side of (4.3.23) as

$$-\tfrac{1}{2}\alpha \mathrm{e}^{j(\alpha+\omega t)}\left(H_0^{(2)}(\alpha)-jH_1^{(2)}(\alpha)\right)\int_{-l}^{l}\frac{\psi_1(\xi)}{\left(l^2-\xi^2\right)^{1/2}}\mathrm{d}\xi. \tag{4.3.24}$$

The second term of (4.3.23) becomes

$$-\frac{j\alpha}{\pi}\mathrm{e}^{j(\alpha+\omega t)}l\int_{-l}^{l}\frac{\left\{UV(\xi)+\omega j\int_{-l}^{\xi}V(\sigma)\mathrm{d}\sigma\right\}}{\left(l^2-\xi^2\right)^{1/2}}$$

$$\times\left\{\int_1^{\infty}\mathrm{e}^{-j\alpha\zeta}\frac{\left(\zeta^2-1\right)^{1/2}}{(\xi+l\zeta)}\mathrm{d}\zeta\right\}\mathrm{d}\xi. \tag{4.3.25}$$

From this expression we consider again the second term. In this term we carry out a partial integration with respect to ζ and use the identity

$$\frac{-l}{\left(l^2-\xi^2\right)^{1/2}}\frac{\partial}{\partial\zeta}\frac{\left(\zeta^2-1\right)^{1/2}}{(\xi+\zeta l)}=\frac{1}{\left(\zeta^2-1\right)^{1/2}}\frac{\partial}{\partial\xi}\frac{\left(l^2-\xi^2\right)^{1/2}}{(\xi+\zeta l)}, \tag{4.3.26}$$

then after some rearrangements, we can write (4.3.25) as

$$-\frac{j\alpha U}{\pi l}\mathrm{e}^{j(\alpha+\omega t)}\int_{-l}^{l}\frac{V(\xi)}{\left(l^2-\xi^2\right)^{1/2}}$$

$$\times\left[\int_1^{\infty}\mathrm{e}^{-j\alpha\zeta}\left(\frac{l\zeta}{\left(\zeta^2-1\right)^{1/2}}-\frac{\xi}{\left(\zeta^2-1\right)^{1/2}}\right)\mathrm{d}\zeta\right]\mathrm{d}\xi$$

$$=\frac{\alpha U}{2l}\mathrm{e}^{j(\alpha+\omega t)}\int_{-l}^{l}\frac{V(\xi)}{\left(l^2-\xi^2\right)^{1/2}}\left\{jlH_1^{(2)}(\alpha)+\xi H_0^{(2)}(\alpha)\right\}\mathrm{d}\xi. \tag{4.3.27}$$

Taking the sum of (4.3.24) and (4.3.27) we find for (4.3.23)

$$\Psi(t)=\tfrac{1}{2}\alpha \mathrm{e}^{j(\alpha+\omega t)}\left[-\left\{H_0^{(2)}(\alpha)-jH_1^{(2)}(\alpha)\right\}\int_{-l}^{l}\frac{\psi_1(\xi)}{\left(l^2-\xi^2\right)^{1/2}}\mathrm{d}\xi\right.$$

$$\left.+\frac{U}{l}\int_{-l}^{l}\frac{V(\xi)}{\left(l^2-\xi^2\right)^{1/2}}\left\{jlH_1^{(2)}(\alpha)+\xi H_0^{(2)}(\alpha)\right\}\mathrm{d}\xi\right]. \tag{4.3.28}$$

Then from (4.3.15), (4.3.21) and (4.3.28) the former unknown constant A follows

$$A = -\frac{1}{\pi}\int_{-l}^{l} \frac{\psi_1(\xi)}{\left(l^2-\xi^2\right)^{1/2}} \mathrm{d}\xi$$

$$+\frac{U}{\pi l}\int_{-l}^{l} \frac{V(\xi)}{\left(l^2-\xi^2\right)^{1/2}} \frac{\left\{ jlH_1^{(2)}(\alpha)+\xi H_0^{(2)}(\alpha)\right\}}{\left\{ H_0^{(2)}(\alpha)-jH_1^{(2)}(\alpha)\right\}} \mathrm{d}\xi. \tag{4.3.29}$$

The Hankel functions we used in the above are related to the Bessel functions by

$$H_\nu^{(2)}(x)=J_\nu(x)-jY_\nu(x), \qquad \nu=0,1,2,\ldots. \tag{4.3.30}$$

4.4. Thrust and efficiency

In section 4.2 we introduced the imaginary unit i with $(i)^2=-1$, in the complex space domain. In section 4.3 we introduced the imaginary unit j with $(j)^2=-1$, in connection with the displacement (4.3.16) of the profile. These different imaginary units have no "interaction", we left unaltered the product ij. The reason that j could be introduced is that the theory is linear. In this section however we consider "quadratic" quantities which are $O(\varepsilon^2)$, such as the thrust and the lost energy, therefore we now consider real motions of the profile, given by

$$\tilde{h}(x,t)=\operatorname*{Re}_{j}\left(h(x)\,\mathrm{e}^{j\omega t}\right)=h_1(x)\cos\omega t-h_2(x)\sin\omega t. \tag{4.4.1}$$

In order to know with respect to which imaginary unit or units the real part of an expression has to be taken we attach to the symbol Re the index i or j or both indices. In the following when necessary for clearness quantities which belong to $\tilde{h}(x,t)$ are provided with a tilde.

The pressures at upper and lower side of the profile follow from (4.2.5)

$$\tilde{p}(x,\pm 0,t)=-\mu\tilde{\varphi}(x,\pm 0,t)\overset{\text{def}}{=}-\mu\operatorname*{Re}_{ij} f^{\pm}(z,t), \qquad |x|<l, \tag{4.4.2}$$

where $f(z,t)$ is given by (4.3.10), $h(x,t)$ by (4.3.16) and $A(t)$ follows from (4.3.17) by (4.3.29). In connection with (4.3.3) we have

$$\left\{\left(\frac{z-l}{z+l}\right)^{1/2}\right\}^{\pm}=\pm i\left(\frac{l-x}{l+x}\right)^{1/2},$$
$$\left\{\left(z^2-l^2\right)^{1/2}\right\}^{\pm}=\pm i\left(l^2-x^2\right)^{1/2},\qquad |x|<l. \tag{4.4.3}$$

Then because both $\psi_1(x, t)$ and $A(t)$ are real with respect to i, we find

$$\tilde{\varphi}(x,\pm 0,t)=\pm\tilde{A}(t)\left(\frac{l-x}{l+x}\right)^{1/2}\pm\frac{1}{\pi}\int_{-l}^{l}\frac{\tilde{\psi}_1(\xi,t)}{\left(l^2-\xi^2\right)^{1/2}}$$
$$\times\left\{\frac{\left(l^2-\xi^2\right)^{1/2}}{(\xi-x)}+\left(\frac{l-x}{l+x}\right)^{1/2}\right\}\mathrm{d}\xi,\qquad |x|<l, \tag{4.4.4}$$

where

$$\tilde{\psi}_1(\xi,t)=\operatorname*{Re}_j\psi_1(\xi,t),\qquad \tilde{A}(t)=\operatorname*{Re}_j A(t). \tag{4.4.5}$$

The thrust $T(t)$ per unit of span belonging to the profile motion (4.4.1), reckoned positive in the negative x direction becomes

$$T(t)=-\int_{-l}^{l}[\tilde{p}]_-^+\frac{\partial\tilde{h}}{\partial x}(x,t)\,\mathrm{d}x+\tfrac{1}{4}\mu\pi\tilde{\Gamma}_l^2(t), \tag{4.4.6}$$

where $[\tilde{p}]_-^+=(\tilde{p}(x,+0,t)-\tilde{p}(x,-0,t))$ and the second term at the right hand side is the suction force (1.11.14), $\tilde{\Gamma}_l(t)$ being the coefficient of the square root singularity of the vorticity at the leading edge (1.11.1). The coefficient $\tilde{\Gamma}_l(t)$ can be calculated as follows. First we observe that the pressure jump over the profile equals

$$[\tilde{p}]_-^+=-\mu U\tilde{\Gamma}_b(x,t), \tag{4.4.7}$$

where $\tilde{\Gamma}_b(x, t)$ is by definition the bound vorticity of the profile. The total vorticity $\tilde{\Gamma}_{\text{tot}}(x, t)$ of the profile can be written as

$$\tilde{\Gamma}_{\text{tot}}(x,t)=\tilde{\Gamma}_b(x,t)-\frac{1}{U}\int_{-l}^{x}\frac{\partial\tilde{\Gamma}_b}{\partial t}\left(\xi,t-\frac{(x-\xi)}{U}\right)\mathrm{d}\xi. \tag{4.4.8}$$

However

$$\lim_{x\to-l}\int_{-l}^{x}\frac{\partial\tilde{\Gamma}_b}{\partial t}\left(\xi,t-\frac{(x-\xi)}{U}\right)\mathrm{d}\xi=0, \tag{4.4.9}$$

because $\partial/\partial t\,\tilde{\Gamma}_b(x, t)$ has at most a square root singularity at $x=-l$, which is absolutely integrable. This means that the singular behaviour of $\tilde{\Gamma}_b(x, t)$ and $\tilde{\Gamma}_{tot}(x, t)$ for $x \to -l$ is the same. Hence using (4.2.14), (4.4.7) and (4.4.2) we find

$$\tilde{\Gamma}_l(t) = 2 \lim_{x \to -l} \frac{(l+x)^{1/2}}{U} \tilde{\varphi}(x, +0, t)$$

$$= \frac{2(2l)^{1/2}}{U} \left\{ \tilde{A}(t) + \frac{1}{\pi} \int_{-l}^{l} \frac{\tilde{\psi}_1(\xi, t)}{(l^2 - \xi^2)^{1/2}} \mathrm{d}\xi \right\}. \tag{4.4.10}$$

When we denote by $k(x, t)$ the external force field parallel to the y axis, exerted by the profile at the fluid, the power $P(t)$ per unit of span needed to maintain the motion becomes

$$P(t) = \int_{-l}^{l} k(x, t) \frac{\partial \tilde{h}}{\partial t}(x, t) \mathrm{d}x = \int_{-l}^{l} [\tilde{p}]_-^+ \frac{\partial \tilde{h}}{\partial t}(x, t) \mathrm{d}x. \tag{4.4.11}$$

For the kinetic energy $\mathrm{d}E/\mathrm{d}t$ lost per unit of time in the fluid we find from (4.4.6) and (4.4.11)

$$\frac{\mathrm{d}E}{\mathrm{d}t} = P(t) - UT(t), \tag{4.4.12}$$

where the second term at the right hand side is the useful work per unit of time. From (1.8.9) however we derive the representation

$$\frac{\mathrm{d}E}{\mathrm{d}t} = \int_{-l}^{l} k(x, t) v_y(x, 0, t) \, \mathrm{d}x. \tag{4.4.13}$$

The questions arises; are (4.4.12) and (4.4.13) equal? In order to make this clear we use a continuity argument based on the last two paragraphs of section 1.11. When the vorticity of the profile is changed in a very small one sided $\tilde{\varepsilon}$ neighbourhood of the leading edge (figure 1.11.3a), so that it is finite at the edge we have practically the same situation, but we lost the suction force in (4.4.6). This force now is distributed over a small neighbourhood of the leading edge and is taken care of by the integral, by making $\tilde{\varepsilon}$ sufficiently small, to any desired degree of accuracy. Hence we can compare (4.4.12) and (4.4.13) neglecting the suction force. However the boundary condition at the profile reads

$$v_y = U \frac{\partial h}{\partial x} + \frac{\partial h}{\partial t}, \tag{4.4.14}$$

from which the equality follows.

Introducing the mean values of the thrust and the power over one period by

$$\bar{T}=\frac{\omega}{2\pi}\int_0^{2\pi/\omega}T(t)\,\mathrm{d}t,\qquad \bar{P}=\frac{\omega}{2\pi}\int_0^{2\pi/\omega}P(t)\,\mathrm{d}t,\tag{4.4.15}$$

we find for the efficiency of the propulsion

$$\eta_T=\frac{U\bar{T}}{\bar{P}}=1-\mathrm{O}(\varepsilon^0).\tag{4.4.16}$$

Finally for the sake of completeness we write the formula for the lift delivered by the profile

$$L(t)=-\int_{-l}^{l}[\tilde{p}]_-^+\,\mathrm{d}x.\tag{4.4.17}$$

4.5. Theoretical and experimental results

In order to illustrate the previous theory of small-amplitude propulsion which describes the two dimensional case of regime i a, we show some theoretical results which are compared with experimental ones as

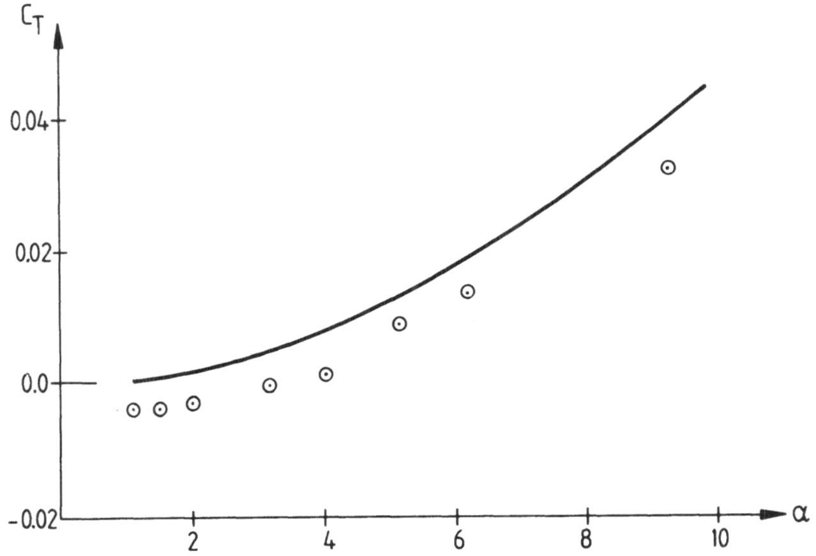

Fig. 4.5.1. Thrust coefficient C_T as a function of reduced frequency α, heaving motion ($C_0=\frac{1}{24}$, $C_1=C_2=0$, $\sigma=0$) of rigid plate, — theory, ⊙ experiment.

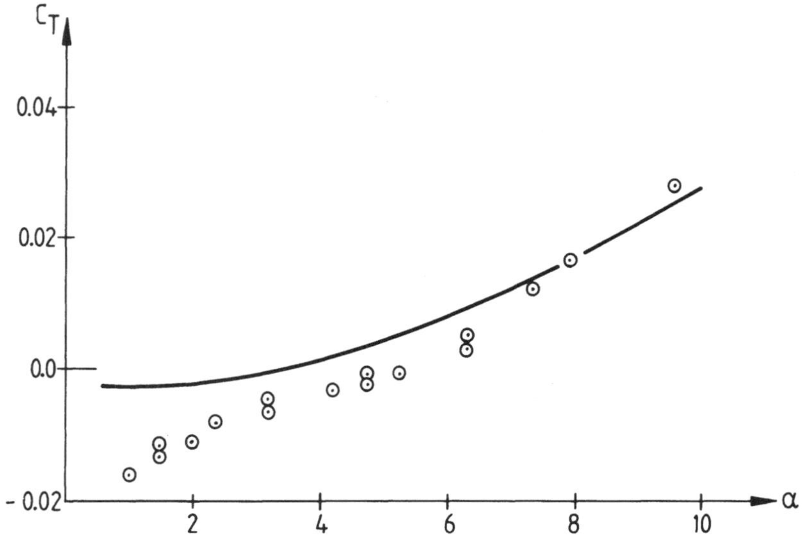

Fig. 4.5.2. Thrust coefficient C_T as a function of reduced frequency α, wave motion with constant amplitude ($C_0 = \frac{1}{12}$, $C_1 = C_2 = 0$, $\sigma = \pi$). — theory, ⊙ experiment.

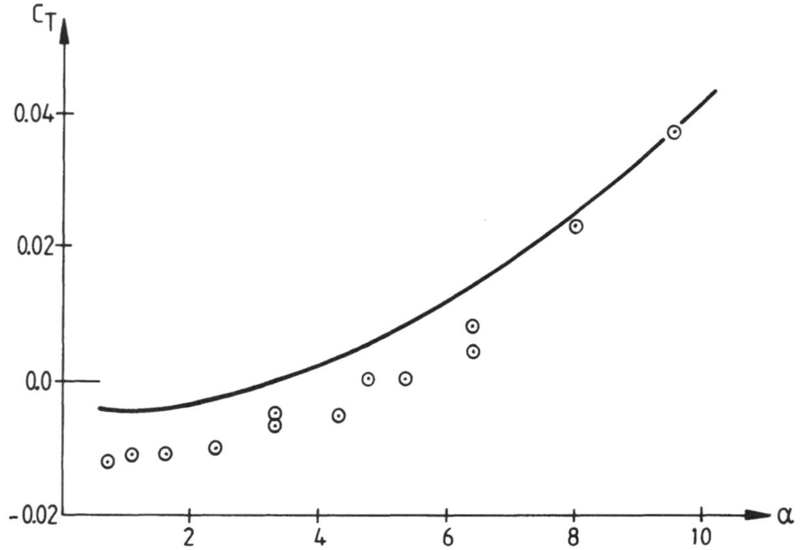

Fig. 4.5.3. Thrust coefficient C_T as a function of reduced frequency α, wave motion with quadratic amplitude ($C_0 = 0.023$, $C_1 = 0.042$, $C_2 = 0.034$, $\sigma = \pi$), — theory, ⊙ experiment.

described by Kelly, Rentz and Siekmann in [33]. The motion of the profile (4.4.1) is given by

$$\tilde{h}(x,t) = \operatorname*{Re}_{j}\left\{\left(C_0 + C_1 x + C_2 x^2\right) e^{-j\alpha x} e^{j\omega t}\right\}, |x| \leqslant 1. \tag{4.5.1}$$

In figures 4.5.1–4.5.3 is drawn the thrust coefficient, defined by

$$C_T = \frac{\bar{T}}{\pi\mu U^2}, \tag{4.5.2}$$

versus the dimensionless reduced frequency $\alpha = \omega/U$ (half length of profile $l = 1$).

First in figure 4.5.1 the purely having motion of a rigid metal plate is considered. Besides a small systematic deviation caused by the resistance induced by the turbulent boundary layer of the slightly viscous fluid, the trend of the experiments is in good agreement with the theory.

In figure 4.5.2 are given the results for a flexible metal profile. The deformation is of constant amplitude and its wavelength equals the chord length. For values $\alpha < 3$, there is some discrepancy between theory and measurements. For that region the wave speed is less than the free stream velocity U. It was observed in the experiments by means of dye flow patterns, that flow separation occurred in the lee of the peaks of waves of the plate.

Flow separation did not occur in the case of a wave motion of the plate, with quadratic amplitude, at least for values $\alpha > 1$. This is shown in figure 4.5.3.

Numerical results for a flexible profile which consists of an elastic plate with bending stiffness and inertia, clamped at its nose and driven by an external force action, are given by Szeless in [60]. For more information with respect to other profile motions and about the efficiency of the small-amplitude propulsion we refer to Wu [74] where also many references are given. For the influence of finite thickness of a profile on its propulsive capacity we mention again [62].

4.6. Elastically coupled profiles

In this section we give some numerical results of the small-amplitude propulsion by means of two profiles one behind the other, regime i b of section 4.1. We have a Cartesian coordinate system (x, y) embedded in an inviscid and incompressible fluid which has an incoming velocity U in the positive x direction. The problem is two dimensional. For $a_1 \leqslant x \leqslant b_1$ and for $a_2 \leqslant x \leqslant b_2 (b_1 < a_2)$ we have two flat and infinitely thin profiles, which are assumed to move in an ε neighbourhood of the x axis. The profiles are coupled by two rigid arms (b_1, s_1) and (s_1, a_2) which are connected to each other at $x = s_1$ by means of a linear elastic hinge. The motion of the first profile is prescribed while the second profile follows passively, hence its motion has to be calculated. The theory is linearized and the free vorticity is assumed to be at the x axis.

This problem is not essentially more difficult than the one of the single profile as discussed in sections 4.2–4.4, although it is much more complicated. For the theory we refer to Sparenberg and de Vries [58]. As we remarked already in section 4.1, the position of the second profile and its mechanical properties have to be determined such that its shed free vorticity is opposite to the vorticity shed by the first one. By this the second profile extracts kinetic energy out of the fluid put into it by the first one.

The numerical results are based on the choice

$$\mu = 1, \qquad U = 1, \qquad a_1 = 0, \qquad b_1 = 1. \tag{4.6.1}$$

The motions of the profiles are written as

$$y = h_l(x, t) = A_l \cos(\omega t + \alpha_l) + B_l x \cos(\omega t + \beta_l), \qquad l = 1, 2, \tag{4.6.2}$$

where $l = 1$, (2), refers to the first (second) profile and we can assume without loss of generality $\alpha_1 = 0$. The results are scaled to a total mean value of thrust $T = 1$. Other quantities which determine the second profile are; the place of its centre of gravity $x = s_z$ its mass m and its moment of inertia I around the centre of gravity. The strength of the elastic hinge is denoted by ν. In order to move the first profile we need at its leading edge an external force of amplitude K_1 and an external

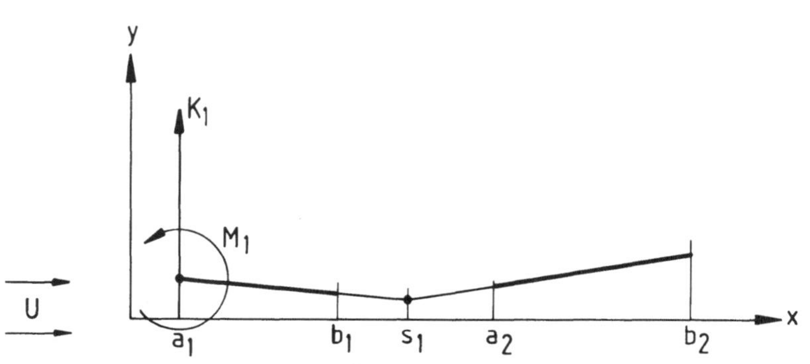

Fig. 4.6.1. Two flat infinitely thin profiles, coupled by an elastic hinge.

moment of amplitude M_1. Here the quantities m, I, ν, K_1, and M_1 are per unit of span. Further in table 4.6.1 the following symbols are used;
T_1 = thrust of first profile,
Z_l = part of thrust of lth profile delivered by its leading edge suction force,
E = kinetic energy lost per unit of span and per unit of time.

The following requirements seem to be reasonable for the functioning of a profile combination, as a propulsive device;

1) moderate energy loss E,
2) moderate suction forces, hence moderate values of Z_1 and Z_2,
3) well balanced distribution of thrust over the two profiles, for instance not too small values of $(b_2 - a_2)$ and T_1,
4) a slot which is not too wide, hence $(a_2 - b_1)$ not too large.

As a result of the testing by means of the computer of about 2.10^6 configurations and the use of an optimization procedure, we give in table 4.6.1 five profile combinations which satisfy to a certain extent the above mentioned demands. For the frequency ω has been chosen the fixed value $\omega = \pi$. Although in general smaller values of the frequency gave better results with respect to the requirements 1), 2) and 3), it turned out that the slots between the two profiles became rather wide.

We now give an example of the calculation from table 4.6.1 of a more or less realistic propulsion device. We assume the following system of units; time, 1 second; length, 1 meter; mass, 1 kilogram mass; force, 0.1 kilogram force.

Suppose we want to construct a propulsion device that moves with a velocity $\tilde{U} = 1.5$, delivers a mean value of thrust $\tilde{T} = 20$ (per unit span) and of which the first profile has a chord length $\tilde{l} = 0.3$. Here and in the following a "~" denotes a quantity with dimension. For the density of

Table 4.6.1.
Profile combinations, $\omega = \pi$, $(T = 1)$.

a_2	$b_2 - a_2$	T_1	Z_1	Z_2	E	A_1	B_1	A_2	α_2	B_2
2.337	0.539	0.514	0.260	0.450	2.500	0.473	0.629	3.328	276.5	1.195
2.359	0.503	0.523	0.260	0.450	2.500	0.486	0.652	2.778	274.8	1.000
2.213	0.500	0.400	0.049	0.450	2.500	0.344	0.597	2.931	276.1	1.109
2.345	0.500	0.504	0.260	0.403	2.500	0.466	0.617	3.869	279.5	1.404
2.266	0.500	0.422	0.260	0.450	1.921	0.420	0.551	3.521	282.1	1.308

β_2	K_1	M_1	β_1	s_1	m	I	ν	s_z
104.7	0.665	1.602	236.1	1.969	0.028	0.068	0.607	2.658
103.5	0.725	0.848	235.3	1.836	0.093	0.015	0.280	2.661
105.0	0.903	1.878	239.4	1.839	0.012	0.069	0.665	2.513
107.6	0.805	0.886	236.4	2.060	0.062	0.028	0.280	2.645
110.3	0.703	1.013	237.3	2.037	0.035	0.026	0.280	2.566

water we take $\tilde{\mu} = 10^3$. We choose from table 4.6.1 for instance the system with the smallest energy loss E, hence the one for which $E = 1.921$. Then we find from dimension analysis [58], for the quantities of interest which determine the desired propulsion device,

$$\begin{aligned}
&\tilde{A}_1 = \left(\frac{\tilde{T}\tilde{l}}{\tilde{\mu}}\right)^{1/2} \frac{A_1}{\tilde{U}} = 0.0217; && \tilde{\omega} = \frac{\tilde{U}}{\tilde{l}}\omega = 15.71; \\
&\tilde{B}_1 = \frac{\tilde{A}_1}{\tilde{l}} \frac{B_1}{A_1} = 0.095; && \tilde{\beta}_1 = \beta_1 = 237.3°; \\
&\tilde{m} = \tilde{\mu}\tilde{l}^2 m = 3.15; && \tilde{I} = \tilde{\mu}\tilde{l}^4 I = 0.211; \\
&\tilde{a}_2 = \tilde{l}a_2 = 0.680; && (\tilde{b}_2 - \tilde{a}_2) = \tilde{l}(b_2 - a_2) = 0.15; \qquad (4.6.3) \\
&\tilde{s}_1 = \tilde{l}s_1 = 0.611; && \tilde{s}_z = \tilde{l}s_z = 0.770; \\
&\tilde{\nu} = \tilde{\mu}\tilde{l}\tilde{U}^2 \nu = 189; && \eta_T = \left(1 + \frac{\tilde{A}_1}{\tilde{l}} \frac{E}{A_1}\right)^{-1} = 0.751; \\
&\tilde{K}_1 = \tilde{\mu}\tilde{U}^2 \tilde{A}_1 \frac{K_1}{A_1} = 81.72; && \tilde{M}_1 = \tilde{\mu}\tilde{l}\tilde{U}^2 \tilde{A}_1 \frac{M_1}{A_1} = 35.33.
\end{aligned}$$

These formulas can be used to calculate the properties of other propulsion devices.

4.7. The Voith-Schneider propeller

In technics there is up to now one important realization of an unsteady large amplitude propulsion device namely the Voith-Schneider propeller of which a scheme is drawn in figure 4.7.1. Under a ship we imagine a horizontal circular disk which rotates about a vertical axis l through its centre C. The rotational velocity of the disk is ω. On the disk are mounted several vertical wing like blades. These blades can perform oscillatory motions about vertical pivotal axes, denoted in figure 4.7.1 by $l_1, \ldots, l_4$ which cut the circular disk in the pivotal points $Q_1, \ldots, Q_4$ respectively.

We now discuss the cylindrical surfaces described by the pivotal axes. We use a Cartesian coordinate system (x, y, z) in rest with respect to the fluid. The circular disk lies in the (x, z) plane and at time $t = 0$ the y axis coincides with the axis of rotation l of the propeller which has a velocity U in the positive x direction. We consider the path in the (x, z) plane of the pivotal point Q_4 which is assumed at $t = 0$ to be on the positive z axis. We find

$$x = R(\nu\omega t + \sin \omega t), \qquad z = R \cos \omega t, \qquad \nu = \frac{U}{\omega R}, \tag{4.7.1}$$

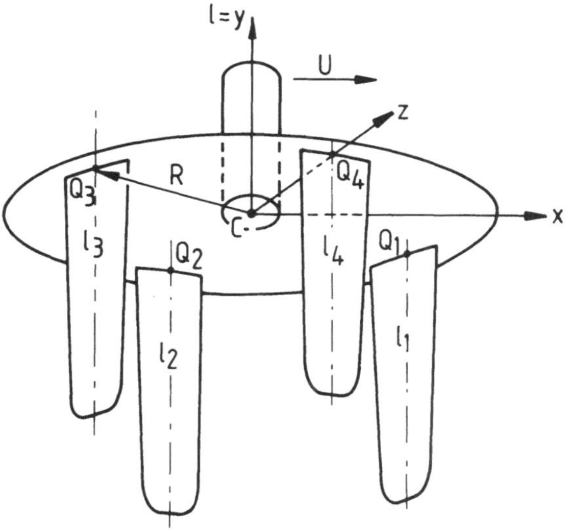

Fig. 4.7.1. Scheme of a four bladed Voith-Schneider propeller.

where R is the distance from the axes $l_1, \ldots, l_4$ to l. This path is a cycloid of which the character is drawn in figure 4.7.2 for two values of ν. When $\nu > 1$ the blades carry out an asymmetrical swimming motion of a fish tail; when $0 < \nu < 1$, the cycloid intersects itself. When $\nu < 0.212$ more than one intersection occurs. In practice we have $\nu < 1$. In order that the propeller provides a mean force in the x direction it is necessary that the blades have an appropriate angle of incidence during their motion. This is accomplished by making them execute a suitable periodic oscillatory motion about their pivotal axes $l_1, \ldots, l_4$. The way in which the motion of the blades about these axes is controlled mechanically will not be discussed here, we refer to Mueller [46]. We only mention that, by turning the whole machinery, we can also turn the direction of the thrust. By this it is possible to steer a ship provided with this type of propeller, hence a rudder becomes superfluous.

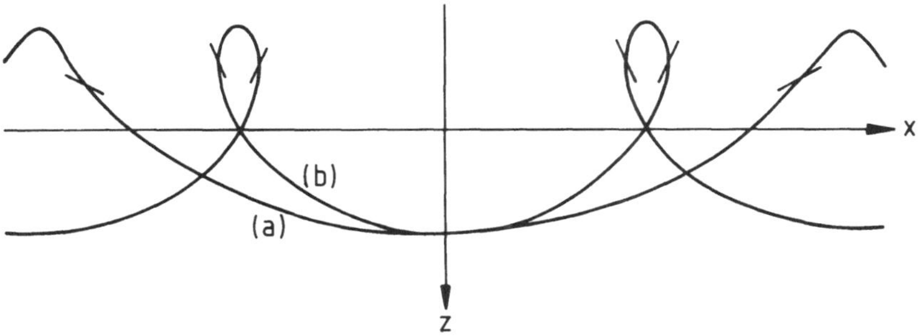

Fig. 4.7.2. Two types of cycloids, (a) $1 < \nu$, (b) $0.212 < \nu < 1$.

When the chordlength of the wings is not too large with respect to the radii of curvature of the cycloids it is probably possible to describe the working of the Voith-Schneider propeller by means of regime ii of section 4.1. Instead of one wing W moving in a neighbourhood of one reference surface H, we have in this case four wings and four reference surfaces, however this is not essential. In order to allow for the bottom of the ship, we can take in figure 4.1.3, the span of W two times the span of the wings of the propeller. Then the plane through the midspan points, of the model, is a plane which by symmetry is not passed through by fluid particles, hence it can represent the bottom. Under the assumption (regime ii) that the free vorticity shed by the Voith-Schneider propeller remains at the place where it is formed it is not difficult, be it complicated, to set up an unsteady lifting surface theory for the blades. This will not be pursued here.

When the cord length of the wings is larger, such that the regime ii can not be applied, we can use regime iii, which is discussed more extensively in the next section. Of course both approaches hold under the condition that the propeller is lightly loaded.

In section 6.10 we discuss the optimization of a Voith-Schneider propeller with many blades.

4.8. Large amplitude unsteady propulsion, rigid profile

We consider a rigid flat profile without thickness of chordlength $2l$. Along the profile we introduce a length parameter s, reckoned from the midpoint M, hence at the trailing edge A, $s=-l$ and at the leading edge B, $s=l$. At the profile we choose a point R with $s=b$, this point moves along some prescribed periodic line represented by the one valued function $y=f(x)$ with period $2h$,

$$f(x+2h)=f(x). \tag{4.8.1}$$

The velocity of the point R along $y=f(x)$ can be prescribed arbitrarily in a periodic way.

First we describe the "base motion" of the profile $(A-B)$ which is, as we defined in section 4.1, the motion for which the circulation around the profile remains zero. Because this unsteady propeller comes under regime iii, we have to consider the direction m and the line r belonging to the flat profile. The direction m is clearly parallel with the profile. Now we have to find a point Q of the line r. We look for such a point on $(A-B)$ itself and denote its parameter value by $s=a$. When we rotate the profile around Q with rotational velocity ω, the velocities of the points of $(A-B)$ are in the direction of the normal $\boldsymbol{n}$ and amount to

$$v_n(s)=\omega(s-a). \tag{4.8.2}$$

The vorticity $\Gamma(s)$, needed on the profile in order that the fluid flows along it, satisfies

$$v_n(s)=\frac{1}{2\pi}\oint_{-l}^{l}\frac{\Gamma(\sigma)}{(s-\sigma)}\,\mathrm{d}\sigma. \tag{4.8.3}$$

The solution of this equation which satisfies the Kutta condition at the trailing edge $s=-l$, hence with $\Gamma(-l)=0$, is (appendix A, or [47])

$$\Gamma(s)=2\omega(a+l-s)\left(\frac{l+s}{l-s}\right)^{1/2}. \tag{4.8.4}$$

The condition that the total circulation is zero yields

$$\int_{-l}^{l}\Gamma(s)\,\mathrm{d}s=\pi\omega l(l+2a)=0, \tag{4.8.5}$$

hence

$$a=-\tfrac{1}{2}l. \tag{4.8.6}$$

This means that Q is the well known three quarter chord point. Hence when a flat profile moves with its three quarter chord point along an arbitrary line L and is tangent to L, its total circulation is identically zero. This can be realized in our case in the following way. The point $R(s=b)$ was assumed to move along $y=f(x)$. We start with the profile

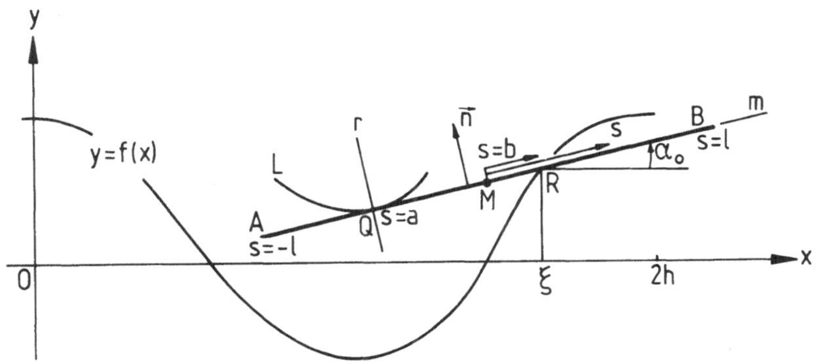

Fig. 4.8.1. Profile $(A-B)$ moving with point R along $y=f(x)$, base motion.

in some position, for instance the one of figure 4.8.1. At the three quarter point Q we attach a little wheel with its axis of rotation along the line r. This wheel rolls on the (x, y) plane and makes that the profile cannot move with its point Q sideways, but only in the momentary direction $A - B$. In other words when the point R moves along $y = f(x)$, the profile behaves as a bicycle of which the front-wheel at R is guided along $y = f(x)$ and the back-wheel at Q follows and describes the demanded line L. If a profile carries out this motion its circulation will remain zero. The line L we obtain in this way depends on the initial position of the profile. If we can find an initial position such that L is periodic in the x direction, we have a simple characterization of the base motion.

We now will discuss this line L in more detail. When we denote the coordinates of the point R by $(\xi, f(\xi))$, ξ is a parameter for L and we denote the coordinates of Q by $x_Q = x_Q(\xi)$, $y_Q = y_Q(\xi)$. The condition of Q moving along the unknown line L, while $(A - B)$ is tangent to L, is

$$\frac{y_Q'(\xi)}{x_Q'(\xi)} = \frac{(f(\xi) - y_Q(\xi))}{(\xi - x_Q(\xi))}, \tag{4.8.7}$$

and the condition that the distance between Q and R is $(\frac{1}{2}l + b)$ is

$$\{\xi - x_Q(\xi)\}^2 + \{f(\xi) - y_Q(\xi)\}^2 = (\tfrac{1}{2}l + b)^2. \tag{4.8.8}$$

Differentiating (4.8.8) with respect to ξ and eliminating $y_Q(\xi)$ and $y_Q'(\xi)$ from this equation and from (4.8.7) and (4.8.8) yields

$$(\tfrac{1}{2}l + b)^2 x_Q' - (\xi - x_Q)^2 - f'(\xi)(\xi - x_Q)\left\{(\tfrac{1}{2}l + b)^2 - (\xi - x_Q)^2\right\}^{1/2} = 0. \tag{4.8.9}$$

Also we have the periodicity condition for L,

$$x_Q(0) = 2h + x_Q(2h). \tag{4.8.10}$$

Solution of (4.8.9) under the condition (4.8.10) yields $x_Q = x_Q(\xi)$ and by (4.8.8) we obtain $y_Q = y_Q(\xi)$. Then the line L, which will be denoted by $y = g(x)$, is determined. Now we can characterize the base motion of the profile by its sliding tangently along L, touching L at its three quarter chord point Q hence for $s = -\frac{1}{2}l$. Also we can calculate the angle $\alpha_0 = \alpha_0(x)$ which the profile makes with the x axis, when the point Q has the coordinates $(x, g(x))$. From now on we assume $L(y = g(x))$ and $\alpha_0(x)$ to be known.

Because the periodic solution of (4.8.9) is essential for the theory, we will interrupt the train of thoughs and sketch a proof due to *F*. Takens of its existence. Consider in figure 4.8.2 the line segment $(Q-R)$ in an arbitrary position, however with the point $Q(s=-\frac{1}{2}l)$ to the left of the y axis. The point $R(s=b)$ moves along $y=f(x)$, starting at $(0, f(0))$ and ending one period further at $(2h, f(2h))$. The point Q follows the point R with a velocity which is directed along $(Q-R)$. Then after this $(Q-R)$ assumes a new position $(Q^{(1)}-R^{(1)})$. This can be done for all possible starting positions of Q at the semi circle I, the end positions $Q^{(1)}$ are lying at the semi circle II. It is easily seen that the vertical segments (Q_1-R) and (Q_2-R) obtain after one period new positions $(Q_1^{(1)}-R^{(1)})$, $(Q_2^{(1)}-R^{(1)})$ which lie to the left of the vertical line $x=2h$. Hence the semi circle *I* is mapped on part of the semi circle II. Besides we know that the points $Q^{(1)}$ at II have the same ordering as the corresponding points Q at *I*. Suppose this is not true then, when we let all segments $(Q-R)$ with Q at *I* move at the same time, two of them have to coincide at some moment and afterwards they would carry out the same motion. This shows that the ordering is preserved. Now we can repeat the mentioned procedure any number of times by letting the point R move along $y=f(x)$ to points $(2nh, f(2nh))$ $n=2, 3, \dots$. Then the points Q_1 and Q_2 obtain positions $Q_1^{(n)}$ and $Q_2^{(n)}$ after each period. There are two possibilities, either the points $Q_1^{(n)}$ and $Q_2^{(n)}$ tend to each other or they remain apart and tend to other positions in a monotonic way. In both cases the limit points yield starting positions of $(Q-R)$ which are reproduced after one period.

We now return to our hydrodynamic problem. In order to obtain a mean value of thrust $\overline{T}$ which is assumed to be small of $O(\varepsilon)$, we have to perturb the base motion, described by L or $\alpha_0(x)$, by an added motion which is $O(\varepsilon)$. First we discuss the equations for the wake consisting of H^* and H as defined in section 4.1 regime iii, where we have to put the shed free vorticity.

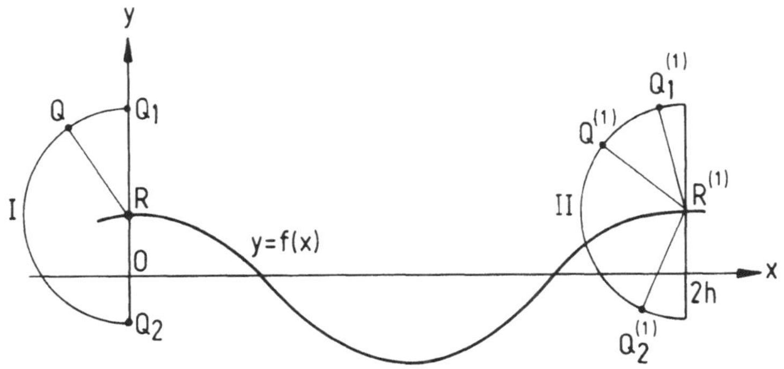

Fig. 4.8.2. Existence proof of periodic solution of (4.8.9).

The vorticity $\Gamma_0(s, t)$ at the profile when carrying out the base motion is by (4.8.4)

$$\Gamma_0(s,t)=\omega(l-2s)\left(\frac{l+s}{l-s}\right)^{1/2},$$
$$\omega=\frac{\mathrm{d}}{\mathrm{d}t}\alpha_0(x_Q)=\alpha_0'(x_Q)\frac{\mathrm{d}x_Q}{\mathrm{d}t}, \tag{4.8.11}$$

where $\mathrm{d}x_Q/\mathrm{d}t$ follows from the prescribed velocity of the point R. By means of the law of Biot and Savart we can determine the velocity field induced by the motion and can calculate the trajectories passed through by the fluid particles. In the following we refer the base motion of the profile to the line L or $y = g(x)$, along which the profile slides with its three quarter point Q, remaining tangent to it. The trajectories $(x(t), y(t))$ of the fluid particles have to satisfy the two coupled ordinary differential equations

$$\dot{x}(t)=v_x(x,y,t)$$
$$=-\frac{\omega}{2\pi}\int_{-l}^{l}\frac{(l-2s)\left(\frac{l+s}{l-s}\right)^{1/2}\left\{y-y_Q-(s+\tfrac{1}{2}l)\sin\alpha\right\}\mathrm{d}s}{R^2}, \tag{4.8.12}$$

$$\dot{y}(t)=v_y(x,y,t)$$
$$=\frac{\omega}{2\pi}\int_{-l}^{l}\frac{(l-2s)\left(\frac{l+s}{l-s}\right)^{1/2}\left\{x-x_Q-(s+\tfrac{1}{2}l)\cos\alpha\right\}\mathrm{d}s}{R^2}, \tag{4.8.13}$$

where

$$R^2=\left\{x-x_Q-(s+\tfrac{1}{2}l)\cos\alpha\right\}^2+\left\{y-y_Q-(s+\tfrac{1}{2}l)\sin\alpha\right\}^2. \tag{4.8.14}$$

We next formulate the concept wake of the base motion. A point (x^*, y^*) belongs at time t^* to the wake, if there exists a solution $(\tilde{x}(t), \tilde{y}(t))$ of (4.8.12) and (4.8.13) with

$$x^*=\tilde{x}(t^*),\ y^*=\tilde{y}(t^*), \tag{4.8.15}$$

and if there exists a $t' \leqslant t$ such that $\tilde{x}(t')$ and $\tilde{y}(t')$ are the coordinates of the trailing edge at the moment t'. Hence as has been said already the wake consists of those fluid particles which once have passed along the profile performing its base motion and left the profile at its trailing edge.

The numerical calculation of the wake is not difficult. We consider a fluid particle which at $t = t_j$ leaves the trailing edge and calculate its motion by (4.8.12) and (4.8.13). This can be done for a large number of times $j = 1, 2, 3, \ldots$, the particles which left at these times the trailing edge then form the wake H^*. In the neighbourhood of the trailing edge H^* is still in motion, however when the profile moves on the particles will ultimately come to rest and form far behind the profile the "ultimate wake" H which again is periodic.

It is interesting to look at the trajectory of an arbitrary fluid particle under influence of the base motion in case of the flat plate. An example is given in figure 4.8.3. The sharp cusps occur at the moment that the three quarter chord point Q of the infinitely thin flat profile passes a point of inflection of the line L, then the fluid comes to rest and starts to move in the inverse direction. Of course the particles which have passed along the profile have the same type of motion. However it is clear that the wake as a line of different fluid particles is smooth.

From the above it follows that besides the line $L(y = g(x))$ and the angle $\alpha_0(x)$, we can also calculate the time dependent part H^* of the wake and its ultimate periodic shape H far behind the profile.

Now as we mentioned, we have to choose an added motion of the profile in the neighbourhood of the base motion, which deviates only by an amount of $O(\varepsilon)$ from it. For instance we can define the added motion by a small periodic rotation $\alpha_1 = \alpha_1(x_Q)$ of the profile around a point S with $s = c$, (figure 4.8.4). Of course more general added motions are possible.

Suppose we have made a choice for $\alpha_1(x_Q)$ then we will need at the profile besides $\Gamma_0(s, t)$ (4.8.11) which is of $O(\varepsilon^0)$, a small extra amount of vorticity $\Gamma_1(s, t)$ of $O(\varepsilon)$, in order to let the fluid pass along it. Using

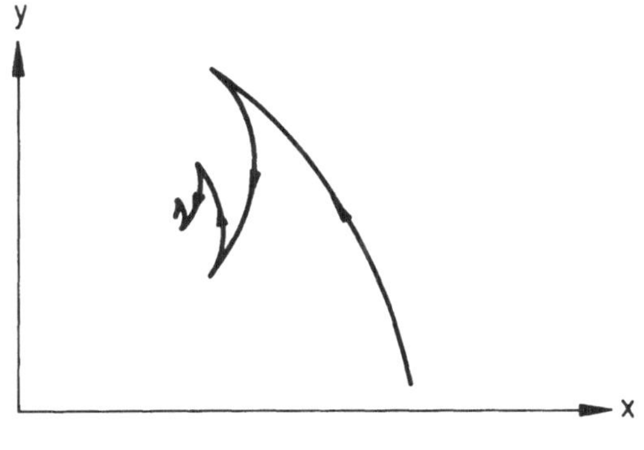

Fig. 4.8.3. Impression of the trajectory of an individual fluid particle, base motion.

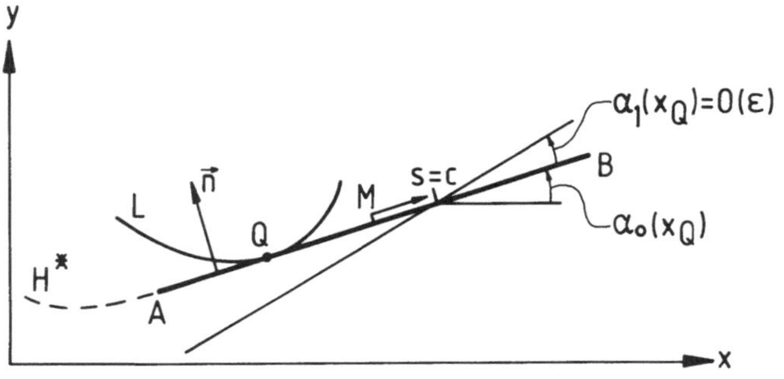

Fig. 4.8.4. The added motion as rotation $\alpha_1(x_Q)$ about $S(s=c)$.

the law of Biot and Savart we have the following singular integral equation for $\Gamma_1(s, t)$

$$\frac{1}{2\pi}\oint_{-l}^{l}\frac{\Gamma_1(\sigma, t)}{(s-\sigma)}\mathrm{d}\sigma = -\alpha_1 V_Q + (s-c)\frac{\mathrm{d}}{\mathrm{d}t}\alpha_1 - \tilde{v}_n, \tag{4.8.16}$$

where V_Q is the velocity of Q along L and v_n is the still unknown normal component of the velocity at the profile induced by the free vorticity at the wake H^*. Because our theory is linearized with respect to the small added motion, it is allowed within the accuracy of the theory to satisfy (4.8.16) at the profile $(A-B)$ carrying out the base motion. By this we have a smooth joining of the shed vorticity layer at H^* and the profile representation. In section (6.8) we will continue this discussion and consider optimum added motions.

4.9. Some additional remarks

For applications of unsteady propulsion it seems that large amplitude oscillation of a hydrofoil is important. A reason is that in case of small amplitude unsteady propulsion the side forces acting at the profile are large with respect to the useful thrust. The side forces are $O(\varepsilon)$ for the regimes i a and i b of section 4.1, while the thrust is only $O(\varepsilon^2)$. For large amplitude linearized theory, regimes ii and iii, the thrust and the side forces are both $O(\varepsilon)$, which is more favourable.

In section 4.1 we had to assume for a rational description of linearized large amplitude propulsion (regime ii), the wing W to be flexible. However, as we mentioned already in section 4.7, when the chord length of W is not too large the profile can be taken rigid without violating too much the assumptions of the linearization procedure. This has been done for instance by Wu [73], by de Graaf [18] and by Chopra [8]. In the work of Katz and Weihs [31] the profile is allowed to be elastic, it has a certain distributed bending stiffness by which it gives passively way to the hydrodynamic pressures. Also in this paper the influence of the transport of the shed free vorticity, by the fluid velocities induced by itself and by the profile motion is considered. These are in fact non linear effects.

In principle there are two different ways of approach to propulsion systems. The first one is to prescribe the motion of the propulsive system or in case of the elastic profile the way in which it is forced. Then the mean value of the thrust and the efficiency can be calculated. This has been carried out in most research pertaining to biology, such as [73], [8] and [31].

The second approach is, when we confine ourselves for instance to regime ii to prescribe the mean value of the thrust and the reference surface H, while we still have a free choice for the profile motions. Here we have the possibility to look for propulsion with an optimum efficiency without specifying the profile motions, because this efficiency depends only on H as will be shown in chapter 5. Then the optimum motion of the wing W is not determined and we can choose motions which meet requirements for a technical realization. The only constraint on the wing motions is that W leaves behind at H the a priori calculated optimum free vorticity. This will be discussed in section 6.8. An analogous reasoning holds for regime iii, however then we have to replace the prescribed surface H by the calculated wake as defined in the previous section.

In optimization theory we will go still one step further. For instance for regime ii we can also optimize the reference surface H itself under some constraints.

It seems to be recommendable in realizations of unsteady propulsion, to use two wings W_1 and W_2 (figure 4.9.1) which carry out motions which are or are nearly reflections of each other with respect to the x axis, in

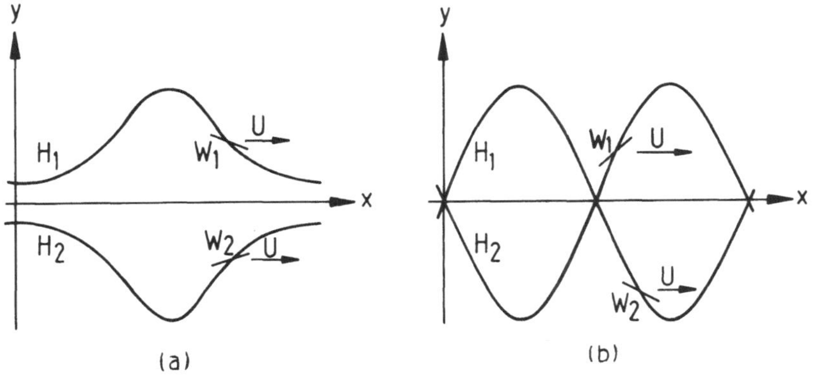

Fig. 4.9.1. Two profiles large amplitude, (a) symmetric motion, (b) nearly symmetric motion.

order to avoid large unsteady moments acting at the propelled body. Then the interaction of the two wings can be important [18]. Some numerical results for the two dimensional interaction in the case of optimum efficiency are given in section 5.8 in connection with a discussion about the qualitative comparison of the efficiencies of different optimum propulsion systems. In section 6.10 we give numerical results of the optimum efficiency of large amplitude propulsion by means of one wing or by means of two wings for the case of finite span.

5. Optimization theory

The theory discussed in this chapter is intended to give insight in the best way of working of a device which produces lift, thrust or any other prescribed force action. We assume the fluid to be inviscid and incompressible. First we have to define what will be called the best way of working of a device. We restrict ourselves to the minimization of a simple cost function, namely the kinetic energy losses per unit of time.

For instance consider a screw propeller with a given diameter, number of blades, velocity of advance and rotational velocity which has to yield a prescribed thrust. The question is: what has to be the circulation distribution along the blades in order that the kinetic energy left behind is as small as possible. This problem has been solved by Betz in the classical paper [5].

In this chapter we will derive necessary conditions for the optimum working of more general devices, including unsteady ones. The motion of the devices is assumed to be periodic. We demand a non zero mean value with respect to time of the force action, otherwise the kinetic energy left behind can be made zero and we have a trivial optimum.

The constraints on the force actions can be rather general. For instance it can be demanded, that a wing carrying out a flapping motion, delivers a prescribed mean value of lift as well as a prescribed mean value of thrust. Then we can ask for the motion which optimizes both force actions at the same time. It will turn out that the free vorticity left behind by the optimum system can be characterized in a simple way.

It is also allowed that the fluid through which the device moves is disturbed by a velocity field of $O(\varepsilon)$. When the situation is favourable energy can be extracted from this field in a useful way, by which the device needs less energy from outside to perform its task. A disturbance velocity field however can also be unfavourable as we will discuss. Then the device needs more energy to generate the prescribed force action than in an undisturbed fluid.

The theory developed here leaves out of consideration an important property of a fluid with respect to optimization namely its viscosity. When viscosity is neglected it will be seen that by increasing the size of a propeller its efficiency can be raised. This is the reason that in our theory we have to make a choice of for instance the diameter of a screw propeller. Such a choice does not need to be made in realistic optimization problems when the fluid is viscous. Then the diameter can be chosen so that the potential theoretical increase of the efficiency caused by an increase of the diameter will be annihilated by a decrease of the efficiency caused by the friction losses due to the fastly moving tips. Although we agree that viscosity is very important in optimization theory we will confine ourselves, in order to avoid mixing difficulties such as unsteady boundary layers and the occurrence of turbulence, to a linearized theory for incompressible and also inviscid fluids.

Essential in the considerations of this chapter is that the force action of the device is $O(\varepsilon)$, such as happens to be the case for the screw propeller or the unsteady propellers with finite amplitude of $O(\varepsilon^0)$ (for instance regimes ii and iii of section 4.1).

In this theory we will neglect forces of $O(\varepsilon^2)$ with respect to the desired force action of $O(\varepsilon)$. The forces of $O(\varepsilon^2)$ are due to leading edge suction and to second order errors of first order forces caused by the assumption that the blade vorticity as well as the trailing vorticity are assumed to be situated at the reference surfaces. For a screw propeller the leading edge suction forces are not important because they are nearly perpendicular to the direction of thrust. When however we consider a shrouded propeller, the suction forces acting at the leading edge of the shroud point in the direction of the thrust and can, in practice, be a non negligible part of it. For the question of the possibility of calculating the efficiency in spite of the errors of $O(\varepsilon^2)$ in the thrust of a propeller we refer to the last paragraph but one of section 4.1.

The concept efficiency only applies to lifting surfaces which perform useful work. For instance it applies to a screw propeller but not to a wing which has to generate a lift force perpendicular to its translational velocity.

5.1. Lifting surface system

We have a Cartesian coordinate system (x, y, z) embedded in an inviscid and incompressible fluid. The fluid is at rest at infinity with respect to the coordinate system. Consider m immaterial geometric reference surfaces H_l

$$H_l(x, y, z) = 0, \qquad l = 1, \ldots, m, \tag{5.1.1}$$

with

$$H_l(x + b, y, z) = H_l(x, y, z), \tag{5.1.2}$$

hence these surfaces are periodic with period b in the x direction and are

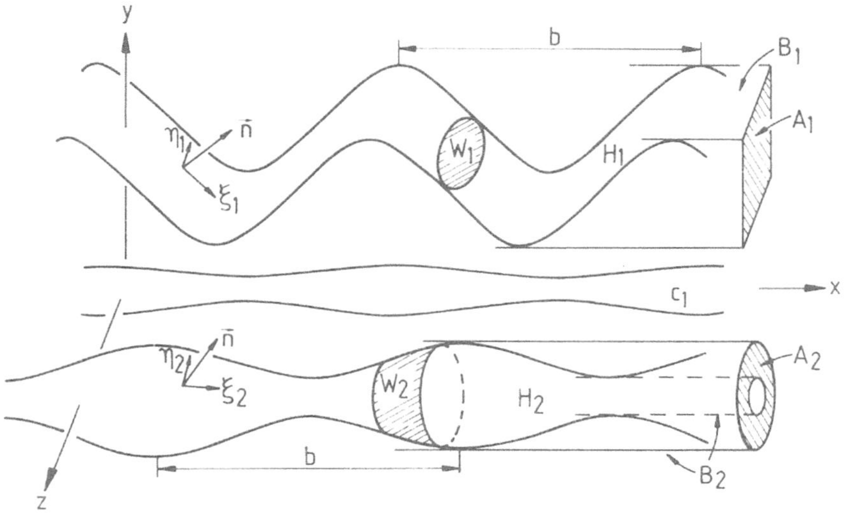

Fig. 5.1.1. A lifting surface system $m = 2, j = 1$.

assumed to be sufficiently smooth. On each surface we have an orthogonal coordinate system (ξ_l, η_l) (figure 5.1.1) in such a way that an increase in ξ_l by a number b_l, while η_l remains constant, makes that we obtain an equivalent point of H_l with respect to its periodicity. The regions of ξ_l and η_l are given by

$$-\infty < \xi_l < +\infty, \qquad \eta_{0,l} \leq \eta_l \leq \eta_{1,l}, \qquad l = 1, \ldots, j, \tag{5.1.3}$$

$$-\infty < \xi_l < +\infty, \qquad \eta_{0,l} \leq \eta_l < \eta_{1,l}, \qquad l = j+1, \ldots, m, \tag{5.1.4}$$

where the half open intervals in (5.1.4) belong to closed surfaces as for instance H_2 in figure 5.1.1. The lines

$$\eta_l = \text{const.}, \tag{5.1.5}$$

will form a one parameter family of curves on H_l, such that through each point of H_l passes one and only one such a line.

In order to introduce a + and a − side on H_l we consider the unit vectors $\boldsymbol{e}_\xi$ and $\boldsymbol{e}_\eta$, tangent to H_l and in the positive directions of ξ_l and η_l respectively. Then we agree that the unit normal $\boldsymbol{n} = \boldsymbol{e}_\xi * \boldsymbol{e}_\eta$ points from the negative side H_l^- to the positive side H_l^+.

Next we have possibly flexible lifting surfaces of finite extent. When these surfaces move exactly along the H_l, the fluid will not be disturbed when it was undisturbed before. Then we have the base motion as discussed in section 4.1. When the surfaces have to perform force actions they will carry out an added motion by moving in an ε neighbourhood of the reference surfaces H_l. Because our theory is linearized, we identify the lifting surfaces as locations of vorticity with their projections on the H_l, hence with their "planforms" W_l.

The force actions of W_l are represented by force fields $\boldsymbol{k}_l$ of $O(\varepsilon)$ which are in the direction of the normal $\boldsymbol{n}$

$$\boldsymbol{k}_l = k_l(\xi_l, \eta_l, t)\boldsymbol{n}_l(\xi_l, \eta_l), \qquad l = 1,\ldots,m, \tag{5.1.6}$$

where k_l are periodic scalar functions (time period τ)

$$k_l(\xi_l + b_l, \eta_l, t + \tau) = k_l(\xi_l, \eta_l, t). \tag{5.1.7}$$

These force fields give rise to pressure differences

$$(p_l^+ - p_l^-) = [p]_{-l}^{+} = k_l(\xi_l, \eta_l, t). \tag{5.1.8}$$

Because our theory will be linear, the free vorticity $\boldsymbol{\gamma}_l$ which is shed by the force fields $\boldsymbol{k}_l$ is assumed to remain where it is formed hence at the surfaces H_l.

In the fluid we admit impermeable rigid bodies which are also periodic with period b in the x direction. These bodies can be simply connected, hence stretch from $x = -\infty$ towards $x = +\infty$ (C_1, figure 5.5.1) or consist of periodically placed disconnected parts. When the lifting surfaces move through the fluid these bodies are boundaries on which the normal component of the induced velocities has to vanish.

We now define a lifting surface system, as the periodic reference surfaces H_l, with the force fields $\boldsymbol{k}_l$, together with the impermeable bodies C_j. Sometimes we denote a system simply by W_l.

Finally we introduce the working region and the working area of a lifting surface system. The working region is the three dimensional region of space enclosed by the most narrow geometric cylindrical surfaces with generators parallel to the x axis which enclose the H_l. The cross section of the cylinders will be called the working area of the system. In figure 5.1.1 the cylindrical surfaces are denoted by B_1 and B_2, where B_2 consists of two parts, one part outside H_2 and one part inside H_2. The working area consists of the regions denoted by A_1 and A_2, where A_2 is multiply connected.

5.2. Energy extraction out of a disturbed fluid

Consider an inviscid and incompressible fluid disturbed by a time independent velocity field $\boldsymbol{v}_0^*$ of $O(\varepsilon)$

$$\boldsymbol{v}_0^*(x, y, z) = \left(v_{0x}^*(x, y, z), v_{0y}^*(x, y, z), v_{0z}^*(x, y, z)\right). \tag{5.2.1}$$

The velocity field is supposed to be periodic in the x direction

$$\boldsymbol{v}_0^*(x+b, y, z) = \boldsymbol{v}_0^*(x, y, z). \tag{5.2.2}$$

We want to discuss the optimum energy extraction out of this fluid by means of a lifting surface system.

In order to keep the discussion as simple as possible we start with one flexible lifting surface moving through the fluid in an ε neighbourhood of a reference surface H, with a constant velocity U in the positive x direction. The discussion we give holds analogously for more lifting surfaces, while also one or more bodies of the type mentioned before are allowed to be present.

The wing has to extract by a suitable motion as much energy out of the velocity field $\boldsymbol{v}_0^*$ as possible. The questions we will consider are how much energy can be extracted by W out of the fluid when a linearized theory is valid and how can this energy be characterized. First we give a mechanical reasoning and then show by an analytical verification that the results are correct.

When the wing moves along the reference strip it can extract energy only by forces of $O(\varepsilon^2)$ in the direction of its motion. These forces originate from the interaction of the bound vorticity of $O(\varepsilon)$ of the wing and the component of the velocity field $\boldsymbol{v}_0^*$ of $O(\varepsilon)$ normal to H. Hence in the linearized theory the normal component of $\boldsymbol{v}_0^*$ is the only thing that matters to the wing. Therefore we replace the velocity field $\boldsymbol{v}_0^*$ by another one $\boldsymbol{v}_0$ which has the same normal component at H and of which the vorticity $\boldsymbol{\gamma}_0$ is entirely confined to H. In this case $\boldsymbol{\gamma}_0$ has two components

$$\boldsymbol{\gamma}_0 = \left(\gamma_{0\xi}(\xi, \eta), \gamma_{0\eta}(\xi, \eta)\right). \tag{5.2.3}$$

This can always be achieved by solving the Neumann problem for a

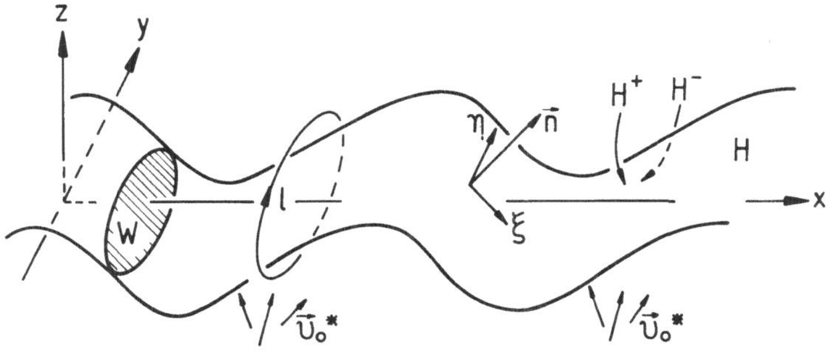

Fig. 5.2.1. Wing moving through disturbed fluid.

velocity potential Φ_0 with outside H, $\Delta\Phi_0 = 0$ and at H

$$\frac{\partial \Phi_0}{\partial n}(\xi, \eta) = \boldsymbol{v}_0^* \cdot \boldsymbol{n}, \tag{5.2.4}$$

where $\boldsymbol{n}$ is the unit normal at H. This potential can be made "uniquely valued" by demanding that the circulation round any contour l (figure 5.2.1) around H is zero. Then the velocity field $\boldsymbol{v}_0$ becomes

$$\boldsymbol{v}_0 = \text{grad}\, \Phi_0. \tag{5.2.5}$$

The kinetic energy of the fluid per period in the x direction has changed, because in general the original field $\boldsymbol{v}_0^*$ is different from the new one $\boldsymbol{v}_0$. It is the kinetic energy of the field $\boldsymbol{v}_0$ which can be extracted by a flexible wing. The reason is that such a wing, when it deforms suitably, can shed at H any vorticity for which the total circulation around H is zero. Hence it can annihilate $\boldsymbol{\gamma}_0$ (5.2.3) and bring to rest the velocity field $\boldsymbol{v}_0$.

The energy belonging to $\boldsymbol{v}_0$ is smaller than or at most equal to the kinetic energy of the original field $\boldsymbol{v}_0^*$. This is clear from the mechanical point of view. The wing W cannot extract more energy out of the fluid than is present in it.

We still give two somewhat different formulations of the principal result. First, the optimum wing W has to leave behind free vorticity which far behind W, induces a velocity field of which the normal component at H is opposite to the normal component of $\boldsymbol{v}_0^*$. Second, the optimum wing has to leave behind that vorticity on H which is needed when H is, as a rigid and impervious surface of zero thickness, embedded in the velocity field $\boldsymbol{v}_0^*$.

Next we give an analytical discussion of the subject. We suppose that the kinetic energy E_0^* of the original field $\boldsymbol{v}_0^*$ per period b in the x direction, is finite

$$E_0^* = \tfrac{1}{2}\mu \iint_{-\infty}^{\infty} \int_0^b (\boldsymbol{v}_0^*)^2 \, \mathrm{d}x \, \mathrm{d}y \, \mathrm{d}z < \infty. \tag{5.2.6}$$

We want to utilize this kinetic energy by a suitable action of the wing, hence by adding a periodic velocity field grad Φ_1 of which the vorticity is confined to H. Note that we are looking again far behind the wing. Then we have to minimize

$$\tilde{E} = \tfrac{1}{2}\mu \iint_{-\infty}^{\infty} \int_0^b (\boldsymbol{v}_0^* + \text{grad}\, \Phi_1)^2 \, \mathrm{d}x \, \mathrm{d}y \, \mathrm{d}z. \tag{5.2.7}$$

Assuming that Φ_1 is the optimum potential we change Φ_1 into $\Phi_1 +$

$\delta\Phi_1$, where $\delta\Phi_1$ is a periodic disturbance potential of which the vorticity is also confined to H. Then we find for the first variation $\delta\tilde{E}$ of $\tilde{E}$

$$\delta\tilde{E} = \mu \iint_{-\infty}^{\infty} \int_0^b (\boldsymbol{v}_0^* + \text{grad}\ \Phi_1) \cdot \text{grad}\ \delta\Phi_1 \, \mathrm{d}x \, \mathrm{d}y \, \mathrm{d}z. \tag{5.2.8}$$

Because the velocity field is without divergence we write

$$\delta\tilde{E} = \mu \iint_{-\infty}^{\infty} \int_0^b \text{div}\{(\boldsymbol{v}_0^* + \text{grad}\ \Phi_1)\delta\Phi_1\} \mathrm{d}x \, \mathrm{d}y \, \mathrm{d}z. \tag{5.2.9}$$

The boundary of the region of integration consists of H and the planes $x = 0$ and $x = b$. When we convert the volume integral into an integral over the boundary, it is seen by the periodicity of $(\boldsymbol{v}_0^* + \text{grad}\ \Phi_1)\delta\Phi_1$ that the planes $x = 0$ and $x = b$ do not contribute. Hence there remains an integration over $H(b)$ which is one period of H with $0 \leq x \leq b$. We find

$$\delta\tilde{E} = -\mu \iint_{H(b)} \{(\boldsymbol{v}_0^* + \text{grad}\ \Phi_1) \cdot \boldsymbol{n}\} [\delta\Phi_1]_-^+ \, \mathrm{d}S, \tag{5.2.10}$$

where $[\delta\Phi_1]_-^+$ is the jump of $\delta\Phi_1$ over H, $\mathrm{d}S$ an element of area and we used the facts that $(\boldsymbol{v}_0^* + \text{grad}\ \Phi_1) \cdot \boldsymbol{n}$ is continuous over H and that $\boldsymbol{n}$ points from H^- towards H^+.

The quantity $[\delta\Phi_1]_-^+$ is still arbitrary, we choose it such that it is zero at $H(b)$ with the exception of a small neighbourhood of an arbitrary point P of $H(b)$. This is most easily effectuated by taking for $\delta\Phi_1$ the potential of a small vortex ring of strength $\delta\gamma$ at H around P (figure 5.2.2) and about points which are equivalent with P with respect to the periodicity of H.Hence because in the optimum case $\delta\tilde{E} = 0$, it follows from (5.2.10)

$$\frac{\partial\Phi_1(P)}{\partial n} = -\boldsymbol{v}_0^*(P) \cdot \boldsymbol{n}(P), \qquad P \in H(b). \tag{5.2.11}$$

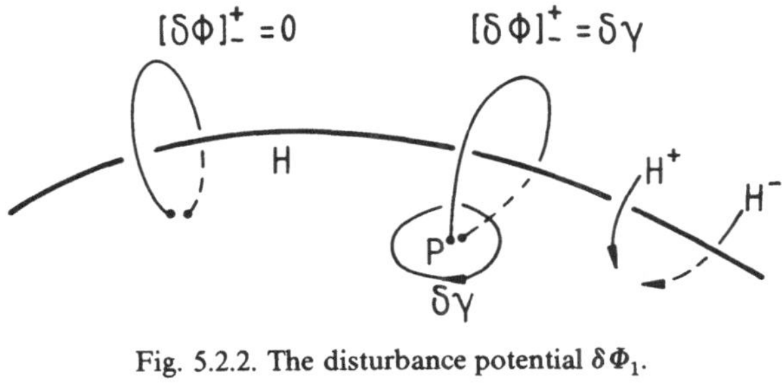

Fig. 5.2.2. The disturbance potential $\delta\Phi_1$.

From (5.2.4) we find

$$\Phi_1 = -\Phi_0. \tag{5.2.12}$$

In the previous mechanical reasoning it was argued that the maximum amount of energy which could be extracted out of the fluid was the kinetic energy $E(\Phi_0)$ belonging to Φ_0. Hence by (5.2.12) we have to prove

$$E_0^* - \tilde{E} = E(\Phi_0) = \tfrac{1}{2}\mu \iint_{-\infty}^{\infty} \int_0^b (\text{grad}\, \Phi_1)^2 \,\mathrm{d}x \,\mathrm{d}y \,\mathrm{d}z. \tag{5.2.13}$$

From (5.2.6) and (5.2.7) it follows that instead of (5.2.13) we have to show

$$-\iint_{-\infty}^{\infty} \int_0^b \boldsymbol{v}_0^* \cdot \text{grad}\, \Phi_1 \,\mathrm{d}x \,\mathrm{d}y\, z = \iint_{-\infty}^{\infty} \int_0^b (\text{grad}\, \Phi_1)^2 \,\mathrm{d}x \,\mathrm{d}y \,\mathrm{d}z. \tag{5.2.14}$$

The left hand side can be written as

$$-\iint_{-\infty}^{\infty} \int_0^b \text{div}(\Phi_1 \boldsymbol{v}_0^*) \,\mathrm{d}x \,\mathrm{d}y \,\mathrm{d}z = \iint_{H(b)} \boldsymbol{v}_0^* \cdot \boldsymbol{n} [\Phi_1]_-^+ \,\mathrm{d}S \tag{5.2.15}$$

and the right hand side as

$$\iint_{-\infty}^{\infty} \int_0^b \text{div}(\Phi_1 \,\text{grad}\, \Phi_1) \,\mathrm{d}x \,\mathrm{d}y \,\mathrm{d}z = -\iint_{H(b)} \frac{\partial \Phi_1}{\partial n} [\Phi_1]_-^+ \,\mathrm{d}S. \tag{5.2.16}$$

Hence equality follows from (5.2.11) and we find

$$E(\Phi_0) = \tfrac{1}{2}\mu \iint_{H(b)} \boldsymbol{v}_0^* \cdot \boldsymbol{n} [\Phi_1]_-^+ \,\mathrm{d}S. \tag{5.2.17}$$

From (5.2.13) it follows that

$$E_0^* \geqslant E(\Phi_0), \tag{5.2.18}$$

because $\tilde{E} \geqslant 0$. This inequality is closely related to Kelvin's minimum energy principle ([3], page 384).

Next we ask for the maximum amount of kinetic energy per period b, which can be extracted out of the velocity field $\boldsymbol{v}_0^*$ by means of an unlimited number of wing systems. The only condition on these systems is that their reference surfaces H_l have to be situated within a geometrical cylinder G

$$G:\ -\infty < x < +\infty, \qquad (y, z) \in A, \tag{5.2.19}$$

where A is some closed region in the (y, z) plane. In the optimum case as we have seen in the first part of this section, the free vorticity shed by the flexible wings moving along the H_l, is such that the H_l can be considered as rigid impervious surfaces. Hence by crowding the cylinder G more and more with reference surfaces, we can cut G into smaller and smaller disconnected regions, say into cubes of volume $\tilde{\varepsilon}^3$ where $\tilde{\varepsilon}$ is the small length of an edge.

Consider such a small cube in which we possibly have continuous free vorticity $\boldsymbol{\gamma}^*$ which belongs to the disturbance velocity field $\boldsymbol{v}_0^*$. Then it follows from the π theorem of dimensional analysis [7] that the kinetic energy contained in this sequestered cube becomes $O(\tilde{\varepsilon}^5)$ when $\tilde{\varepsilon} \to 0$. This means that the total kinetic energy for larger and larger numbers of H_l in the cylinder G tends to zero, because the number of cubes is only of $O(\tilde{\varepsilon}^{-3})$. In other words by the wings W_l in the optimum case, for larger and larger numbers of suitably chosen H_l, "all" the kinetic energy will be annihilated in the cylinder G.

However by the shed free vorticity of the W_l or what is the same by the impervious H_l, the component of the resulting velocity field, normal to the boundary ∂G of G has to vanish. The field outside the cylinder induced by the shed vorticity has a uniquely valued potential Φ_1 which has clearly to satisfy

$$\frac{\partial \Phi_1}{\partial n} = -\boldsymbol{v}_0^* \cdot \boldsymbol{n}, \qquad (x, y, z) \in \partial G. \tag{5.2.20}$$

This potential again follows from the solution of a Neumann problem.

From the foregoing it follows that the extractable kinetic energy E has the value

$$E = \tfrac{1}{2}\mu \left\{ \iiint_{V_0(b)} (\boldsymbol{v}_0^*)^2 \, \mathrm{dVol} - \iiint_{V_0(b)} (\boldsymbol{v}_0^* + \operatorname{grad} \Phi_1)^2 \, \mathrm{dVol} + \iiint_{V_i(b)} (\boldsymbol{v}_0^*)^2 \, \mathrm{dVol} \right\}, \tag{5.2.21}$$

where

$$\begin{aligned} V_0(b) &: 0 \leqslant x \leqslant b, \qquad (x, y, z) \notin G, \\ V_i(b) &: 0 \leqslant x \leqslant b, \qquad (x, y, z) \in G. \end{aligned} \tag{5.2.22}$$

Following the same reasoning as we used in (5.2.13)...(5.2.16) we can write (5.2.21) as

$$E = \tfrac{1}{2}\mu \left\{ \iint_{\partial G(b)} \boldsymbol{v}_0^* \cdot \boldsymbol{n} \Phi_1 \, \mathrm{d}S + \iint_{V_i(b)} \boldsymbol{v}_0^{*2} \, \mathrm{dVol} \right\}, \tag{5.2.23}$$

where $\boldsymbol{n}$ is the unit normal at ∂G pointing out of G and

$$\partial G(b): 0 \leqslant x \leqslant b, (x, y, z) \in \partial G. \tag{5.2.24}$$

It can be proved analogously as we did before, that

$$\Phi_0 = -\Phi_1, \tag{5.2.25}$$

is the potential of the velocity field $\boldsymbol{v}_0$ outside G, by which we can replace $\boldsymbol{v}_0^*$ outside G, without changing the maximum amount of extractable kinetic energy.

5.3. The variational problem for lifting surface systems

Our object will be the optimization of lifting surface systems as defined in section 5.1. The reference surfaces H_l and the impermeable rigid periodical bodies C_j are prescribed, while the force fields $k_l(\xi_l, \eta_l, t)$ (5.1.6) which represent the lifting surfaces may be varied. For simplicity we start with the case that only one reference surface H (coordinates ξ, η), is present. The fluid is assumed to be disturbed by a time independent velocity field $\boldsymbol{v}_0^*$ of $O(\varepsilon)$, which is periodic with period b in the x direction.

When the wing of finite extent has passed along, it has left behind free vorticity $\boldsymbol{\gamma}$ at H by which it alters the kinetic energy which before the passing of W, belonged to $\boldsymbol{v}_0^*$ alone. The resulting kinetic energy is wasted and should be made as small as possible. We introduce the velocity potential $\Phi(x, y, z)$ which belongs to the velocity field induced by $\boldsymbol{\gamma}$. This potential is independent of time because the wing is supposed to be already at a large distance. The kinetic energy E left behind per period can be written as

$$E = \tfrac{1}{2}\mu \iint_{-\infty}^{\infty} \int_0^b \{\boldsymbol{v}_0^* + \operatorname{grad} \Phi\}^2 \mathrm{d}x \, \mathrm{d}y \, \mathrm{d}z. \tag{5.3.1}$$

We remark that in this expression the behaviour of the wing is represented only by the potential Φ of the disturbance velocities caused by the shed vorticity $\boldsymbol{\gamma}$. From this it follows that any wing which sheds the same free vorticity, has the same losses of kinetic energy. Hence we can for instance replace the lifting surfaces, with respect to linearized optimization theory, also by lifting lines with suitable circulation distributions. This sometimes facilitates thinking about a problem (sections 5.5 and 5.7).

The energy E (5.3.1) has to be minimized under some constraints. For instance we can demand that the mean value of the thrust with respect to time, reckoned positive in the positive x direction, must have some prescribed value $\overline{T}$. This condition can by (5.1.6) be written as

$$\frac{1}{\tau}\int_0^{\tau}\iint_H k(\xi,\eta,t)\cos_{(n,x)}(\xi,\eta)\,\mathrm{d}S\,\mathrm{d}t=-\overline{T}, \tag{5.3.2}$$

where $\cos_{(n,x)}$ is the cosine of the angle between the normal $\boldsymbol{n}$ at H and the positive x direction and $\mathrm{d}S$ is an element of area of H. In (5.3.2) only the area of the lifting surface, which is a finite part of H, contributes to the integral.

Using the periodicity of the external force field and of the reference surface H we can write (5.3.2) as

$$\frac{1}{\tau}\int_0^{\tau}\sum_{m=-\infty}^{+\infty}\iint_{H(b)} k(\xi,\eta,t+m\tau)\cos_{(n,x)}(\xi,\eta)\,\mathrm{d}S\,\mathrm{d}t=-\overline{T}, \tag{5.3.3}$$

where $H(b)$ is a period of H, hence

$$\frac{1}{\tau}\int_{-\infty}^{+\infty}\iint_{H(b)} k(\xi,\eta,t)\cos_{(n,x)}(\xi,\eta)\,\mathrm{d}S\,\mathrm{d}t=-\overline{T}. \tag{5.3.4}$$

Only a finite interval of time contributes to this integral, namely the interval during which W passes $H(b)$. First we carry out the integration with respect to t for a fixed point (ξ,η) at H. We connect H^+ and H^- by a contour h (figure 5.3.1) and calculate the increase of the circulation along h by formula (1.3.18), where here the force field is concentrated at H and directed perpendicular to it. This increase in circulation is equal to the jump $[\Phi]_-^+$ of the potential Φ across H. Then we can write (5.3.4) as

$$\frac{\mu}{\tau}\iint_{H(b)}[\Phi]_-^+(\xi,\eta)\cos_{(n,x)}(\xi,\eta)\,\mathrm{d}S=\overline{T}, \tag{5.3.5}$$

the arguments of $[\Phi]_-^+$ are ξ and η because we consider limit values of Φ at H.

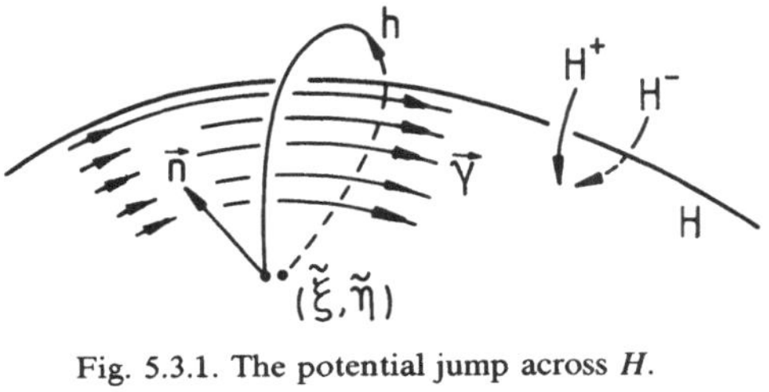

Fig. 5.3.1. The potential jump across H.

The jump $[\Phi]_-^+$ in the potential at a point $(\tilde{\xi}, \tilde{\eta})$ at H is also equal to the amount of shed vorticity $\boldsymbol{\gamma}$ which lies between $(\tilde{\xi}, \tilde{\eta})$ and that boundary of H which is enclosed by h. The direction of this vorticity is coupled with a right hand screw to the orientation of the contour h from the H^- side towards the H^+ side ($[\Phi]_-^+ < 0$, $k > 0$).

Formula (5.3.5) represents a constraint on the potential functions $\Phi(x, y, z)$ which are admitted in the variational problem. By replacing for instance the function $\cos_{(n,x)}(\xi, \eta)$ by $\cos_{(n,y)}(\xi, \eta)$, which is the cosine between the normal $\boldsymbol{n}$ at H and the positive y direction, we put a demand on $\Phi(x, y, z)$ so that a mean force in the y direction is delivered. More generally we can replace $\cos_{(n,x)}(\xi, \eta)$ in (5.3.5) by any periodic $g(\xi, \eta)$ and also we can put a number of constraints simultaneously at Φ

$$\iint_{H(b)} [\Phi]_-^+(\xi, \eta) g_\nu(\xi, \eta)\, \mathrm{d}S - F_\nu = 0, \qquad \nu = 1, \ldots, M, \tag{5.3.6}$$

where g_ν are prescribed periodic functions and the F_ν prescribed constants.

Next we consider the case that we have m lifting surfaces, hence m reference surfaces H_l, $l = 1, \ldots, m$. Then we have to replace (5.3.6) by

$$\sum_{l=1}^{m} \iint_{H_l(b)} [\Phi]_-^+|_l(\xi_l, \eta_l) g_{l\nu}(\xi_l, \eta_l)\, \mathrm{d}S - F_\nu = 0, \qquad \nu = 1, \ldots, M, \tag{5.3.7}$$

where $[\Phi]_-^+|_l$ denotes the values of $(\Phi^+ - \Phi^-)$ at H_l. By choosing the functions $g_{l\nu}(\xi_l, \eta_l) \equiv 0$ for a number of values of l the corresponding wings are not subject to constraint ν. Hence we can also put constraints on the resultant force action of selected groups of wings.

We assume the constraints not to be contradictory. Consider for instance a screw propeller with given angular velocity ω and velocity of advance U. Then it is easily seen that the thrust T and the moment M around the axis of rotation satisfy in the linearized theory

$$TU = M\omega. \tag{5.3.8}$$

Hence T and M cannot be prescribed independently of each other. Equation (5.3.8) follows from an energy consideration because both T and M are $O(\varepsilon)$ while the lost kinetic energy is $O(\varepsilon^2)$. Another way of deriving (5.3.8) is by means of a consideration of the blade geometry.

5.4. Necessary condition for an optimum

In this section we will give necessary conditions for the minimization of (5.3.1) with respect to Φ. We consider as in section 5.1, m reference surfaces H_l $(l=1,\ldots,m)$ and a number n of periodic bodies C_j $(j=1,\ldots,n)$. On the function Φ we put M constraints (5.3.7). Then instead of E we consider the functional

$$\tilde{E}=E+\sum_{\nu=1}^{M}\lambda_\nu\left\{\sum_{l=1}^{m}\iint_{H_l(b)}[\Phi]_-^+|_l g_{l\nu}(\xi_l,\eta_l)\,\mathrm{d}S-F_\nu\right\}, \tag{5.4.1}$$

where the λ_ν are Lagrange multipliers.

Suppose $\Phi(x,y,z)$ is the optimum potential we are looking for. Then we replace Φ by $\Phi+\delta\Phi$, where the perturbation potential $\delta\Phi$ has to be periodic with period b in the x direction and has to satisfy the condition of tangential flow to the periodic bodies C_j. We start with the first variation δE of E (5.3.1) which reads (see (5.2.9))

$$\delta E=\mu\iint_{-\infty}^{\infty}\int_0^b\operatorname{div}\{(\boldsymbol{v}_0^*+\operatorname{grad}\Phi)\delta\Phi\}\,\mathrm{d}x\,\mathrm{d}y\,\mathrm{d}z. \tag{5.4.2}$$

In (5.4.2) the integration is carried out over the slab $0\le x\le b$, $-\infty<y$, $z<+\infty$, minus the region cut out by the bodies C_j.

The boundary of the region of integration consists of $H_l(b)$ $(l=1,\ldots,m)$, the planes $x=0$ and $x=b$ minus the parts where the C_j cut these planes and the relevant parts of the boundaries ∂C_j of the C_j. When we convert the volume integral in (5.4.2) into an integral over the boundary, it is seen that besides the planes $x=0$ and $x=b$ also the boundaries ∂C_j yield no contribution, because the flow is tangent to them. Hence there remain the integrations over the $H_l(b)$, we find

$$\delta E=-\mu\sum_{l=1}^{m}\iint_{H_l(b)}\{(\boldsymbol{v}_0^*+\operatorname{grad}\Phi)\cdot\boldsymbol{n}\}|_l[\delta\Phi]_-^+|_l\,\mathrm{d}S, \tag{5.4.3}$$

because $(\boldsymbol{v}_0^*+\operatorname{grad}\Phi)\cdot\boldsymbol{n}$ is continuous at the H_l and $\boldsymbol{n}$ points from H_l^- towards H_l^+.

In the case of an optimum the first variation $\delta\tilde{E}$ of $\tilde{E}$ (5.4.1) has to vanish hence with (5.4.3)

$$\delta\tilde{E}=-\mu\sum_{l=1}^{m}\iint_{H_l(b)}\{(\boldsymbol{v}_0^*+\operatorname{grad}\Phi)\cdot\boldsymbol{n}\}|_l[\delta\Phi]_-^+|_l\,\mathrm{d}S$$
$$+\sum_{\nu=1}^{M}\lambda_\nu\sum_{l=1}^{m}\iint_{H_l(b)}g_{l\nu}(\xi_l,\eta_l)[\delta\Phi]_-^+|_l\,\mathrm{d}S=0. \tag{5.4.4}$$

As in section 5.2 the jump $[\delta\Phi]_-^+$ can be chosen to be zero everywhere, with the exception of the neighbourhood of some arbitrary point P at $H_l(b)$ (figure 5.2.2), then we find from (5.4.4)

$$\left.\frac{\partial \Phi}{\partial n}\right|_l (\xi_l, \eta_l) = -\boldsymbol{v}_0^* \cdot \boldsymbol{n}|_l (\xi_l, \eta_l) + \frac{1}{\mu} \sum_{\nu=1}^{M} \lambda_\nu g_{l\nu}(\xi_l, \eta_l), \qquad l = 1, \ldots, m. \tag{5.4.5}$$

This relation is a necessary condition for the normal component $\partial\Phi/\partial n$ of the velocity induced at each point of the reference surfaces H_l by the free vorticity shed by the optimum wings. Hence we have to solve for the calculation of Φ a Neumann problem (also $\partial\Phi/\partial n = 0$ at ∂C_j), while the Lagrange multipliers λ_ν are still unknown. Having solved the problem we can determine te λ_ν by the conditions (5.3.7).

From the necessary conditions (5.4.5) for an optimum we can draw two conclusions. First, we could have replaced the original disturbance field $\boldsymbol{v}_0^*$ by the field $\boldsymbol{v}_0$ as defined in section 5.2, hence with its vorticity γ_0 confined to H_l and with a potential Φ_0 defined outside H_l. Second, we can divide the action of the lifting surface system into two parts, in other words, the potential Φ can be split into two parts

$$\Phi = \Phi_1 + \Phi_2 \tag{5.4.6}$$

which we will discuss separately.

In the first place the lifting surfaces W_l have to "clean" their reference surfaces H_l from the vorticity γ_0 belonging to the modified disturbance velocity field $\boldsymbol{v}_0$ of which the kinetic energy can be extracted. The corresponding potential Φ_1 follows from

$$\left.\frac{\partial \Phi_1}{\partial n}\right|_l (\xi_l, \eta_l) = -\boldsymbol{v}_0^* \cdot \boldsymbol{n}|_l (\xi_l, \eta_l) = -\left.\frac{\partial \Phi_0}{\partial n}\right|_l (\xi_l, \eta_l), \; l = 1, \ldots, m, \tag{5.4.7}$$

where Φ_0 and Φ_1 have the same meaning as in section 5.2 and $\Phi_1 = -\Phi_0$.

In performing a motion suitable for this task, force actions have to be exerted on the fluid and their reactions act at the W_l. These reactions of the fluid can already "partly fulfill" the constraints (5.3.7) put on the lifting surface system.

The second part Φ_2 of Φ belongs to that part of the motion of the W_l, which is carried out along their "cleaned" reference surfaces H_l. These motions have to supply the deficiency in the demanded force actions (5.3.7)

$$\left.\frac{\partial \Phi_2}{\partial n}\right|_l (\xi_l, \eta_l) = \frac{1}{\mu} \sum_{\nu=1}^{M} \lambda_\nu g_{l\nu}(\xi_l, \eta_l), \qquad l = 1, \ldots, m. \tag{5.4.8}$$

The λ_ν have to be chosen such that $\Phi = \Phi_1 + \Phi_2$ satisfies (5.3.7).

Now consider the kinetic energy E_p put into the fluid per period of length b of the motion by an optimum lifting surface system W_l. This energy is the difference between the kinetic energy in the fluid over one period far downstream of W_l and the kinetic energy which was present in the fluid at that period long before W_l arrived, hence

$$E_p = \tfrac{1}{2}\mu \iint_{-\infty}^{\infty} \int_0^b (\boldsymbol{v}_0^* + \operatorname{grad} \Phi_1 + \operatorname{grad} \Phi_2)^2 \, \mathrm{d}x \, \mathrm{d}y \, \mathrm{d}z$$

$$- \tfrac{1}{2}\mu \iint_{-\infty}^{\infty} \int_0^b (\boldsymbol{v}_0^*)^2 \, \mathrm{d}x \, \mathrm{d}y \, \mathrm{d}z. \tag{5.4.9}$$

It is not difficult to show by using formulas analogous to (5.2.13) that we can write this energy also in the form

$$E_p = \tfrac{1}{2}\mu \iint_{-\infty}^{\infty} \int_0^b (\operatorname{grad} \Phi_2)^2 \, \mathrm{d}x \, \mathrm{d}y \, \mathrm{d}z - \tfrac{1}{2}\mu \iint_{-\infty}^{\infty} \int_0^b (\boldsymbol{v}_0)^2 \, \mathrm{d}x \, \mathrm{d}y \, \mathrm{d}z. \tag{5.4.10}$$

In words, for the calculation of E_p we may replace from the beginning the original disturbance field $\boldsymbol{v}_0^*$ by the modified field $\boldsymbol{v}_0$. Hence all fields $\boldsymbol{v}_0^*$ with the same normal component at H_l are equivalent with respect to the optimization problem.

From this it follows that after the energy extraction by Φ_1, we can consider the creation of the potential Φ_2 to be carried out in an undisturbed fluid.

Finally we discuss the force actions needed to induce Φ_1. Suppose these are all in the desired direction, then the $|\lambda_\nu|$ can be smaller than they have to be without the disturbance field $\boldsymbol{v}_0^*$. This is favourable for two reasons. First, all energy is extracted from the field $\boldsymbol{v}_0$ and becomes available. Second, the energy needed for the smaller values of $|\lambda_\nu|$ is less than the energy needed to satisfy the constraints in an unperturbed fluid. In the other case, when the force actions needed to create Φ_1 are or are partly opposite to the desired ones, the $|\lambda_\nu|$ or some of them have to be larger than in the undisturbed case. Then although energy is extracted from the field $\boldsymbol{v}_0$, the extra energy needed for these larger values of $|\lambda_\nu|$ can become so large that it nullifies or makes negative the profit of the energy extraction. In the latter case the disturbance field $\boldsymbol{v}_0^*$ is unfavourable. This will be shown to be possible by means of the simple example discussed in the next section.

5.5. Optimum wing in up- or downward flow

In order to demonstrate the use of the optimization condition (5.4.5), we consider an optimum wing moving with constant velocity U in the positive x direction along a flat strip H; $-\infty < x < +\infty$, $|y| \leq a$, $z = 0$ (figure 5.5.1)). The coordinate system (ξ, η) of H can be identified in this case with (x, y). The disturbance velocity field $\boldsymbol{v}_0^*$ is constant and has the form

$$\boldsymbol{v}_0^* = (0, 0, v_{0z}^*), \qquad v_{0z}^* = O(\varepsilon). \tag{5.5.1}$$

When the wing has passed and is far away in the positive x direction, all phenomena are independent of x. Hence, for the period b in the x direction we can take any number, for instance the value $b = 1$. We demand that the wing delivers a lift force L in the positive z direction. This example will show the effect of the direction of $\boldsymbol{v}_0^*$ as mentioned in the last paragraph of the preceding section.

First we calculate the vorticity $\boldsymbol{\gamma}_0$ at H which belongs to the modified disturbance field $\boldsymbol{v}_0$. The vorticity is, because $\boldsymbol{v}_0^*$ is independent of the x and the y coordinate, parallel to the x axis and will be denoted by a scalar function $\gamma_0(y)$. The potential Φ_0 of $\boldsymbol{v}_0$ for $(x, y, z) \notin H$, is independent of x and satisfies

$$\frac{\partial \Phi_0}{\partial z} = v_{0z}^*, \qquad |y| \leqslant a, \qquad z = 0. \tag{5.5.2}$$

Hence we have for $\gamma_0(y)$ by (1.11.2) the singular integral equation

$$-\frac{1}{2\pi} \oint_{-a}^{a} \frac{\gamma_0(\eta)}{(\eta - y)} \mathrm{d}\eta = v_{0z}^*, \qquad |y| \leq a, \tag{5.5.3}$$

where $\gamma_0(y)$ is positive with a right hand screw in the positive x direction.

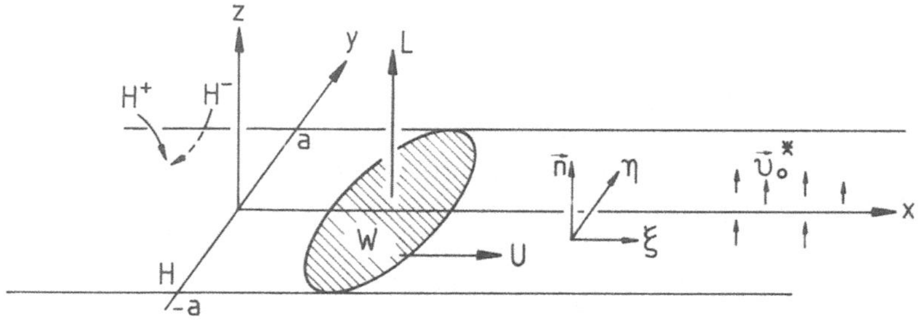

Fig. 5.5.1. Wing moving along a flat strip.

Further because Φ_0 has to be uniquely valued the circulation around H has to be zero

$$\int_{-a}^{a} \gamma_0(y)\,\mathrm{d}y = 0, \tag{5.5.4}$$

and we tolerate for $\gamma_0(y)$, in connection with the integrability of the kinetic energy, singularities at $y = \pm a$ of the square root type. The solution of this equation is discussed in [47] page 249 and in appendix A formula (A.3.14) with $c_2 = 0$, we find

$$\gamma_0(y) = -2v_{0z}^{*}y(a^2-y^2)^{-1/2}, \qquad |y| \le a, \qquad z = 0. \tag{5.5.5}$$

In this case the necessary condition (5.4.5) for the potential Φ of the flow field far behind the wing, becomes

$$\frac{\partial \Phi}{\partial z} = -v_{0z}^{*} + \lambda, \qquad |y| \le a, \qquad z = 0, \tag{5.5.6}$$

where λ is the only Lagrange multiplier which occurs in this problem because we need only one function $g_{l\nu}$ namely $\cos_{(n,y)} \equiv 1$. The vorticity which belongs to this equation is, because also here the right hand side is a constant, by (5.5.5)

$$\gamma(y) = -2(-v_{0z}^{*} + \lambda)y(a^2-y^2)^{-1/2}. \tag{5.5.7}$$

From (5.5.7) it follows

$$[\Phi]_{-}^{+}(y) = 2(-v_{0z}^{*} + \lambda)\int_{-a}^{y} \frac{\eta\,\mathrm{d}\eta}{(a^2-\eta^2)^{1/2}} = -2(-v_{0z}^{*} + \lambda)(a^2-y^2)^{1/2}. \tag{5.5.8}$$

Because $[\Phi]_{-}^{+}(y)$ equals the circulation of the wing W at the place y, we find that the well known elliptic circulation distribution has appeared.

In order to calculate λ we use (5.3.7) which reduces in this simple case, to

$$\int_{-a}^{a} [\Phi]_{-}^{+}(y)\,\mathrm{d}y = \frac{L}{\mu U}. \tag{5.5.9}$$

This formula follows more directly from (5.3.5) by putting $\tau = U^{-1}$ and replacing $\cos_{(n,x)}$ by $\cos_{(n,y)} \equiv 1$.

Hence we have

$$-2\mu U(-v_{0z}^* + \lambda)\int_{-a}^{a} (a^2 - y^2)^{1/2}\,\mathrm{d}y = -\mu\pi U a^2(-v_{0z}^* + \lambda) = L, \tag{5.5.10}$$

or

$$\lambda = \frac{-L}{\mu\pi U a^2} + v_{0z}^*. \tag{5.5.11}$$

The kinetic energy per unit of length in the x direction, which remains in the fluid after the wing has passed by (when we have again changed from $\boldsymbol{v}_0^*$ to $\boldsymbol{v}_0$), equals

$$-\tfrac{1}{2}\mu\int_{-a}^{a} [\Phi + \Phi_0]_-^+ \frac{\partial}{\partial n}(\Phi + \Phi_0)\,\mathrm{d}y = -\tfrac{1}{2}\mu\int_{-a}^{a} -2\lambda(a^2 - y^2)^{1/2}\cdot\lambda\,\mathrm{d}y$$

$$= \tfrac{1}{2}\pi\mu a^2\left(-\frac{L}{\mu\pi U a^2} + v_{0z}^*\right)^2. \tag{5.5.12}$$

The energy necessary to move the wing over one unit of length in the x direction is this amount minus the energy per unit of length belonging to the modified velocity field $\boldsymbol{v}_0$ already present in the fluid. This becomes

$$\tfrac{1}{2}\left(\frac{L^2}{\mu\pi U^2 a^2} - 2\frac{L v_{0z}^*}{U}\right). \tag{5.5.13}$$

When there is no disturbance field v_0^* this energy has the value

$$\frac{L^2}{2\mu\pi U^2 a^2}. \tag{5.5.14}$$

Hence when $v_{0z}^* > 0$ the energy needed to move the wing becomes smaller and inversely.

We now compare these results with the remarks of the last paragraph of the previous section, especially the influence of the direction of the force actions connected to the part Φ_1 of Φ, needed to annihilate the vorticity γ_0 of $\boldsymbol{v}_0$. In order to calculate these force actions we replace the wing by a lifting line parallel to the y axis, with bound vorticity $\Gamma_1(y)$ (figure 5.5.2). As has been argued before, this does not reduce generality when we consider linearized optimization theory. The bound vorticity

$\Gamma_1(y)$ has in order to extract the kinetic energy of the velocity field $\boldsymbol{v}_0$, to shed free vorticity $\gamma_1(y)$ opposite to $\gamma_0(y)$, clearly

$$\Gamma_1(y) = -2v^*_{0z}(a^2 - y^2)^{1/2}, \tag{5.5.15}$$

reckoned positive with a right hand screw in the positive y direction. Far behind the lifting line the free vorticity γ_1 creates the potential $\Phi_1 = -\Phi_0$. In figure 5.5.2 the situation is drawn when Γ_1 is slightly above γ_0. When the bound vorticity Γ_1 moves with the velocity U in the positive x direction, it delivers for the case $v^*_{0z} > 0$ of figure 5.5.2, two force components. First, a force $\tilde{T}$ in the positive x direction of magnitude

$$\tilde{T} = -\mu\tfrac{1}{2}v^*_{0z}\int_{-a}^{a} \Gamma_1(y)\,\mathrm{d}y = \tfrac{1}{2}\pi\mu a^2 v^{*2}_{0z} = \mathrm{O}(\varepsilon^2), \tag{5.5.16}$$

by which energy is extracted out of the fluid. Here we have to use $\frac{1}{2}v^*_{0z}$ in the calculation of $\tilde{T}$ because behind $\Gamma_1(y)$ the vorticity has disappeared by the energy extraction. Second, a lift force $\tilde{L}$ in the positive z direction of magnitude

$$\tilde{L} = -\mu U\int_{-a}^{a} \Gamma_1(y)\,\mathrm{d}y = \mathrm{O}(\varepsilon), \tag{5.5.17}$$

which contributes to the prescribed lift force L. It is even possible that $\tilde{L}$ is larger than L, then it is clear that we do not need to extract all the kinetic energy out of the modified disturbance velocity field $\boldsymbol{v}_0$. This can happen to a seagull flying along a range of dunes when a transverse wind is blowing.

When $v^*_{0z} < 0$ also $\Gamma_1(y)$ changes sign, hence $\tilde{T}$ (5.5.16) remains positive however $\tilde{L}$ (5.5.17) becomes negative, this is the unfavourable situation. Then the energy, needed to deliver the lift force $L - \tilde{L}$ in the

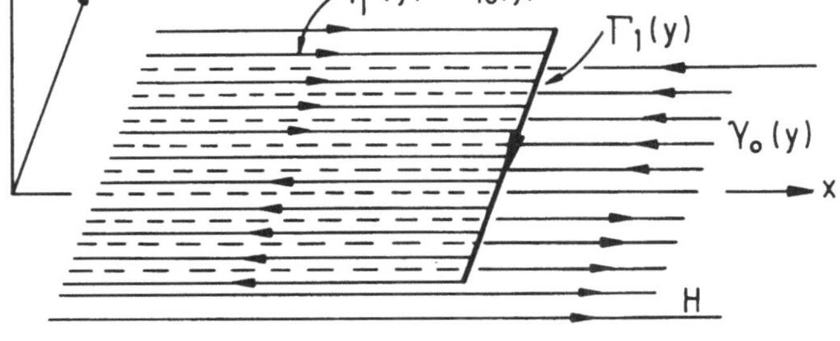

Fig. 5.5.2. Extraction of kinetic energy out of $\boldsymbol{v}_0$; $v_{0z} > 0$.

now undisturbed fluid, minus $\tilde{T}U$; is larger than the energy needed to deliver L in the undisturbed fluid.

Analogous phenomena can occur for instance with respect to a screw propeller working in a disturbed fluid behind a ship. Here the fluid is dragged along by the hull and has a velocity in the desired direction of the thrust, this is favourable.

5.6. Classes of lifting surface systems, quality number

Up to now we have introduced lifting surface systems and have discussed their optimization. The reference surfaces H_l were given and we determined the circulation distributions or the force fields k_l (5.1.6) of the wings W_l moving along them, in such a way that certain constraints (5.3.7) are satisfied and that the kinetic energy left behind per period is minimum. Now we will give more freedom to the admitted lifting surface systems, to this end we introduce "classes" of these systems. We use the notations and definitions of section 5.1. It is again allowed that the fluid is disturbed in advance by a velocity field $\mathbf{v}_0^*(x, y, z)$ of $O(\varepsilon)$, periodic in the x direction.

A class of periodically moving lifting surface systems consists of all those systems which meet a certain number of conditions. There are four principal properties which systems belonging to one and the same class have in common.

They will have:

1. the same mean velocity of advance,
2. the same period,
3. admitted reference surfaces which can be embedded in the same cylindrical region, called the working region of the class, outside of which there are the same rigid bodies,
4. the same prescribed mean force actions.

Additional conditions with respect to the geometry of the reference surfaces can be imposed. For instance we can demand that these surfaces make an angle with the x axis which is smaller than or equal to a prescribed value. This means that the surfaces H are not allowed to be too steep.

We will consider a "best lifting surface system of a class". This is a system which leaves behind the greatest lower bound of the kinetic energy losses of all systems of that class.

Not each class has one or more of these wing systems. Assume for simplicity that the fluid is undisturbed, hence $\mathbf{v}_0^* = 0$ and consider the class of screw propellers with prescribed diameter, velocity of advance, thrust and number of blades, while the pitch of the helicoidal reference surface is left free. As we will see later on, the kinetic energy can be lowered when it is possible to increase the rotational velocity. Hence no finite rotational velocity can appear as the result of the optimization process and hence there is no best screw propeller of this class. This is the reason that in the definition of best lifting surface system of a class, we used the concept "greatest lower bound of the kinetic energy losses" because that always exists. Of course this example only holds in our linearized theory and with the neglect of viscosity and cavitation.

We next introduce the concept "ideal lifting surface system connected to a class C". To this end we consider an other class C_0 which has in common with C the four principal properties mentioned in the second paragraph of this section, further there are no conditions on C_0. A best lifting surface system of C_0 is called an ideal lifting surface system connected to C or pure and simple an ideal system, when C or C_0 is known. Note that the class C_0 can be more comprehensive than C because we can have, additional geometrical conditions on the reference surfaces H_l of C.

Suppose we take some set of reference surfaces H_l which are compatible with C_0, hence they are within the working region and have the correct period. Then we can find an optimum wing system W_l for these H_l by the method described before. This system belongs to C_0. We cannot expect that W_l is an ideal system, because in general we will need other reference surfaces. Choose another set $\tilde{H}_l$ which is also compatible with C_0 and look at the combined set $H_l \cup \tilde{H}_l$. Again we can find an optimum wing system by our previous method which now belongs to $H_l \cup \tilde{H}_l$. This system also belongs to C_0 and again it does not need to be ideal with respect to this class. Now we can formulate the following criterion:

When for each combination of a fixed set H_l with any other set $\tilde{H}_l$ no vorticity is needed in the optimum case on the set $\tilde{H}_l$, it follows that the set H_l itself is able to yield an ideal system.

We will discuss this subject in some more detail. In order to find an ideal lifting surface system we follow the same method as in the last part of section 5.2, where we considered the energy extraction by means of lifting surfaces W_l moving along reference surfaces H_l which have to remain within the geometrical cylinder G (5.2.19). Also now we have a periodic initial disturbance field $\boldsymbol{v}_0^*$ in space. Then we crowd the cylinder G more and more with periodic reference surfaces H_l and optimize each time the lifting surface systems under the constraint of some prescribed force action. By (5.4.6) we can split the originating potentials Φ into Φ_1 and Φ_2, where Φ_1 belongs to vorticity which annihilates the normal component of $\boldsymbol{v}_0^*$ at the H_l. Hence the effect of Φ_1 outside G is that in the limit of very many H_l it compensates the normal component of $\boldsymbol{v}_0^*$ at the boundary ∂G of G. Then the potential Φ_2, which belongs to wing systems with the same reference surfaces H_l, takes care of the prescribed force actions. As we discussed in section 5.4 we can consider Φ_2 to be created by the W_l but now moving in an undisturbed fluid. The force action which has to be delivered by Φ_2 depends also on the force action already delivered by Φ_1. We return to this in section 5.7 and in chapter 6.

From the foregoing it follows that the action of an ideal lifting surface system connected to class C can be split into two parts. First find Φ_1 which is created by "infinitely" many H_l and which brings to rest the flow in the cylinder G. Next find a best lifting surface system of C_0 in an undisturbed fluid.

The potential Φ_1 is the same as the one mentioned in (5.2.25). The vorticity belonging to Φ_2 and the potential Φ_2 itself will be discussed in the next section for the special case of a propulsion device (5.7.4) (5.7.5)). However it can be seen easily how the method can be generalized to other force actions.

We now discuss an interesting quantity of an optimum lifting surface system W_l which belongs to a certain class C, namely its quality number. We denote again by E_p the amount of kinetic energy put into the fluid per period of length b by W_l. This amount is given in (5.4.9) or more explicitly in (5.4.10). By E_i we denote the amount of kinetic energy put into the fluid per period of length b by an ideal lifting surface system connected to the class C. This amount is given by

$$E_i = \tfrac{1}{2}\mu \iint_{-\infty}^{\infty} \int_0^b (\operatorname{grad} \Phi_2)^2 \,\mathrm{d}x\,\mathrm{d}y\,\mathrm{d}z - \tfrac{1}{2}\mu \iiint_{V_0(b)} (\tilde{\boldsymbol{v}}_0)^2 \,\mathrm{dVol}$$

$$- \tfrac{1}{2}\mu \iiint_{V_i(b)} (\boldsymbol{v}_0^*)^2 \,\mathrm{dVol}, \tag{5.6.1}$$

where $V_0(b)$ and $V_i(b)$ are defined in (5.2.22).

It can happen theoretically that E_p or E_i or both are negative which means that more energy can be extracted out of the field $\boldsymbol{v}_0$ than is needed for the force action. In practical situations of for instance ship propulsion however this will hardly occur and we assume $E_p > 0$ and $E_i > 0$. Note that in the case of the already mentioned seagull flying along a range of dunes or of a sailplane flying in thermals this need not to be true. Then we define the quality number q by

$$q = \frac{E_i}{E_p} \leq 1, \qquad E_p > 0, \qquad E_i > 0, \tag{5.6.2}$$

the inequality follows directly from the meaning of E_p and E_i.

As an illustration of the use of q we consider a class C of propellers. With respect to the efficiency η_T and the value q of an optimum propeller chosen from C, we can roughly speaking distinguish between four cases

$$\eta_T \approx 1, q \approx 1; \quad \eta_T \approx 1, q < 1; \quad \eta_T < 1, q \approx 1; \quad \eta_T < 1, q < 1. \tag{5.6.3}$$

In our linearized theory, for the type of propulsion we consider in this chapter, we have $\eta_T = 1 - O(\varepsilon)$. Hence it seems that always $\eta_T \approx 1$ however the linearized theory is most probably not devoid of sense for let us say η_T is about 0.7. For this kind of values we write $\eta_T < 1$.

The efficiency is about one in the first two cases (5.6.3), hence the propeller cannot be improved substantially. In the third case q is nearly one, this means that also in this case the propeller cannot be improved significantly because its energy loss per period is nearly equal to the energy loss of an ideal propeller connected to class C ($E_p \approx E_i$). In the fourth case $\eta_T < 1$ and $q < 1$ we can state that the propeller can be improved from the potential theoretical point of view when we possibly enlarge the class to which it had to belong, by demanding only the four principal properties 1,...,4 given in the second paragraph of this section. In section 6.4 we give an application of the quality number in relation with the influence of endplates on the performance of a screw propeller.

Using the quality number we can write the efficiency of a propeller as

$$\eta_T = \frac{b\bar{T}}{b\bar{T} + E_p} = \left(1 + \frac{E_i}{qb\bar{T}}\right)^{-1}, \tag{5.6.4}$$

where $\bar{T}$ is the mean value of the thrust. Note that E_i is proportional to $\bar{T}^2$.

5.7. The ideal propeller

Consider the circular cylindrical region G (figure 5.7.1)

$$G:\ -\infty < x < +\infty, \qquad y^2 + z^2 \leq R^2, \tag{5.7.1}$$

which is assumed to be the working region of a class of lifting surface systems as discussed in the previous section. The reference surfaces H_l will have a period b in the x direction, the lifting surfaces have a mean velocity of advance U and have to deliver together a mean value of thrust $\bar{T}$. The fluid is undisturbed, hence $\mathfrak{v}_0^* = 0$. We will construct an ideal propeller for this class. That the cross section is circular is only to fix attention, it will become clear that it is allowed to possess any shape. The results we find in this section yield the potential Φ_2 in E_i (5.6.1) which then can be considered to be known for propellers. There is one complication, as we already mentioned in section 5.6. The thrust $\bar{T}$ which has to be delivered by Φ_2 depends on the thrust created already by Φ_1 when the maximum amount of kinetic energy of the disturbance field $\mathfrak{v}_0^*$ is extracted.

A set H_l of two reference surfaces H_1 and H_2 is sufficient for our purpose. The surface H_1 is the cylinder which is the boundary of the working region.

$$H_1: y^2 + z^2 - R^2 = 0. \tag{5.7.2}$$

The surface H_2 is more complicated. First, it is periodic in the x direction with period b and axisymmetric with respect to the x axis. Second, it contracts up to the x axis, then expands up to the surface H_1, then contracts again and so on.

The condition for the potential $\Phi(x, y, z)$ belonging to the free vorticity shed by the optimum propeller follows from (5.4.5). Because $v_0^* = 0$ and we have only to deliver a thrust, we find

$$\left.\frac{\partial \Phi}{\partial n}\right|_l (\xi_l, \eta_l) = \lambda \cos_{(n,x)}(\xi_l, \eta_l), \qquad l = 1, 2. \tag{5.7.3}$$

We have two reference surfaces H_1 and H_2 and only one condition for both together. Hence (5.7.3) follows from (5.4.5) for $m = 2$, $M = 1$, and one function $g_{l\nu}$ namely $\cos_{(n,x)}$. Therefore only one Lagrange multiplier λ appears. Condition (5.7.3) can be interpreted as follows. The potential Φ can be created by translating the two reference surfaces H_1 and H_2, as rigid and impermeable surfaces with the velocity λ in the positive x direction. The vorticity needed on the impermeable H_1 and H_2 is the free vorticity left behind by the optimum propeller. The constant λ follows from the condition that the mean value of the thrust has to be $\bar{T}$.

It is clear that when H_1 and H_2 are placed in a flow parallel with the x axis, no velocity inside H_1, hence inside the working region will occur. This means that H_2 will not carry any free vorticity inside H_1. Outside H_1, we have the undisturbed parallel flow, hence the vorticity on H_1, is constant, circular and perpendicular to the x axis.

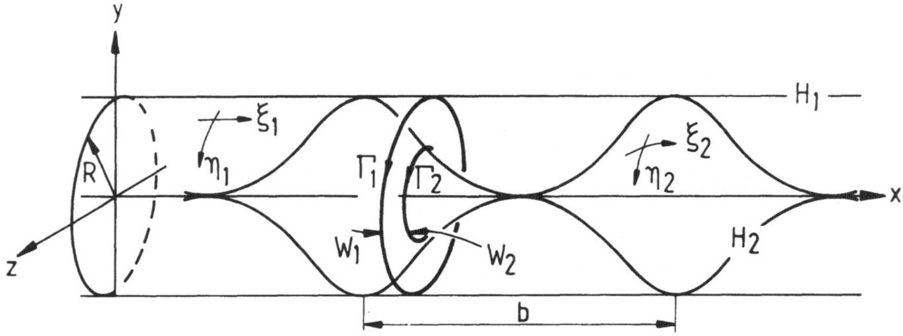

Fig. 5.7.1. A possible "realization" of an ideal thrust producing system.

We now consider two flexible ring wings W_1 and W_2, moving in the neighbourhood of H_1 and H_2 respectively. Without restricting generality, we represent them by two circular bound vortices Γ_1 and Γ_2. For simplicity we assume that the velocity of Γ_1 is U and that the velocity of Γ_2 along H_2 is such that it remains in the plane of Γ_1. Clearly quite different bound vorticity configurations can be used to represent the propulsion system.

The thrust is entirely delivered by Γ_2 because Γ_1 cannot produce a force of $O(\varepsilon)$ in the x direction. The process is as follows. The strength of Γ_1 increases linearly with time, hence it sheds the optimum constant circular vorticity at H_1. This however cannot go on indefinitely because then the strength of Γ_1 would increase beyond all bounds and the propulsion system would not be periodic. It is here that Γ_2 appears on the stage. First it cannot leave behind any vorticity inside H_1. Hence its strength has to be constant at parts of H_2 which are not at H_1. However when Γ_2 has to deliver a thrust it must have opposite signs at the opposite slopes which H_2 forms with the x axis. When Γ_2 arrives at the contact circle of H_2 with H_1, it changes sign instantaneously and leaves behind a concentrated free vortex at H_1. However when Γ_1 changes its magnitude at exactly the same place it cancels this concentrated free vortex. By this change Γ_1 need not to grow beyond all bounds. The bound vorticity Γ_2 can however also change its sign inside H_1 at the points where H_2 touches the x axis, there its radius and hence its length is zero and no free vorticity is left behind. In figure 5.7.2 we have drawn the strength of Γ_1 and Γ_2 as a function of x. We note that it is not allowed to leave behind concentrated free vorticity, because as we remarked already at the end of section 1.10, the kinetic energy around a concentrated vortex is infinite. Hence the efficiency of the propeller would theoretically be zero.

From the construction it follows that this propulsion device is ideal in

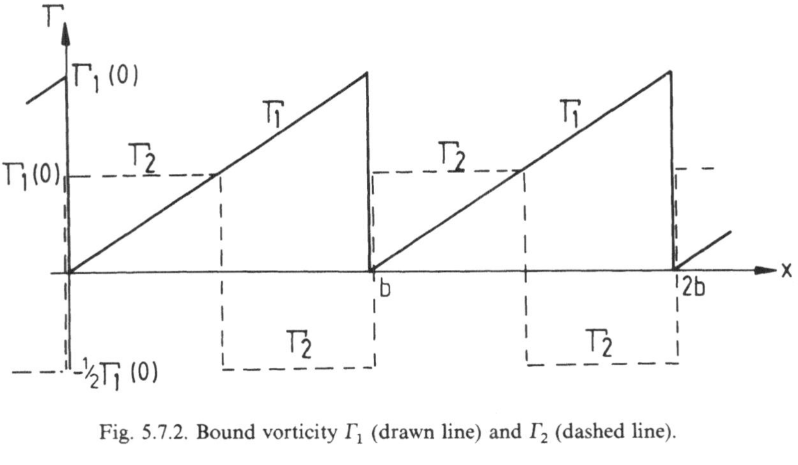

Fig. 5.7.2. Bound vorticity Γ_1 (drawn line) and Γ_2 (dashed line).

the sense of the previous section. When any other reference surface is added inside H_1, it will not get any vorticity on it by the optimization procedure. Hence the free vorticity behind the supplemented propeller can be chosen the same as before. This means that the efficiency of the original one cannot be raised. We remark that the surface H_2 is not unique each reference surface which divides the interior of H_1 into disconnected parts, will do.

When we look at the free vorticity which is left behind by our ideal propeller, we see that it is the same vorticity as is left behind by an actuator disk with constant load, velocity of advance U, working region (5.7.1), and thrust $\bar{T}$, described by a linearized theory (section 2.2). This means that both systems must have the same efficiency. However the system discussed here has the highest possible efficiency in comparison with each system of lifting surfaces of the considered class. Hence also the efficiency (2.2.17) of the actuator disk yields an upperbound for any conceivable propeller of the class under consideration. This is the reason that the actuator disk with constant load acting in an undisturbed fluid sometimes is called "the ideal propeller".

The quality number q, as introduced in the previous section, is in the case of a fluid without disturbance velocity, $\boldsymbol{v}_0^* = 0$, independent of the prescribed thrust $\bar{T}$. In this case both E_i and E_p (5.6.2) are proportional to $\bar{T}^2$. This is not true when $\boldsymbol{v}_0^* \neq 0$. Then E_i and E_p possess parts which differ from each other, originating from the extraction of kinetic energy out of $\boldsymbol{v}_0^*$.

In connection with the actuator disk theory (2.2.7) it is seen that the velocity field far behind the just mentioned propeller follows from

$$\operatorname{grad} \Phi_2(x, y, z) = \left(-\frac{\bar{T}}{\mu U \pi R^2}, 0, 0 \right), \quad y^2 + z^2 < R^2, \tag{5.7.4}$$

$$\operatorname{grad} \Phi_2(x, y, z) = (0, 0, 0), \quad y^2 + z^2 > R^2. \tag{5.7.5}$$

Note that in this discussion no essential use has been made of the fact that the working region G of the propeller has a circular cross section. We could have defined the working region as well as we have done in (5.2.19). Then for H_1 we choose the boundary ∂G of G and for H_2 any surface inside H_1 which touches H_1 periodically along closed lines and which contracts periodically upto a point. In this latter case (5.7.4) and (5.7.5) change into

$$\operatorname{grad} \Phi_2(x, y, z) = \left(-\frac{\bar{T}}{\mu U |A|}, 0, 0 \right), \quad (y, z) \in A, \tag{5.7.6}$$

$$\operatorname{grad} \Phi_2(x, y, z) = (0, 0, 0), \quad (y, z) \notin A, \tag{5.7.7}$$

where $|A|$ is the area of A. This is the function grad Φ_2 which has to be used in (5.6.1) in case of a propulsion device. However as we observed before, $\bar{T}$ has to be chosen such that together with the thrust delivered already by Φ_1, the demanded total thrust is obtained. An example of this procedure is given in section 6.4.

For the case of a propulsion device working in an undisturbed fluid, $\boldsymbol{v}_0^* = 0$, we can go one step further with (5.6.4),

$$E_i = \frac{\bar{T}^2 b}{2\mu U^2 |A|}, \qquad \boldsymbol{v}_0^* = 0, \tag{5.7.8}$$

and

$$\eta_T = \left(\frac{b\bar{T}}{b\bar{T} + E_p}\right) = \left(1 + \frac{\bar{T}}{2\mu q U^2 |A|}\right)^{-1}, \qquad \boldsymbol{v}_0^* = 0. \tag{5.7.9}$$

We conclude this section with an example of an other type of ideal lifting surface system for a class with the working region G of (5.7.1). This system is demanded to yield a mean lift force L in the y direction. It follows from the criterion in italics of section 5.6 that in the case of $\boldsymbol{v}_0^* = 0$ we look for the ring wing moving along the cylinder

$$-\infty < x < \infty, y^2 + z^2 = R^2. \tag{5.7.10}$$

We find its optimum shed free vorticity by placing this cylinder as an impervious and rigid surface in a parallel flow of magnitude λ but now in the y direction. This follows from (5.4.5) with only one function $g_{l\nu}$ namely $\cos_{(n,y)}$. Then inside the cylinder the fluid is at rest and we can easily calculate the optimum shed vorticity.

Consider a simple ring wing which consists of a part of finite length of a circular cylinder. The profiles are straight lines of equal length parallel to the axis of the cylinder. The leading edge and the trailing edge are both circles. When we place this wing under a small angle of incidence in a parallel flow then, by using the theory given in [4] and the theory given here, it is not difficult to prove that the optimum circulation distribution occurs.

5.8. Comparison of the quality number of optimum propellers by inspection

In this section we compare optimum propulsion systems which have prescribed reference surfaces. We will show that sometimes it is possible, by simple inspection of the shape of the reference surfaces, to judge which one can be expected to yield a propeller with the highest efficiency. We will formulate a rather vague but useful, statement for this comparison, which is related to the criterion in italics of section 5.6, it says:

Given two propulsion systems which have the same mean velocity of advance U, the same working region, the same mean value of thrust $\bar{T}$ and length period b. Then the propeller of which the reference surfaces, considered to be impervious and rigid, hamper most in the working region a flow parallel to the mean velocity of advance, will have the highest efficiency.

The mentioned parallel flow is the flow of velocity λ, discussed below formula (5.7.3).

The examples we discuss are two dimensional. Each propulsion device consists of two flexible profiles W_1 and W_2, moving in an ε neighbourhood of the two reference surfaces H_1 and H_2 respectively, which are given by

$$H_l\colon\ y = \pm 0.15 \mp h \pm \delta \pm 0.15 \cos 2\pi x \pm h \cos 4\pi x, \qquad l = 1, 2, \quad (5.8.1)$$

where the upper signs belong to $l = 1$ and the lower signs to $l = 2$. The cross section of the H_l are drawn in figure 5.8.1 for two specific choices of h. The parameter δ is still variable. The two figures 5.8.1*a* and *b* are clearly related. The surface H_1 of figure 5.8.1*b* originates from H_1 of figure 5.8.1*a* by a reflection and a translation. A method to calculate the optimum shed free vorticity at the surfaces H_1 and H_2 will be discussed in section 6.8.

We now discuss the quality number q (5.6.2) of these two types of propellers. As their working area per unit of span we take the two parts S_1 and S_2 covered by the amplitude of H_1 and H_2, separated by a gap of width 2δ. Hence this area equals 0.6. The calculated values of q are given

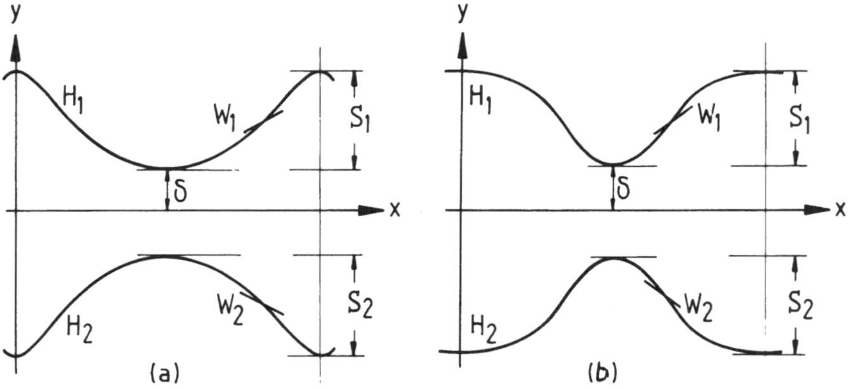

Fig. 5.8.1. Two different arrangements of geometrically congruent shapes of reference surfaces, (*a*) $h = 0.0375$, (*b*) $h = -0.0375$.

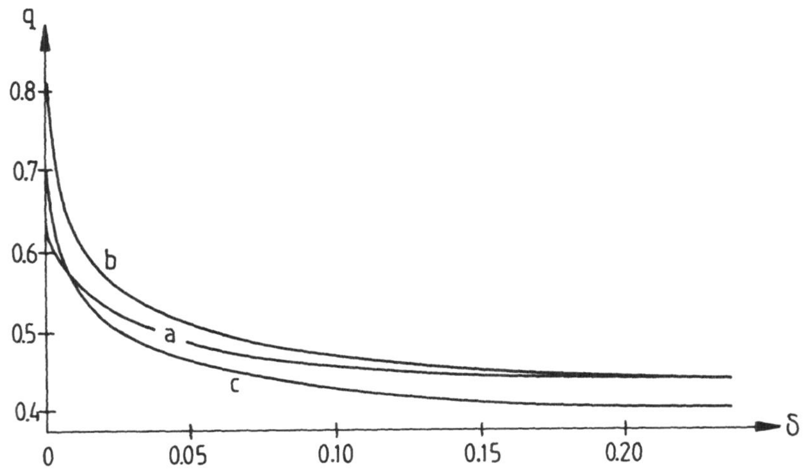

Fig. 5.8.2. The quality number q as a function of the minimum distance 2δ of the two wings, for three different sets of H_1 and H_2.

in figure 5.8.2, as a function of the minimum distance 2δ between the two wings W_1 and W_2 for the two cases a and b of figure 5.8.1 and also for the case c where H_1 and H_2 have a simple cosine shape with $h=0$ in (5.8.1). Note that in all three cases the working area per unit of span is the same.

First we discuss the cases a and b with $h=0.0375$ and $h=-0.0375$ respectively and try to predict which one has the highest quality number. It is clear that for large values of δ there is no interaction between the wings W_1 and W_2 and between the free vortex sheets H_1 and H_2 in each case a and b. Further the geometry of the surfaces H_1 and H_2 in both cases is the same, hence there will be no difference in their quality numbers q for $\delta \to \infty$. When δ becomes smaller interaction will occur. Then the parallel flow of velocity λ is hampered in S_1 besides by H_1 also by H_2 and analogously for S_2. Hence it is clear that for smaller values of δ in the optimum case q increases for configuration a as well as for b. The question is for which of these the value of q increases most. For $\delta=0$, the working regions S_1 and S_2 have a boundary in common and then in case b more "fluid is enclosed" in between H_1 and H_2 than in case a. Hence in case b the parallel flow seems to be hampered more than in case a and it can be expected that q for $\delta=0$, is larger in case b than in case a, moreover it is likely that this holds for all values of δ. These conclusions are confirmed by the calculated results given in figure 5.8.2.

Next we consider case c ($h=0$) with respect to cases a and b. For large values of δ it is conceivable that the more sharply indented H_1 and H_2 of cases a and b hamper more the parallel flow than the gently waving pure cosine will do, although they have the same amplitude. This is in agreement with figure 5.8.2, where for large values of δ the quality number q of case c is less than the value of q for the cases a and b. When $\delta = 0$ however the two pure cosines of case c "enclose more fluid" than the surfaces H_1 and H_2 in case a and less than the surfaces belonging to case b. So it seems predictable that for sufficiently small values of δ the quality number of case c is in between those of case a and b. Also this is confirmed by the results given in figure 5.8.2 [18].

The optimum efficiency of these propulsion systems follows from their quality number q as given in figure 5.8.2. Using (5.7.9) with $S = S_1 + S_2 = 0.6$, hence now for a non circular working area, we obtain

$$\eta_T = \left(1 + \frac{\bar{T}}{1.2\mu q U^2}\right)^{-1}, \tag{5.8.2}$$

where $\bar{T}$ is the mean value of the thrust per unit of span and U the mean velocity of advance of the wings W_1 and W_2 in the x direction.

It is seen from figure 5.8.2 that there is a sharp increase in the quality number for smaller values of δ. We emphasize that the numerical results apply to the two dimensional case, with other words to wings which have infinite span. When the span of the wings is finite and of the order of magnitude of the amplitude of the motion $(0.6 + 2\delta)$ in figure 5.8.1, the optimization problem is essentially three dimensional and it is no longer clear what happens for $\delta \to 0$. It is of interest to investigate such situations in connection with the application of sculling propulsion to real ships. We will return to this subject in section 6.10.

In the same way we can compare other propulsion systems. For instance, consider two optimum ship screws, with the same values of U, diameter, hub radius, $\bar{T}$ and rotational velocity, but differing in their number of blades. Then the one with the highest number of blades will have the highest value of q and hence the highest efficiency (5.7.9). Analogously, from two optimum screw propellers differing only in their rotational velocity the one with the highest rotational velocity will have the highest value of q and hence the highest efficiency. Of course in the latter example the blades have in each case to be in the neighbourhood of the appropriate helicoidal surfaces.

Again we have to realize that viscosity effects of the fluid have been left out of consideration.

5.9. Energy extraction, translating rigid wing

Since a screw propeller generally works in the disturbed flow behind a ship, it seems of interest to investigate the possibilities of energy extraction out of an inhomogeneous inflow at the propeller "disk". Up to now we have assumed that the lifting surfaces which extracted energy out of a

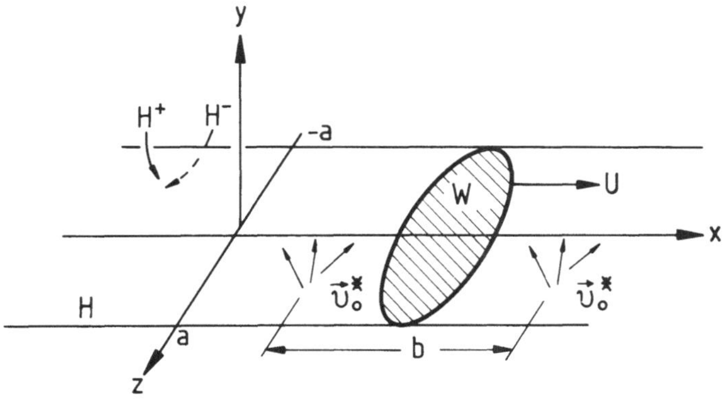

Fig. 5.9.1. Energy extraction by means of rigid wing W.

disturbance field were flexible and moving along arbitrarily curved reference surfaces. A screw propeller generally has undeformable blades connected rigidly to the hub. In order to make the considerations not unnecessarily complicated we discuss in this section a rigid planar wing W moving through a disturbed fluid. The same reasoning can be used for the non planar blades of the ship screw moving along helicoidal surfaces.

The reference strip H along which the wing W moves with velocity U, is defined by

$$H: -\infty < x < +\infty, \qquad y = 0, \qquad |z| \leq a. \tag{5.9.1}$$

As before we have a periodic disturbance field $\boldsymbol{v}_0^*$, which we replace by $\boldsymbol{v}_0$ having the same normal component at H and of which the vorticity $\gamma_0(x, z) = (\gamma_{0x}(x, z), \gamma_{0y}(x, z))$ is confined to H. This vorticity field can be expanded into a Fourier series

$$\gamma_{0x}(x, z) = \alpha_{x0}(z) + \sum_{n=1}^{\infty} \left\{ \alpha_{xn}(z) \cos \frac{2\pi n x}{b} + \beta_{xn}(z) \sin \frac{2\pi n x}{b} \right\}, \tag{5.9.2}$$

$$\gamma_{0z}(x, z) = \alpha_{z0}(z) + \sum_{n=1}^{\infty} \left\{ \alpha_{zn}(z) \cos \frac{2\pi n x}{b} + \beta_{zn}(z) \sin \frac{2\pi n x}{b} \right\}. \tag{5.9.3}$$

A vorticity field is free of divergence div $\gamma_0(x, z) = 0$, hence

$$\frac{\partial}{\partial z} \alpha_{z0}(z) = 0, \ \frac{\partial}{\partial z} \alpha_{zn}(z) = \frac{2\pi n}{b} \beta_{xn}(z), \ \frac{\partial}{\partial z} \beta_{zn}(z) = -\frac{2\pi n}{b} \alpha_{xn}(z). \tag{5.9.4}$$

Because outside H there is no vorticity we assume

$$\alpha_{z0} = 0, \qquad \alpha_{zn}(\pm a) = 0, \qquad \beta_{zn}(\pm a) = 0, \tag{5.9.5}$$

otherwise $\gamma_0(x, z)$ would have concentrated vorticity in the x direction at the edges of the strip and $\boldsymbol{v}_0^*$ would be infinite at those lines. This we do not consider by reasons of simplicity.

The velocity potential Φ_0 of the modified disturbance velocity field $\boldsymbol{v}_0$ is likewise expanded in a Fourier series

$$\Phi_0(x, y, z) = \Phi_{00c}(y, z) + \sum_{n=1}^{\infty} \left\{ \Phi_{0nc}(y, z) \cos\frac{2\pi nx}{b} + \Phi_{0ns}(y, z) \sin\frac{2\pi nx}{b} \right\}. \tag{5.9.6}$$

The jump $[\Phi_0]_-^+$ of this potential over H can be written as

$$\begin{aligned}
[\Phi_0]_-^+(x, z) &= -\int_z^a \alpha_{x0}(s)\,\mathrm{d}s - \sum_{n=1}^{\infty} \int_z^a \left\{ \alpha_{xn}(s) \cos\frac{2\pi nx}{b} + \beta_{xn}(s) \sin\frac{2\pi nx}{b} \right\} \mathrm{d}s \\
&= -\int_z^a \alpha_{x0}(s)\,\mathrm{d}s + \frac{b}{2\pi} \sum_{n=1}^{\infty} \frac{1}{n} \left\{ -\beta_{zn}(z) \cos\frac{2\pi nx}{b} + \alpha_{zn}(z) \sin\frac{2\pi nx}{b} \right\} \\
&\overset{\text{def}}{=} [\Phi_0]_{-0c}^+(z) + \sum_{n=1}^{\infty} \left\{ [\Phi_0]_{-nc}^+(z) \cos\frac{2\pi nx}{b} + [\Phi_0]_{-ns}^+(z) \sin\frac{2\pi nx}{b} \right\},
\end{aligned} \tag{5.9.7}$$

where we used (5.9.4) and (5.9.5). Then we can write the kinetic energy E, belonging to Φ_0 over one period in the x direction, as

$$\begin{aligned}
E &= -\tfrac{1}{2}\mu \int_{-a}^{a} \int_0^b \left\{ [\Phi_0]_-^+(x, z) \frac{\partial \Phi_0}{\partial y}(x, 0, z) \right\} \mathrm{d}x\,\mathrm{d}z \\
&= -\tfrac{1}{4} b\mu \int_{-a}^{a} \left[2\left\{ [\Phi_0]_{-0c}^+(z) \frac{\partial}{\partial y} \Phi_{00c}(0, z) \right\} + \sum_{n=1}^{\infty} \left\{ [\Phi_0]_{-nc}^+(z) \frac{\partial}{\partial y} \Phi_{0nc}(0, z) \right.\right. \\
&\quad \left.\left. + [\Phi_0]_{-ns}^+(z) \frac{\partial}{\partial y} \Phi_{0ns}(0, z) \right\} \right] \mathrm{d}z \overset{\text{def}}{=} E_0 + \sum_{n=1}^{\infty} (E_{cn} + E_{sn}) \overset{\text{def}}{=} \sum_{n=0}^{\infty} E_n.
\end{aligned} \tag{5.9.8}$$

Next we consider the rigid wing W moving with velocity U along H. We introduce the coordinates ξ and ζ which are in the direction of x and z however of a reference system (ξ, η, ζ) which translates with W. Because $\boldsymbol{\gamma}_0(x, z)$ has period b in the x direction we write for the still unknown vorticity $\boldsymbol{\gamma}_W = (\gamma_{W\xi}, \gamma_{W\zeta})$ on W

$$\gamma_{W\xi}(\xi, \zeta, t) = \tilde{\alpha}_{\xi 0}(\xi, \zeta) + \sum_{n=1}^{\infty} \left\{ \tilde{\alpha}_{\xi n}(\xi, \zeta) \cos\frac{2\pi n U t}{b} + \tilde{\beta}_{\xi n}(\xi, \zeta) \sin\frac{2\pi n U t}{b} \right\}, \tag{5.9.9}$$

$$\gamma_{W\zeta}(\xi, \zeta, t) = \tilde{\alpha}_{\zeta 0}(\xi, \zeta) + \sum_{n=1}^{\infty} \left\{ \tilde{\alpha}_{\zeta n}(\xi, \zeta) \cos\frac{2\pi n U t}{b} + \tilde{\beta}_{\zeta n}(\xi, \zeta) \sin\frac{2\pi n U t}{b} \right\}. \tag{5.9.10}$$

We suppose that W moves just above H, hence just above the vorticity $\boldsymbol{\gamma}_0$ and keep in mind that we have to take the limit of vanishing distance between W and $\boldsymbol{\gamma}_0$. The vorticity $\boldsymbol{\gamma}_W$ caused by W induces a velocity potential $\tilde{\Phi}(\xi, \eta, \zeta, t)$. The pressure jump across W becomes

$$[p]_-^+(\xi, \zeta, t) = -\mu U \left[\frac{\partial \tilde{\Phi}}{\partial \xi}\right]_-^+ (\xi, 0, \zeta, t) - \mu \left[\frac{\partial \tilde{\Phi}}{\partial t}\right]_-^+ (\xi, 0, \zeta, t). \tag{5.9.11}$$

The potential induced by the already existing vorticity $\boldsymbol{\gamma}_0$ at H, being just below the wing, does not contribute to a jump of the pressure across W.

Extraction of energy out of the fluid is effectuated by the component in the x direction of the force exerted by the fluid at W. Suppose the shape of W is given by

$$y = f(\xi, \zeta) = \mathrm{O}(\varepsilon), \qquad (\xi, \zeta) \in B, \tag{5.9.12}$$

where B is the planform of the wing. Then the contribution of an elementary area $\mathrm{d}\xi\, \mathrm{d}\zeta$ of W to the mean value with respect to time, of the force in the x direction becomes

$$\frac{U}{b} \int_0^{b/U} \left\{ [p]_-^+(\xi, \zeta, t) \frac{\partial f}{\partial \xi}(\xi, \zeta)\, \mathrm{d}\xi\, \mathrm{d}\zeta \right\} \mathrm{d}t. \tag{5.9.13}$$

It is clear from (5.9.9) and (5.9.11) that we can develop $[p]_-^+$ in a Fourier series with respect to t. It follows from (5.9.13) that only a non zero value can be contributed to the mean force in the x direction, by the time independent part of $[p]_-^+$. With other words by the time independent part of $[\partial\tilde{\Phi}/\partial\xi]_-^+$, this reads by (5.9.9)

$$-\int_\zeta^a \frac{\partial}{\partial\xi}\tilde{\alpha}_{\xi 0}(\xi, s)\,\mathrm{d}s = \int_\zeta^a \frac{\partial}{\partial s}\tilde{\alpha}_{\zeta 0}(\xi, s)\,\mathrm{d}s = -\tilde{\alpha}_{\zeta 0}(\xi, \zeta), \tag{5.9.14}$$

where we used again div $\boldsymbol{\gamma}_w = 0$. Hence we can write (5.9.13) as

$$\frac{\mu U^2}{b}\int_0^{b/U}\left\{\tilde{\alpha}_{\zeta 0}(\xi, \zeta)\frac{\partial f}{\partial\xi}(\xi, \zeta)\,\mathrm{d}\xi\,\mathrm{d}\zeta\right\}\mathrm{d}t = \mu U\tilde{\alpha}_{\zeta 0}(\xi, \zeta)\frac{\partial f}{\partial\xi}(\xi, \zeta)\,\mathrm{d}\xi\,\mathrm{d}\zeta. \tag{5.9.15}$$

When for one reason or another the wing W can leave behind vorticity of which the x component has one of the strengths

$$-\alpha_{x0}(z), \qquad -\alpha_{xn}(z)\cos\frac{2\pi nx}{b}, \qquad -\beta_{xn}(z)\sin\frac{2\pi nx}{b}, \tag{5.9.16}$$

then the corresponding "parcel" of kinetic energy E_0, E_{cn} or $E_{sn}(n \neq 0)$ would disappear from E (5.9.8).

The parcel E_0 is changed by free vorticity shed by W, which is parallel to the x axis and independent of the x coordinate. Such free vorticity equals $\tilde{\alpha}_{\xi 0}(\xi, \zeta)$ (5.9.9) at the trailing edge. By a suitable camber distribution the wing W is able to extract E_0 completely by means of the inner parts of its planform. It is also possible that suction forces at the leading edge extract parts of E_0.

The parcels E_{cn} and $E_{sn}(n \neq 0)$ of the kinetic energy can only be influenced by shed vorticity which is periodic in the x direction, hence which in the (ξ, η, ζ) system is created by vorticity at W which is periodic in time. The corresponding periodic pressure jumps however yield by (5.9.13) no mean value of a force in the x direction at inner parts of the planform B. This means that these energy parcels can only be influenced when suction forces at the leading edge of W occur. We will discuss this somewhat further.

The action of the wing W is split up into separate parts by considering separate wings with the same planform as W however working in differently disturbed fluids.

i) Wing $W_0(=W)$ moves through a fluid which has at H the vortex field $\alpha_{x0}(z)$ (5.9.2) in the x direction, W_0 has a camber such that it sheds free vorticity which annihilates $\alpha_{x0}(z)$.
ii) Wings $W_n(n = 1, 2, \ldots)$ are the flat "impermeable projections" of W

at the $y = 0$ plane and move through a fluid which has at H the vortex field (5.9.2), (5.9.3)

$$\left\{ \alpha_{xn}(z) \cos\frac{2\pi nx}{b} + \beta_{xn}(z) \sin\frac{2\pi nx}{b}, \right.$$

$$\left. \alpha_{zn}(z) \cos\frac{2\pi nx}{b} + \beta_{zn}(z) \sin\frac{2\pi nx}{b} \right\}, \tag{5.9.17}$$

they shed free vorticity which we denote by

$$\left\{ \alpha^*_{xn}(z) \cos\frac{2\pi nx}{b} + \beta^*_{xn}(z) \sin\frac{2\pi nx}{b}, \right.$$

$$\left. \alpha^*_{zn}(z) \cos\frac{2\pi nx}{b} + \beta^*_{zn}(z) \sin\frac{2\pi nx}{b} \right\}. \tag{5.9.18}$$

Wing W_0 extracts the kinetic energy E_0 (5.9.8) entirely, while the wings W_n extract a certain amount of energy E_n^* out of $(E_{cn} + E_{sn})$. The wings $W_n(n = 1, 2 \ldots)$ have to extract, because they are flat, energy by means of their suction forces at the leading edge, however this could not be done otherwise as we discussed in relation with (5.9.13). When we now let move the wing $W(= W_0)$ through the total vortex field (5.9.2), (5.9.3) it will, by the linearity of the problem, subtract the same amounts E_n^* from each of the energy parcels E_0, $(E_{cn} + E_{sn})$ as the wings $W_n(n = 0, 1, 2, \ldots)$.

Summarizing: consider a rigid wing W with span $2a$, moving with velocity U along H, while at H we have vorticity (5.9.2), (5.9.3), then from the considerations above we have the following two results:

1) The wing can extract the energy E_0 (5.9.8) wholly by means of its pressure jump at the inner parts of the wing or also partly by this pressure jump and partly by suction forces at the leading edge. When the camber is such that the component $\alpha_{x0}(z)$ (5.9.2) is annihilated then E_0 is extracted completely. If only $\alpha_{x0}(z)$ is present and if the wing is designed such that at the leading edge the flow is smooth, then the extraction of energy occurs entirely at the inner parts of the wing. The complete extraction of E_0 can always be effectuated by means of a wing with finite chord length.

2) The extraction of kinetic energy out of the E_n $(n = 1, 2, \ldots)$ (5.9.8) is determined by the planform of the wing only and is effectuated by suction forces at the leading edge. We do not know if all the energy out of the $E_n(n = 1, 2, \ldots)$ can be extracted by wings of finite chord length. In the next section we make it plausible that it can be done by wings of "infinite chord length". It is also proved there that in the two dimen-

sional case it cannot be done by wings of finite chord length. So it is at least plausible that to extract all the energy $E_n (n = 1, 2, \dots)$, a wing is needed with infinite chord length.

The above considerations hold mutatis mutandis for the rigid blades of a screw propeller, moving along the helicoidal strips. We remark that the friction caused by the viscosity in real fluids will be detrimental for the gain predicted by potential theory for blades with very large chord lengths. Further the extraction by means of suction forces at the sharp leading edge of the blades does not seem to be very reliable, small disturbances can cause separation of the flow by which the magnitude of the suction forces will fall off. Hence the most important energy to recover in this case seems to be E_0, which can be done by blades with finite chord lengths while no suction forces are needed for that purpose.

5.10. Optimum energy extraction by rigid wing

Consider a flat wing W moving along a strip H (figure 5.9.1) of width $2a$. On H we have vorticity $\gamma_0(x, z)$ as in the previous section. In order to fix the attention we assume that the wing is a rectangle of width $2a$ and chord length l with $l \gg 2a$, which moves with a velocity U in the positive x direction.

The component of the velocity induced by $\gamma_0(x, z)$, normal to W has to be cancelled by vorticity γ_W at W. This vorticity will be split up into three parts

$$\gamma_W(x, z, t) = \gamma_{w1}(x, z, t) + \gamma_{W2}(x, z, t) + \gamma_{W3}(x, z, t), \qquad (5.10.1)$$

where for the vorticity at W we now use the coordinates (x, z) instead of those of a reference system which moves with W.

The first part we choose as

$$\gamma_{W1}(x, z, t) = -\gamma_0(x, z), \qquad (x, z) \in W, \qquad (5.10.2)$$

and state that this vorticity for $l \gg 2a$, is shed at the trailing edge of W, hence it annihilates the vorticity γ_0 at H behind the wing. The correctness of this will be checked.

The vorticity γ_{W1} needs a correction for two reasons. First, it is not free of divergence at the leading edge of W. Second, it cannot cancel all the normal velocities induced at W by γ_0, because in front of the leading edge we have still vorticity of which the induced velocities at W are not compensated by those of γ_{W1}. The first short coming will be repaired by the vortex system γ_{W2}, which can be defined in a fixed region near the leading edge and which can be chosen in an infinite number of ways. The second short coming will be corrected by choosing γ_{W3} such that the normal velocities at W induced by γ_{W2} and the vorticity γ_0 at H upstream of the leading edge are compensated. These velocities however tend to zero at least as

$$0(|x-x_l|^{-1}), \qquad (x-x_l) \to -\infty, \tag{5.10.3}$$

where x_l is the x coordinate of the leading edge. This relation can be shown to be true by an asymptotic expansion of the velocities induced by some representative vortex system. We now argue that also γ_{W3} tends to zero for $(x-x_l) \to -\infty$.

Consider a region $\tilde{W}$ of W of which the length is large with respect to the width $2a$ of the strip H. This region is situated at a large distance $\tilde{l} \gg 2a$ from the leading edge as well as from the trailing edge. Suppose that $|\gamma_{W3}|$ assumes finite values larger than c with $c > 0$, on $\tilde{W}$ no matter how far $\tilde{W}$ is situated behind the leading edge. On $\tilde{W}$ the velocities induced by the vorticity γ_0 at H in front of the leading edge and by γ_{W2} have disappeared when $\tilde{l}$ is sufficiently large. This means that the periodic γ_{W3} on the region $\tilde{W}$ is not allowed to induce a non-zero velocity component normal to W, because there γ_{W1} already compensates γ_0. This can happen only when $\gamma_{W3} = 0$ at W. Namely its circulation around $\tilde{W}$ is zero because a wing can create only such vorticity. Hence the potential Φ_{W3} of its induced velocities outside $\tilde{W}$ is a uniquely valued function of which the derivative normal to W is zero and which is zero at infinity $((y^2+z^2) \to \infty)$. This means that $\Phi_{W3} = 0$ and the same holds for γ_{W3}. Hence when the chord length l of the wing W is large enough, it is clear that in the neighbourhood of the trailing edge γ_{W1} dominates by far and this confirms our statement below (5.10.2).

The result is that when the wing is flat and has a very large, in fact an infinite, chord length it annihilates the vorticity γ_0 at H and extracts by its suction forces at the leading edge all the kinetic energy belonging to γ_0 out of the fluid.

The question remains if a flat wing with finite chord length can extract all kinetic energy belonging to vorticity with Fourier coefficients $\alpha_{xn}(z)$, $\beta_{xn}(z)$, $\alpha_{zn}(z)$, $\beta_{zn}(z)$, $n \neq 0$ (5.9.2) (5.9.3) out of the fluid. That this is not plausible can be seen by the following calculation. We consider a two dimensional problem. The Cartesian coordinate system (x, y) is embedded in an incompressible and inviscid fluid. The fluid has an incoming velocity U in the positive x direction (figure 5.10.1). With the fluid is transported a vortex layer $\gamma_0(x, t)$ of strength

$$\gamma_0(x,t) = A\, e^{i\nu(x-Ut)}. \tag{5.10.4}$$

This problem can be considered as a special case of figure (5.9.1) for a

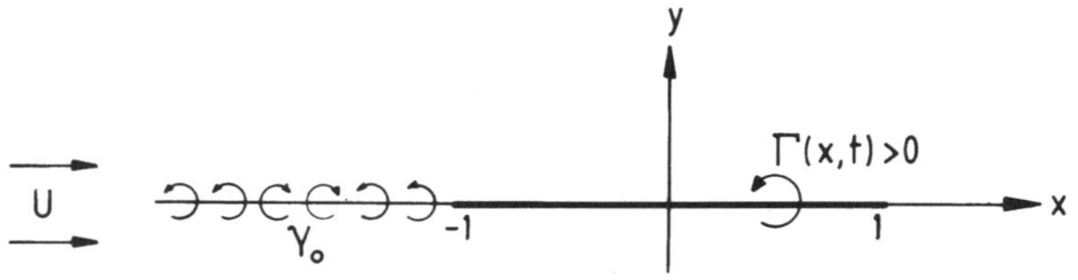

Fig. 5.10.1. Two dimensional profile with incoming vortex layer γ_0.

very wide strip H or for very large values of a. Then we neglect in a finite neighbourhood of the edges of H at $z=a$, $z=-a$, vorticity with a nonzero x component needed to satisfy the condition div $\gamma_0=0$. However the amount of energy added by this vorticity component can be made arbitrarily small with respect to the total amount of energy connected to γ_0, when the width of the strip is sufficiently large.

That our profile has the chord length 2 is not a restriction of generality, it is only a choice of the unit of length. Also that the profile is chosen to be flat is not a restriction, because we already discussed previously that a profile with camber cannot extract more energy out of a periodic incoming flow than a flat profile.

We assume that it is possible for some value of ν to annihilate all the free vorticity (5.10.4) behind the profile and will arrive at a contradiction. If there is no vorticity behind the profile the vorticity $\Gamma(x,t)$ at the profile has to satisfy

$$\oint_{-1}^{1}\frac{\Gamma(\xi,t)}{(\xi-x)}\mathrm{d}\xi=-A\,\mathrm{e}^{-i\nu Ut}\int_{-\infty}^{-1}\frac{\mathrm{e}^{i\nu\xi}}{(\xi-x)}\mathrm{d}\xi,\qquad -1<x<1. \qquad (5.10.5)$$

Writing

$$\Gamma(x,t)=\Gamma(x)\,\mathrm{e}^{-i\nu Ut}, \qquad (5.10.6)$$

we obtain

$$\oint_{-1}^{1}\frac{\Gamma(\xi)}{(\xi-x)}\mathrm{d}\xi=-A\int_{-\infty}^{-1}\frac{\mathrm{e}^{i\nu\xi}}{(\xi-x)}\mathrm{d}\xi,\qquad -1<x<1. \qquad (5.10.7)$$

The solution of this equation which satisfies the Kutta condition at the trailing edge follows from appendix A ((A.3.1) and (A.3.9))

$$\Gamma(x)=\frac{A}{\pi^2}\left(\frac{1-x}{1+x}\right)^{1/2}\oint_{-1}^{1}\left(\frac{1+\xi}{1-\xi}\right)^{1/2}\left\{\int_{-\infty}^{-1}\frac{\mathrm{e}^{i\nu\sigma}}{(\sigma-\xi)}\mathrm{d}\sigma\right\}\frac{\mathrm{d}\xi}{(\xi-x)}. \qquad (5.10.8)$$

The condition that no free vorticity is present behind the profile reads

$$\gamma_0(-1,t)=A\,\mathrm{e}^{i\nu(-1-Ut)}=\frac{1}{U}\frac{\mathrm{d}}{\mathrm{d}t}\int_{-1}^{1}\Gamma(x,t)\,\mathrm{d}x. \tag{5.10.9}$$

By (5.10.6) we write (5.10.9) as

$$\int_{-1}^{1}\Gamma(x)\,\mathrm{d}x=\frac{iA}{\nu}\mathrm{e}^{-i\nu}. \tag{5.10.10}$$

Substituting (5.10.8) into (5.10.10) we find, after a rearrangement of the integrations, that the following condition must hold

$$\frac{1}{\pi^2}\int_{-\infty}^{-1}\mathrm{d}\sigma\,\mathrm{e}^{i\nu\sigma}\int_{-1}^{1}\mathrm{d}\xi\left(\frac{1+\xi}{1-\xi}\right)^{1/2}\frac{1}{(\sigma-\xi)}$$

$$\times\oint_{-1}^{1}\left(\frac{1-x}{1+x}\right)^{1/2}\frac{\mathrm{d}x}{(\xi-x)}=\frac{i}{\nu}\mathrm{e}^{-i\nu}. \tag{5.10.11}$$

Using the integrals ([21] II),

$$\oint_{0}^{\pi}\frac{\cos nx}{\cos x-\cos\alpha}\mathrm{d}x=\pi\frac{\sin n\alpha}{\sin\alpha},\qquad n=0,1,2,\dots, \tag{5.10.12}$$

and

$$\int_{a}^{b}\frac{\mathrm{d}x}{(cx+d)\{(x-a)(b-x)\}^{1/2}}=\frac{\pi}{\{(ac+d)(bc+d)\}^{1/2}}, \tag{5.10.13}$$

$ac+d>0,\ bc+d>0,$

we write condition (5.10.11) as

$$-\int_{-\infty}^{-1}\mathrm{e}^{i\nu\sigma}\left(1+\frac{(1+\sigma)}{(\sigma^2-1)^{1/2}}\right)\mathrm{d}\sigma=\frac{i}{\nu}\mathrm{e}^{-i\nu}. \tag{5.10.14}$$

Multiplication of the real parts of (5.10.14) by $\sin\nu$ and of the imaginary parts by $\cos\nu$ and addition of the two results yields

$$-\int_{-\infty}^{-1}\left(1+\frac{(1+\sigma)}{(\sigma^2-1)^{1/2}}\right)\sin\nu(1+\sigma)\,\mathrm{d}\sigma$$

$$=\int_{0}^{\infty}\left\{1-\left(\frac{\xi}{\xi+2}\right)^{1/2}\right\}\sin\nu\xi\,\mathrm{d}\xi=\frac{1}{\nu}. \tag{5.10.15}$$

By two times partial integration of the second integral we rewrite this condition as

$$\int_0^\infty \sin \nu\xi \frac{(2\xi+1)}{\xi^{3/2}(\xi+2)^{5/2}} \mathrm{d}\xi = 0. \tag{5.10.16}$$

However, because the factor of $\sin \nu\xi$ in the integrand is a monotonicly decreasing function, it follows that (5.10.16) cannot be satisfied. This means that the assumption that the profile extracts all the energy out of the fluid is not correct.

6. Applications of optimization theory

We now discuss a number of specific problems concerned with the optimization of propulsion devices. These problems are such that they can be treated by means of the theory of the previous chapter. Our considerations are directed to the determination of the shed free vorticity of the optimum propellers from which follows their bound vorticity.

The first subject is a screw propeller provided with endplates at the tips of the blades. The number of blades ranges from one upto and including four, these blades are mounted on a two sided infinitely long cylindrical hub and the incoming flow is allowed to be disturbed by an axisymmetric velocity field of $O(\varepsilon)$.

When the endplates are sufficiently long and are lying in the neighbourhood of a cylindrical surface, their tips can coincide and then a ring propeller arises. In the optimum case however the ring is not rotationally symmetric. The angles of incidence have to vary along the ring in an appropriate way from one blade to a neighbouring one, this variation is then repeated between the other blades. Hence there is an essential difference between an optimum ring propeller and a conventional ring propeller or between an optimum ring propeller and a shrouded propeller of which the blade tips move closely along the inner surface of the shroud. In these two cases the ring and the shroud are axisymmetric or anyhow the shape of the shroud is fixed with respect to the ship.

If a shrouded propeller has to be optimized or a ring propeller under the condition that the ring remains rotationally symmetric, a much more complicated problem arises. In this case the local angles of incidence of the blades are free while the ring or the shroud in the linearized theory can be considered as a circle cylinder of finite length. Now it is no longer possible to carry out the calculations for the optimum free vorticity far behind the propeller. The interaction of blades and ring or of blades and shroud has to be considered at the propeller itself. Such a problem will not be considered here.

The second problem is related to the optimization of a yacht sailing close to wind. We calculate the optimum circulation distribution along the sails, under the constraint of a prescribed heeling moment. In this example we do not take into account the action of the keel, which has to be considered as a wing working in the medium water. In fact sail and keel have to be optimized together as one propulsion system. This more complicated problem is discussed by Wiersma in [71]. For more general information about yachts we refer to [44].

As a third application we consider the large amplitude unsteady sculling propulsion of regimes ii and iii of section 4.1. With respect to the velocity field induced by the base motion of regime iii, a somewhat complicated method has to be followed in order to calculate the circulation distribution of the rigid profile from the optimum shed free vorticity.

The optimum Voith-Schneider propeller with many blades is the fourth application. The chapter is concluded with qualitative considerations about the influence of the finite length of a hub on the spanwise circulation distribution of the blades of an optimum screw propeller. This gives information about the essential theoretical shortcomings of the use of the twosided infinitely long cylinder as a model of a hub in optimization theory.

6.1. Optimum screw propeller with endplates

We consider the optimization of a screw propeller of which the blades are provided with endplates, Klaren and Sparenberg [38]. The propeller has N blades mounted on a twosided infinitely long cylinder C of radius r_i, which represents the hub. In figure 6.1.1 we have drawn one blade with endplate and the "hub". The helicoidal reference surfaces along which the blades move are described by

$$\varphi + ax + \frac{2\pi n}{N} = 0, \quad r_i \leq r \leq r_0, \quad a = \frac{\omega}{U}, \quad n = 0, 1, \ldots, N-1, \tag{6.1.1}$$

where U is the velocity of advance of the propeller in the positive x

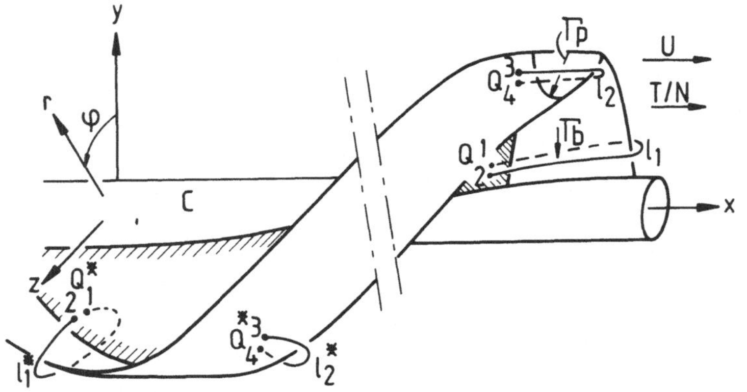

Fig. 6.1.1. One blade with endplate.

direction, ω is the rotational velocity and $2r_0$ is the diameter of the propeller. The endplates move along strips wound helicoidally around the cylinder $r = r_0$,

$$-ax - \frac{2\pi n}{N} - \alpha \leq \varphi \leq -ax - \frac{2\pi n}{N} + \alpha, \quad n = 0, 1, \ldots, N-1, \tag{6.1.2}$$

for some suitable value $0 \leq \alpha \leq \pi/N$. When $\alpha = \pi/N$ these strips cover entirely the cylinder $r = r_0$. In that case as we mentioned in the introduction, the endplates have with respect to optimization the same influence as a ring around the propeller. This optimum ring will not be rotationally symmetric but has changing angles of incidence in between the blades. We note that when $\alpha = \pi/N$ the total span of the endplates can be much smaller than the length of the circumference of the ring. This because the span of the end plates can be chosen perpendicular to the relative flow direction at the tips of the blades. Hence, although with respect to optimization endplates and ring have the same influence when $\alpha = \pi/N$, the wetted area of the endplates can be much smaller than the wetted area of the ring. This is favourable with respect to resistance caused by viscosity.

Equations (6.1.1) and (6.1.2) describe the reference surfaces along which the lifting surfaces are moving. The hub is a body as admitted in section 5.1. Because we have only one constraint on the force action, namely a prescribed thrust, we denote the reference surfaces by one symbol H.

We suppose a disturbance field $\boldsymbol{v}_0^*$ of $O(\varepsilon)$ to be present which has a zero component normal to the hub C. This field is chosen to be independent of the x coordinate because a propeller behind a ship always rotates in the same disturbed flow,

$$\boldsymbol{v}_0^*(r, \varphi) = \left(v_{0x}^*(r, \varphi), v_{0r}^*(r, \varphi), v_{0\varphi}^*(r, \varphi)\right). \tag{6.1.3}$$

It is also assumed that the propeller blades encounter after each revolution when moving in the x direction, the same disturbance velocities. This periodicity with respect to φ has the same meaning for the screw propeller as the periodicity of $\boldsymbol{v}_0^*$ in the x direction in section 5.9 for the rigid wing. Fourier expansions can be made here in the φ direction exactly analogous to the Fourier expansions in the x direction in that section. It follows that the kinetic energy "parcels" in this section belonging to the higher Fourier components can be extracted only by means of suction forces. This seems not very well possible by means of the sharp leading edges of the propeller blades and of the endplates.

Therefore we consider instead of (6.1.3) disturbance fields which only depend on r

$$\boldsymbol{v}_0^*(r) = \left(v_{0x}^*(r), 0, v_{0\varphi}^*(r)\right), \tag{6.1.4}$$

where the velocity in the r direction has to be zero in connection with the equation of continuity (appendix B, (B.2.5)). Because the screw propeller working in a field $\boldsymbol{v}_0^*$ of the form (6.1.4) is a stationary propeller its thrust T is constant and we do not use the symbol $\bar{T}$. The same will be done in other cases when the thrust is not dependent on time.

For the optimization of the propeller we can, as we discussed in section 5.2, replace $\boldsymbol{v}_0^*(r)$ by a field $\boldsymbol{v}_0(x, r, \varphi)$ of which the vorticity $\boldsymbol{\gamma}_0$ is on the reference surface H and on the hub C. This vorticity lies along helicoidal lines. The component of $\boldsymbol{v}_0$ normal to H is the same as the corresponding component of $\boldsymbol{v}_0^*$. The vorticity $\boldsymbol{\gamma}_0$ has to be such that the circulation calculated for a contour encircling H and C is zero, then outside H and C the velocity field $\boldsymbol{v}_0$ has the uniquely valued potential Φ_0. It is the kinetic energy of $\boldsymbol{v}_0(x, r, \varphi)$ which can be extracted entirely by blades with finite chord lengths and even without suction forces at the leading edges.

In order that the propeller in the inhomogeneous field $\boldsymbol{v}_0$ is optimum we have for the potential Φ of the velocity field induced by the shed vorticity far behind the propeller

$$\frac{\partial \Phi}{\partial n} = -\frac{\partial \Phi_0}{\partial n} + \lambda \cos_{(n,x)} \tag{6.1.5}$$

at $H + C$. This equation follows from (5.4.5) with only one Lagrange multiplier for the constraint on the thrust, while the function $g = \cos_{(n,x)}$ follows from a comparison of (5.3.6) and (5.3.5). Our next aim is to calculate the unknown velocity potential Φ.

We introduce a helicoidal coordinate system ρ, ξ, σ by

$$\rho = ar, \quad \zeta = \varphi + ax, \quad \sigma = \varphi; \qquad a = \frac{\omega}{U}. \tag{6.1.6}$$

It follows from appendix B (B.2.12) that the potential equation written in these coordinates assumes the form

$$\left\{\frac{\partial^2}{\partial \zeta^2} + \frac{1}{\rho}\frac{\partial}{\partial \rho}\rho\frac{\partial}{\partial \rho} + \frac{1}{\rho^2}\left(\frac{\partial^2}{\partial \zeta^2} + 2\frac{\partial^2}{\partial \zeta \partial \sigma} + \frac{\partial^2}{\partial \sigma^2}\right)\right\}\Phi = 0. \tag{6.1.7}$$

In the next section we derive some properties of the potential Φ.

6.2. Properties of the potential

First we show that Φ is independent of σ. This is clear for the velocity field grad Φ, because the vorticity behind the blades and endplates is lying along helicoidal lines. However because Φ arises from the velocity field by means of integration, it is not evident for Φ itself. Suppose Φ is given at a point $(\rho_0, \zeta_0, \sigma_0)$ and we want to calculate the potential at a general point $(\rho_1, \zeta_1, \sigma_1)$. Then (figure 6.2.1)

$$\Phi(\rho_1, \zeta_1, \sigma_1) = \Phi(\rho_0, \zeta_0, \sigma_0) + |\boldsymbol{v}(\rho_0, \zeta_0)| \cos\alpha \int_{(\rho_0,\zeta_0,\sigma_0)}^{(\rho_0,\zeta_0,\sigma_1)} |\mathrm{d}\boldsymbol{s}_1| + \int_{(\rho_0,\zeta_0,\sigma_1)}^{(\rho_1,\zeta_0,\sigma_1)} \boldsymbol{v}(\rho, \zeta_0)\cdot \mathrm{d}\boldsymbol{s}_2 + \int_{(\rho_1,\zeta_0,\sigma_1)}^{(\rho_1,\zeta_1,\sigma_1)} \boldsymbol{v}(\rho_1, \zeta)\cdot \mathrm{d}\boldsymbol{s}_3, \tag{6.2.1}$$

where $\alpha = \alpha(\rho_0, \zeta_0)$ is the angle between the velocity $\boldsymbol{v}$ and the helicoidal line through $(\rho_0, \zeta_0, \sigma_0)$, the $\mathrm{d}\boldsymbol{s}_i (i = 1, 2, 3)$ are infinitesimal vectors along the lines of integration. From (6.2.1) it follows that we can write the potential at the point $(\rho_1, \zeta_1, \sigma_1)$ as

$$\Phi(\rho_1, \zeta_1, \sigma_1) = \Phi(\rho_0, \sigma_0, \zeta_0) + (\sigma_1 - \sigma_0)k \cos\alpha + F(\rho_1, \zeta_1), \tag{6.2.2}$$

where k is a nonzero constant and F is independent of σ_1. This latter statement applies because if we had chosen another value of σ_1, then each point of the path of integration of the second and third integral would have moved along a helicoidal line over the same distance which would not have changed the values of these integrals.

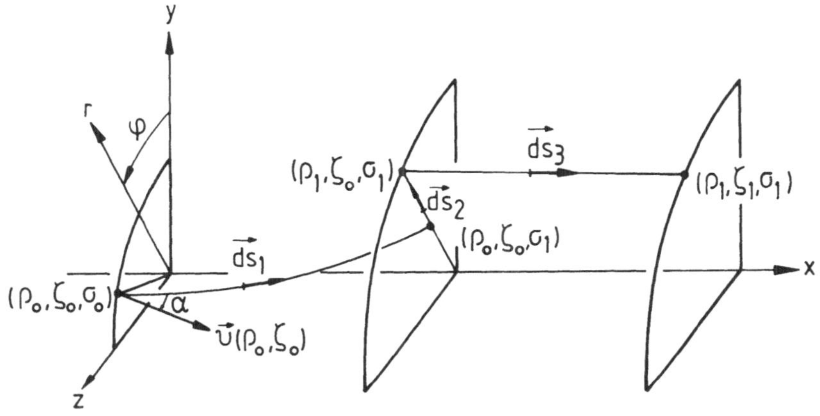

Fig. 6.2.1. Path of integration from $(\rho_0, \zeta_0, \sigma_0)$ towards $(\rho_1, \zeta_1, \sigma_1)$.

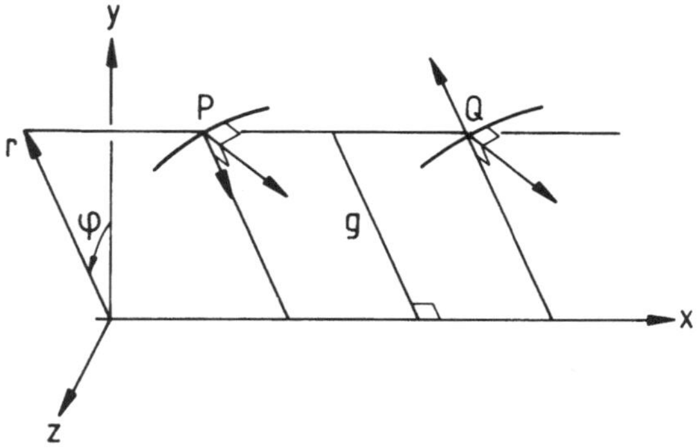

Fig. 6.2.2. Velocities at two points symmetrically placed with respect to a bisector surface.

Formula (6.2.2) shows that when $\cos\alpha \neq 0$, we have a term which is linear in $(\sigma_1 - \sigma_0)$. Because the disturbance velocities induced by the vortex sheets behind the propeller tend to zero for $r \to \infty$ we have

$$\frac{\partial\Phi}{\partial x} = a\frac{\partial\Phi}{\partial\zeta} = a\frac{\partial F}{\partial\zeta} \to 0, \qquad \rho = ar \to \infty. \tag{6.2.3}$$

From (6.1.6) and (6.2.2) in combination with (6.2.3) we find

$$\frac{\partial\Phi}{r\partial\varphi} = \frac{1}{r}\left(\frac{\partial F}{\partial\zeta} + k\cos\alpha\right) \to \frac{k\cos\alpha}{r}, \qquad \rho = ar \to \infty. \tag{6.2.4}$$

Hence because there is no total circulation round the x axis, we know that $\partial\Phi/\partial\varphi$ has to tend to zero more quickly than r^{-1} for $r \to \infty$. From this it follows $\cos\alpha = 0$ or $\alpha = \frac{1}{2}\pi$. This means that all velocities behind the optimum propeller are perpendicular to the helicoidal lines and

$$\Phi = \Phi(\rho, \zeta). \tag{6.2.5}$$

Combining this result with (6.1.7) we obtain that Φ satisfies

$$\left\{\frac{1}{\rho}\frac{\partial}{\partial\rho}\rho\frac{\partial}{\partial\rho} + \left(1 + \frac{1}{\rho^2}\right)\frac{\partial^2}{\partial\zeta^2}\right\}\Phi(\rho, \zeta) = 0. \tag{6.2.6}$$

Before we finally formulate the boundary value problem in helicoidal coordinates we derive some symmetry properties of the optimum potential Φ, hence far behind the propeller. In that region we have helicoidal surfaces (6.1.1), the N strips (6.1.2) and the hub C. Also we imagine N helicoidal bisector surfaces exactly in between the helicoidal reference surfaces. Now we take a generator line g (with x = const., φ = const.) of a bisector surface (figure 6.2.2) and rotate the whole system of vortex sheets, bisector surfaces and hub around this line over an angle of 180°. After that each surface coincides with one or another surface of its own type of the original configuration. When we multiply the vorticity at the rotated vortex sheets and at C by -1, we obtain the original velocity field. Consider two points P and Q, which have equal values of φ and r and which changed position after the rotation. It follows easily from the just mentioned trick, that the velocities in these points possess the symmetry relation shown in figure 6.2.2. The velocity components perpendicular to the r direction are parallel and of equal magnitude, while the components in the r direction are also of equal magnitude but opposite in sense.

When the points P and Q tend to each other and hence to the bisector plane, we see that the radial components have to vanish. This means that, because we already found $\alpha = \frac{1}{2}\pi$ (figure 6.2.1), the whole bisector surface under consideration has the same potential, say zero. Then, however, each bisector surface has potential zero, otherwise we would have non zero velocities for $r \to \infty$.

An analogous reasoning can be given for two points which change position when we rotate the whole system over 180° around a generating line of a helicoidal free vorticity sheet behind a propeller blade. The only difference is that here the velocity components in the r direction do not have to vanish when both points tend to the sheet, because by the presence of the vorticity a discontinuity is tolerated. Hence the helicoidal vortex sheets in general do not have a constant potential. However, on their extension for values with $r > r_0$ the potential is again a constant. This constant is zero because for $r \to \infty$ the fluid is undisturbed and the potential of the bisector surfaces is zero.

From the above we find that the potential $\Phi(\rho, \zeta)$ is an odd function of ζ with respect to the bisector surfaces and the vorticity sheets behind the blades. It is clear that the results we found for Φ also hold for Φ_0.

We now return to condition (6.1.5) which has to be fulfilled at $H + C$. The unit normal $\boldsymbol{n}$ to the helicoidal surfaces (6.1.1) is, using cylindrical coordinates

$$\boldsymbol{n} = (1 + a^2 r^2)^{-1/2}(ar, 0, 1), \tag{6.2.7}$$

hence expressed in helicoidal coordinates the cosine of the angle between this normal and the x axis has the value

$$\cos_{(n,x)}(\rho) = \frac{\rho}{(1 + \rho^2)^{1/2}}. \tag{6.2.8}$$

The derivative of the potential in cylindrical coordinates, normal to the helicoidal sheets becomes

$$\frac{\partial \Phi}{\partial n} = \operatorname{grad} \Phi \cdot \boldsymbol{n} = \frac{1}{(1 + a^2 r^2)^{1/2}} \left(ar \frac{\partial \Phi}{\partial x} + \frac{1}{r} \frac{\partial \Phi}{\partial \varphi} \right). \tag{6.2.9}$$

Changing to helicoidal coordinates (6.1.6) we find

$$\frac{\partial}{\partial n}\Phi(\rho,\zeta)=\frac{a}{\rho(1+\rho^2)^{1/2}}\left((1+\rho^2)\frac{\partial}{\partial\zeta}+\frac{\partial}{\partial\sigma}\right)\Phi(\rho,\zeta)$$

$$=\frac{a}{\rho}(1+\rho^2)^{1/2}\frac{\partial}{\partial\zeta}\Phi(\rho,\zeta). \tag{6.2.10}$$

Substitution of (6.2.8) and (6.2.10) into (6.1.5), yields for the boundary condition at the helicoidal sheets

$$\frac{\partial\Phi}{\partial\zeta}=-\frac{1}{a}\frac{\rho(\rho v_{0x}^*+v_{0\varphi}^*)}{(1+\rho^2)}+\frac{\lambda}{a}\frac{\rho^2}{(1+\rho^2)}\stackrel{\text{def}}{=}h(\rho),\quad ar_i\le\rho\le ar_0, \tag{6.2.11}$$

where we introduced the abbreviation $h(\rho)$.

At the strips (6.1.2) and at the hub we find, because $\boldsymbol{n}\perp\boldsymbol{v}_0^*(r)$ and $\cos_{(n,x)}=0$, that condition (6.1.5) changes into

$$\frac{\partial\Phi}{\partial\rho}=0. \tag{6.2.12}$$

Note that the infinitely long hub has the same effect as an "optimum inner ring" at $r=r_i$, to which the blades are fixed.

Summarizing we conclude that we have to solve the boundary value problem drawn in figure 6.2.3, where Φ has to satisfy (6.2.6).

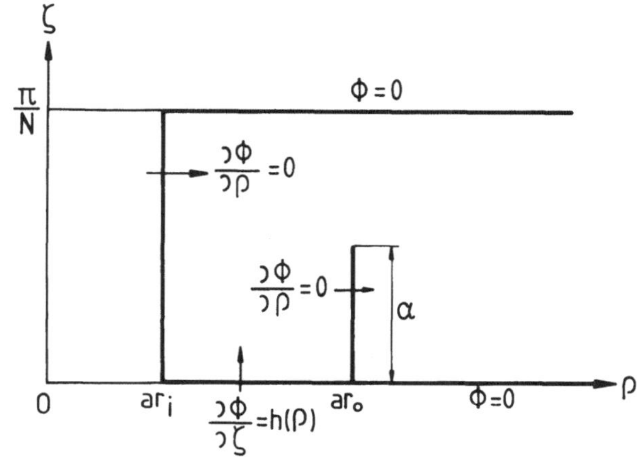

Fig. 6.2.3. Boundary values for the velocity potential Φ.

By the linearity of the theory we can, as we did in (5.4.6), split the potential Φ into two parts

$$\Phi = \Phi_1 + \Phi_2 \overset{\text{def}}{=} \frac{1}{a}\tilde{\Phi}_1 + \frac{\lambda}{a}\tilde{\Phi}_2, \qquad \Phi_1 = \frac{1}{a}\tilde{\Phi}_1 = -\Phi_0, \tag{6.2.13}$$

where we introduced $\tilde{\Phi}_1$ and $\tilde{\Phi}_2$, which satisfy (6.2.6) and are defined in an obvious way by the first equality of (6.2.11).

The splitting (6.2.13) elucidates again the action of an optimum propeller in an inhomogeneous flow. First the field $\boldsymbol{v}_0$ (6.1.4) is annihilated and hence all its energy is extracted by the term $a^{-1}\tilde{\Phi}_1 = -\Phi_0$. This causes in general some thrust. Second, this thrust is completed by the part $\lambda a^{-1}\tilde{\Phi}_2$ to the desired value T by a suitable choice of λ. This latter potential is the potential of an optimum screw propeller working in an homogeneous flow.

It can happen that the thrust caused by the extraction of the kinetic energy of $\boldsymbol{v}_0$ is in the wrong direction. Then the thrust to be delivered by $\lambda a^{-1}\tilde{\Phi}_2$ is larger than T. This happens when $\boldsymbol{v}_0$ is mainly in the direction of the negative x axis. Then we have extra losses which however are reduced to a minimum. When $\boldsymbol{v}_0$ is mainly in the direction of the positive x axis the thrust belonging to $a^{-1}\tilde{\Phi}_1$ will be in the desired direction. Besides that energy is extracted, also a smaller thrust than T has to be delivered by $\lambda a^{-1}\tilde{\Phi}_2$. This will result in a saving of energy. The latter case happens in general behind a ship's hull, where the propeller is working in a fluid which by viscosity effects has obtained a velocity in the direction of the motion of the ship. By this, part of the friction losses of the hull can be regained by the propeller.

6.3. Circulation distribution, thrust

We will give in this section formulas for the optimum spanwise circulation distributions $\Gamma_b(r)$ and $\Gamma_p(r)$ of the blades and of the endplates respectively. Consider around one of the blades a contour l_1, (figure 6.1.1), we reckon $\Gamma_b(r)$ positive when it is coupled with a right hand screw, to the negative r direction. By a continuous deformation of l_1 we can without cutting vortex lines, hence without changing the amount of enclosed vorticity, obtain the contour l_1^*. Thus

$$\Gamma_b(r) = \Phi(Q_2) - \Phi(Q_1) = \Phi(Q_2^*) - \Phi(Q_1^*) = 2\Phi(\rho, 0), \tag{6.3.1}$$

where the points Q_1^* and Q_2^* are far behind the propeller, each at an appropriate side of the helicoidal vortex sheet behind the blade under consideration and have the same value of r as the points Q_1 and Q_2 of l_1.

The value of $\Phi(\rho, 0)$ in (6.3.1) follows from the solution of the boundary value problem of figure 6.2.3.

Analogously we obtain the circulation $\Gamma_p(\zeta)$ around an endplate, reckoned positive when it has (right hand screw) a component in the direction of increasing φ. From figures 6.1.1 and 6.2.3 we see that for a certain value of ζ we have

$$\Gamma_p(\zeta) = \Phi(Q_3) - \Phi(Q_4) = \Phi(Q_3^*) - \Phi(Q_4^*)$$

$$= \Phi(ar_0^+, \zeta) - \Phi(ar_0^-, \zeta), \tag{6.3.2}$$

where $ar_0^+(ar_0^-)$ means $\rho \downarrow (\uparrow) ar_0$. Also this holds far behind the propeller.

From the geometry of the propeller it follows that the demanded thrust T of $O(\varepsilon)$ has to be delivered by the blades, the endplates do not contribute in the linearized theory. Applying Joukowski's theorem (1.9.4) we find

$$T = N\mu\omega \int_{r_i}^{r_0} r\Gamma_b(r)\, \mathrm{d}r = \frac{2N\mu\omega}{a^2} \int_{ar_i}^{ar_0} \rho\Phi(\rho, 0)\, \mathrm{d}\rho. \tag{6.3.3}$$

Due to the splitting (6.2.13) of Φ we write (6.3.3) as

$$T = \frac{2N\mu\omega}{a^3}(M_1 + \lambda M_2), \tag{6.3.4}$$

where

$$M_j = \int_{ar_i}^{ar_0} \rho\tilde{\Phi}_j(\rho, 0)\, \mathrm{d}\rho, \qquad j = 1, 2. \tag{6.3.5}$$

Hence by (6.3.4) the still unknown constant λ becomes

$$\lambda = M_2^{-1}\left\{\frac{a^3 T}{2N\mu\omega} - M_1\right\}, \tag{6.3.6}$$

in which M_1 depends on the disturbance velocity field $\boldsymbol{v}_0^*$.

6.4. Efficiency and quality number

We calculate first the kinetic energy $\bar{E}_p$ which is put into the fluid by the screw propeller per unit of time. The bar above the symbol denotes that the energy loss is per unit of time while in (5.4.10) we used E_p for the energy loss per period of length. The same will hold for $\bar{E}_i$ and E_i (5.6.1). Hence we have to consider as we discussed in section 1.8, the kinetic energy in the region of space K between two planes perpendicular to the x axis at a distance U apart and at a large distance behind the propeller. Also we determine the kinetic energy which was present in that region long before the propeller has passed. The difference between these two energies is the kinetic energy added by the optimum propeller per unit of time. Because we consider a screw propeller with N equal blades, the kinetic energy in K is $2N$ times the kinetic energy in that part K_N of K which is inbetween a helicoidal surface behind one of the blades and a neighbouring bisector surface both extended to infinity.

By (5.4.10) with b replaced by U and carrying out the integration over K_N we find after a partial integration for the difference between the mentioned kinetic energies,

$$\bar{E}_p = -\mu N \frac{\lambda^2}{a^2} \iint_S \tilde{\Phi}_2 \frac{\partial \tilde{\Phi}_2}{\partial n} \mathrm{d}S + \mu N \iint_S \Phi_0 \frac{\partial \Phi_0}{\partial n} \mathrm{d}S, \tag{6.4.1}$$

where S is the helicoidal surface behind one blade over a distance U measured along the x axis. We now discuss the two terms at the right hand side of (6.4.1).

From (6.2.10) and (6.2.11) it follows that we can change the first term into

$$-\mu N \frac{\lambda^2}{a} \iint_S \tilde{\Phi}_2 \frac{\rho}{\left(1+\rho^2\right)^{1/2}} \mathrm{d}S. \tag{6.4.2}$$

The element of area $\mathrm{d}S$ of the helicoidal sheet is by (6.1.1)

$$\mathrm{d}S = \mathrm{d}r\left(\mathrm{d}x^2 + r^2 \mathrm{d}\varphi^2\right)^{1/2} = \left(1 + a^2 r^2\right)^{1/2} \mathrm{d}x\, \mathrm{d}r. \tag{6.4.3}$$

Then by using the definition (6.3.5) we write (6.4.2) as

$$-\mu N \frac{\lambda^2}{a} \int_{r_i}^{r_0} \int_0^U \tilde{\Phi}_2(\rho, 0) \rho \, \mathrm{d}x\, \mathrm{d}r = -\frac{\lambda^2}{a^2} \mu U N M_2. \tag{6.4.4}$$

The second term at the right hand side of (6.4.1) becomes by using (6.2.13)

$$-\frac{\mu U N}{a^2} \int_{ar_i}^{ar_0} \tilde{\Phi}_1(\rho, 0)\left(\rho v_{0x}^*(\rho) + v_{0\varphi}^*(\rho)\right) \mathrm{d}\rho \overset{\mathrm{def}}{=} -\frac{1}{a^2} \mu U N M_3, \tag{6.4.5}$$

where the disturbance velocities are expressed in the variable $\rho = ar$ and we introduced the integral M_3.

Hence we can write (6.4.1) as

$$\bar{E}_p = -\frac{1}{a^2}\mu UN\left(\lambda^2 M_2 + M_3\right). \tag{6.4.6}$$

Next we calculate the kinetic energy added per unit of time to the fluid by an ideal propeller, in the sense of section 5.6, which yields the same thrust T, has the same velocity U and of which the working region is the cylinder G

$$G: \ -\infty \leq x \leq +\infty, \qquad r_i \leq \left(y^2 + z^2\right)^{1/2} \leq r_0. \tag{6.4.7}$$

We use (5.6.1) where again we have to take $b = U$ and because the disturbance field $\boldsymbol{v}_0^*$ is tangent to the boundary ∂G of G we have $\boldsymbol{v}_0 = 0$ hence the second integral is zero in this expression. The Φ_2 which appears in (5.6.1) is discussed in section 5.7 and belongs to the homogeneous slipstream behind the ideal propeller (actuator disk) of which the region A (5.7.6), (5.7.7) is given by the second part of (6.4.7). We have to calculate the thrust T_i which has to be delivered by the actuator disk. This thrust follows from the thrust T_e already created by the extraction of the kinetic energy out of the field $\boldsymbol{v}_0^*$, namely

$$T_i = T - T_e. \tag{6.4.8}$$

So we have to consider T_e. The most simple way in this case is to calculate the amount of momentum in the x direction, which disappears per unit of time when the fluid inside G is brought to a standstill. This amount of momentum is

$$2\pi U\mu \int_{r_i}^{r_0} v_{0x}^*(r) r \, \mathrm{d}r. \tag{6.4.9}$$

Hence we find

$$T_i = T - 2\pi U\mu \int_{r_i}^{r_0} v_{0x}^*(r) r \, \mathrm{d}r. \tag{6.4.10}$$

Using (5.6.1) with $\boldsymbol{v}_0 \equiv 0$ and (5.7.6), the kinetic energy $\bar{E}_i$ added per unit of time to the fluid by the ideal propeller becomes

$$\bar{E}_i = \frac{T_i^2}{2\pi\mu U\left(r_0^2 - r_i^2\right)} - \pi\mu U \int_{r_i}^{r_0} \left(v^{*2}_{0x}(r) + v^{*2}_{0\varphi}(r)\right) r \, \mathrm{d}r. \tag{6.4.11}$$

Then the quality coefficient is by definition

$$q = \bar{E}_i / \bar{E}_p \leq 1. \tag{6.4.12}$$

The efficiency η_T of the screw propeller can now be written as

$$\eta_T = \frac{TU}{TU + \bar{E}_p} = \left(1 + \frac{\bar{E}_i}{qTU}\right)^{-1}. \tag{6.4.13}$$

In the case that the disturbance field v_0^* is identically zero we find by (5.7.9)

$$\eta_T = \left(1 + \frac{T}{2\pi\mu q U^2 (r_0^2 - r_i^2)}\right)^{-1}, \qquad v_0^* \equiv 0. \tag{6.4.14}$$

These optimum efficiencies can be realized theoretically still by all kinds of shapes of blades and endplates and types of profiles. The only thing that matters in this linearized theory, is the spanwise distribution of the total circulation around the blades and the endplates.

6.5. Numerical results

For a numerical method to solve the boundary value problem of figure 6.2.3 we refer to [38]. We only remark that the condition $\partial\Phi/\partial\rho = 0$ at the endplate $r = ar_0$, $0 \le \zeta \le \alpha$, causes difficulties with respect to the accuracy of the solution of a finite difference approximation of the partial differential equation (6.2.6) in the neighbourhood of the point $r = ar_0$, $\zeta = \alpha$ of the endplate.

First we give results in case that the propeller is working in a homogeneous flow, hence when the disturbance field $v_0^*(r)$ is zero. For the relative radius of the hub we have chosen for these and the following numerical results $r_i = 0.2r_0$. The quality number q in this case of an undisturbed flow, is only influenced by three parameters. It depends on the number of blades N, on the number $N\alpha/\pi$ (6.1.2) which is the fraction of the circumference of the screw $2\pi r_0$ covered by the total span of the endplates and on ar_0, the inverse of the advance coefficient.

In table 6.5.1 values of q are given. It follows from this table that an increase in one of the just mentioned three parameters, while the other two remain constant, causes an increase in q. This could be expected, because viscosity effects are neglected in this theory. The number of blades starts with the not realistic value $N = 1$, however it is interesting to see that the quality factor for $N = 1$, differs much in the case of no endplates ($\alpha = 0$) from its value for $N = 4$. When however $N\alpha/\pi = 1$ in both cases ($N = 1$, $N = 4$), this difference becomes small. This holds most strongly for low advance coefficients $ar_0 = 2$, however also for $ar_0 = 4$ the favourable action of the endplates is clear. Summarizing we can say that

Table 6.5.1
Values of quality number q, homogeneous flow.

ar_0	$N\alpha/\pi$	1	2	3	4	N
2	0.0	0.181	0.297	0.367	0.413	
	0.125	0.272	0.368	0.423	0.460	
	0.25	0.346	0.427	0.470	0.499	
	0.5	0.457	0.511	0.538	0.556	
	0.75	0.526	0.558	0.576	0.587	
	1.0	0.548	0.572	0.587	0.597	
3	0.0	0.295	0.456	0.539	0.589	
	0.125	0.399	0.527	0.592	0.631	
	0.25	0.486	0.588	0.637	0.667	
	0.5	0.620	0.678	0.705	0.720	
	0.75	0.702	0.730	0.743	0.751	
	1.0	0.729	0.746	0.755	0.760	
4	0.0	0.398	0.574	0.655	0.701	
	0.125	0.498	0.637	0.700	0.736	
	0.25	0.586	0.693	0.740	0.766	
	0.5	0.719	0.775	0.799	0.811	
	0.75	0.800	0.823	0.832	0.838	
	1.0	0.827	0.838	0.843	0.846	

in the optimum case endplates with respect to potential theory, seem to be effective in case of a small number of blades and of small numbers ar_0.

In figure 6.5.1 graphs are drawn of the spanwise optimum circulation distribution of blades and endplates of a three bladed propeller $N = 3$ for $ar_0 = 3$ and for a number of lengths of the endplates or better for a number of values of $N\alpha/\pi$. The coordinates ρ and ζ are defined in (6.1.6) and the dimensionless thrust coefficient C_T by

$$C_T = \frac{T}{\mu U^2 \pi \left(r_0^2 - r_i^2\right)}. \tag{6.5.1}$$

It is seen from this figure that at the tips of the endplates the strength of the circulation distribution has an infinite derivative with the exception for $N\alpha/\pi = 1$. In this case the vorticity of one tip of an endplate A passes just along the tip of one neighbouring endplate while the other tip of A is just on the verge of the vorticity sheet of the other neighbouring endplate. It is also possible to replace in this case, the separate endplates by one closed ring. The circulation distribution along the endplates or the closed ring becomes practically linear, hence in the optimum cases free vorticity is shed on the cylinder $r = r_0$ with nearly a constant density.

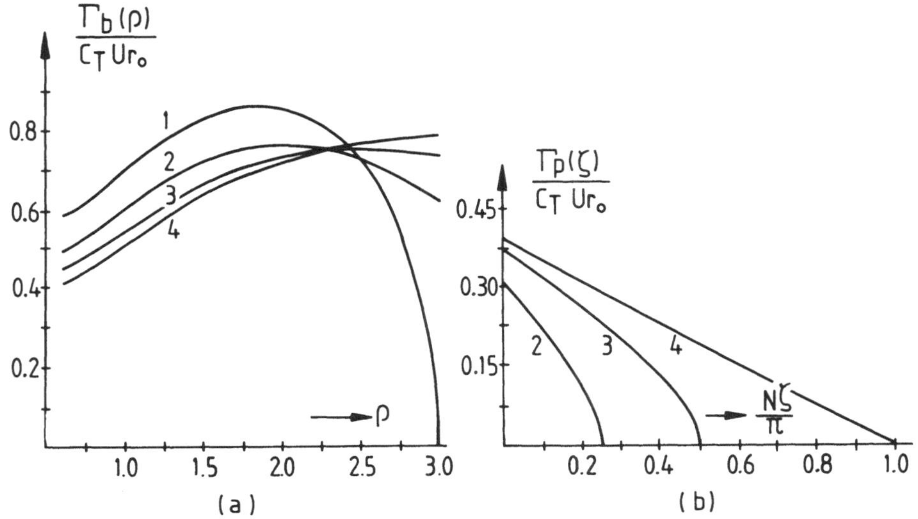

Fig. 6.5.1. Optimum circulation distributions, $N=3$, $ar_0=3$, 1) $N\alpha/\pi=0$, 2) $N\alpha/\pi=0.25$, 3) $N\alpha/\pi=0.5$, 4) $N\alpha/\pi=1$, (a) along blade, (b) along endplate.

Next we consider the case of an inhomogeneous inflow. We restrict ourselves to disturbance velocity fields $\boldsymbol{v}_0^*$ (6.1.4) which only have an axial component. These fields are assumed to have the form

$$v_{0x}^*(r)=\sum_{j=0}^{n}\beta_j\rho^j,\qquad v_{0r}^*(r)=v_{0\varphi}^*(r)=0,\tag{6.5.2}$$

where the coefficients β_j have dimension of velocity. In order to give the results in a convenient way we need a few more definitions.

The solutions of the boundary value problem (figure 6.2.3) with $v_{0x}^*(r)=\rho^j$, $v_{0\varphi}^*=0$, $a=1$ and $\lambda=0$ (6.2.11), will be denoted by $\psi_j(\rho,\zeta)$ $(j=0,1,\ldots,n)$. Hence when v_{0x}^* is given by (6.5.2) we have, by the linearity of the problem, for the functions $\tilde{\Phi}_1$ and $\tilde{\Phi}_2$ (6.2.13)

$$\tilde{\Phi}_1(\rho,\zeta)=\sum_{j=0}^{n}\beta_j\psi_j(\rho,\zeta),\qquad \tilde{\Phi}_2(\rho,\zeta)=-\psi_0(\rho,\zeta).\tag{6.5.3}$$

Substitution of (6.5.3) into (6.3.5) and (6.4.5) leads to

$$M_1=\sum_{j=0}^{n}\beta_jM_{j,0},\qquad M_3=\sum_{j=0}^{n}\sum_{k=0}^{n}\beta_j\beta_kM_{j,k},\tag{6.5.4}$$

where

$$M_{j,k}=\int_{ar_i}^{ar_0}\rho^{k+1}\psi_j(\rho,0)\,\mathrm{d}\rho,\qquad k,j=0,1,\ldots,n.\tag{6.5.5}$$

Finally we introduce the numbers $Q_{j,k}$, which are dimensionless, by

$$Q_{j,k} = -\frac{2N}{a^2\pi\left(r_0^2 - r_i^2\right)} M_{j,k}. \tag{6.5.6}$$

We note that $Q_{j,k}$ is determined by r_i/r_0, N, α and ar_0.

The unknown λ follows from (6.3.6) which can be written as

$$\frac{\lambda}{U} = \frac{1}{Q_{0,0}}\left\{C_T - \sum_{j=1}^{n} \frac{\beta_j}{U} Q_{j,0}\right\}. \tag{6.5.7}$$

A straight forward calculation yields for the quality factor (6.4.12)

$$q = \left[\left\{C_T - \frac{2}{Ua^2\left(r_0^2 - r_i^2\right)} \int_{ar_i}^{ar_0} v_{0x}^* \cdot \rho \, \mathrm{d}\rho\right\}^2 - \frac{2}{U^2a^2\left(r_0^2 - r_i^2\right)} \int_{ar_i}^{ar_0} v^{*2}_{0x} \cdot \rho \, \mathrm{d}\rho\right]$$

$$\times \left\{-\left(\frac{\lambda}{U}\right)^2 Q_{0,0} + \frac{1}{U^2} \sum_{j=0}^{n} \sum_{k=0}^{n} \beta_j \beta_k Q_{j,k}\right\}^{-1}$$

$$\stackrel{\text{def}}{=} \frac{G_1}{G_2}, \tag{6.5.8}$$

where we have introduced the abbreviations G_1 and G_2.

In the case of an inhomogeneous flow $v_{0x}^*(r) \not\equiv 0$ the quality factor depends besides on the three parameters N, $N\alpha/\pi$ and ar_0 mentioned in the first paragraph of this section also on v_{0x}^* and on the thrust T or on the thrust coefficient C_T. We emphasize that the dependence on C_T does not occur in the homogeneous case $v_{0x}^* \equiv 0$, as follows also from (6.5.8).

In Table 6.5.2 values of $Q_{j,k}$ are given for the denoted values of the parameters N, α and ar_0, while j and k range through 0, 1 and 2. This means that in (6.5.2) the summation is carried out for $n = 2$. Table 6.5.2 can be used as follows. Suppose a field $\boldsymbol{v}_0^* = (v_{0x}^*(r), 0, 0)$ is given by means of numerical values of v_{0x}^* for a number of values of r. Draw approximately a second order polynomial "through" these points. Then the coefficients β_0, β_1, β_2 (6.5.2) are known. Further we assume to be given values of N, α and ar_0 which also appear in table 6.5.2, by which we can select the appropriate $Q_{j,k}$. If the thrust T and hence the thrust coefficient C_T is also given, equation (6.5.7) yields λ, and q follows from (6.5.8).

For the efficiency we find

$$\eta_T = \left(1 + \frac{G_1}{2qC_T}\right)^{-1}, \tag{6.5.9}$$

where G_1 is defined in (6.5.8).

Next we discuss shortly the effect of $v_{0x}^*(r)$ (6.5.2) on the optimum circulation distribution. From (6.2.13), (6.3.1) and (6.5.3) we find

$$\Gamma_b(r) = +2\Phi(\rho, 0) = +\frac{2}{a}\left\{\sum_{j=0}^{2} \beta_j \psi_j(\rho, 0) - \lambda\psi_0(\rho, 0)\right\}. \tag{6.5.10}$$

It is not difficult to see that the functions $\psi_j(\rho, 0)$ have the same sign. We consider $\psi_j(\rho, 0)/\psi_0(\rho, 0)$, $j = 1, 2$. Figure 6.5.2 shows that these functions have larger values away from the hub. For positive values of β_j, the incoming flow has larger velocities at the tips of the blades than in the neighbourhood of the hub. From (6.5.10) and figure 6.5.2 it follows that

Table 6.5.2.
Values of $Q_{i,j}$; $r_i/r_0 = 0.2$; (top) $ar_0 = 3$, (bottom) $ar_0 = 4$.

N	2			3		
$N\alpha/\pi$	0	0.5	1	0	0.5	1
$Q_{0,0}$	0.445	0.679	0.748	0.525	0.702	0.756
$Q_{1,0}$	0.860	1.424	1.597	1.036	1.475	1.627
$Q_{2,0}$	1.816	3.251	3.692	2.220	3.363	3.780
$Q_{0,1}$	0.860	1.435	1.597	1.036	1.494	1.626
$Q_{1,1}$	1.753	3.141	3.541	2.166	3.305	3.677
$Q_{2,1}$	3.844	7.380	8.397	4.832	7.802	8.820
$Q_{0,2}$	1.819	3.291	3.686	2.222	3.438	3.773
$Q_{1,2}$	3.847	7.413	8.387	4.835	7.872	8.809
$Q_{2,2}$	8.672	17.764	20.238	11.100	19.027	21.595
N	2			3		
$N\alpha/\pi$	0	0.5	1	0	0.5	1
$Q_{0,0}$	0.559	0.774	0.838	0.638	0.793	0.842
$Q_{1,0}$	1.439	2.139	2.368	1.676	2.191	2.392
$Q_{2,0}$	4.053	6.452	7.269	4.789	6.593	7.365
$Q_{0,1}$	1.440	2.163	2.368	1.676	2.227	2.391
$Q_{1,1}$	3.929	6.292	7.022	4.694	6.534	7.207
$Q_{2,1}$	11.530	19.646	22.241	14.021	20.471	23.056
$Q_{0,2}$	4.056	6.578	7.258	4.790	6.783	7.351
$Q_{1,2}$	11.537	19.797	22.213	14.028	20.715	23.027
$Q_{2,2}$	34.874	63.281	71.873	43.230	66.702	75.586

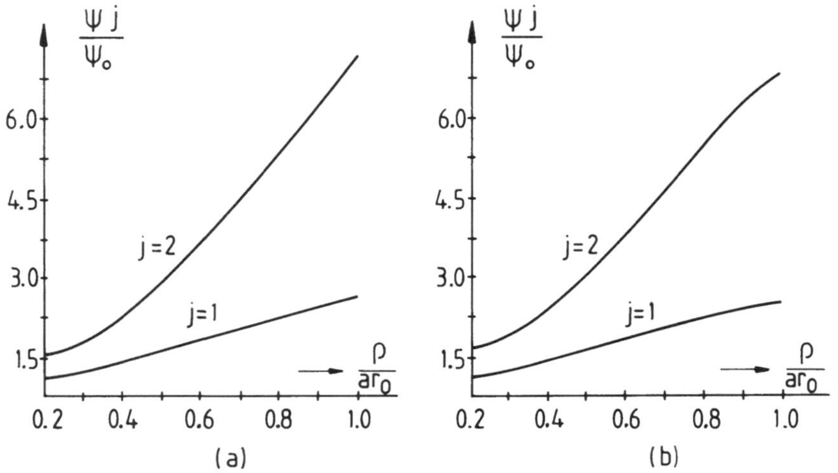

Fig. 6.5.2. The functions $\psi_j(\rho, 0)/\psi_0(\rho, 0)$, $j = 1, 2$; $N = 3$, $ar_0 = 3$, (a) $N\alpha/\pi = 0$, (b) $N\alpha/\pi = 0.5$.

in this case the circulation distribution $\Gamma_b(r)$ has relatively larger values in the neighbourhood of the hub than in the case of homogeneous inflow. This is in agreement with the fact that a propeller will have a higher efficiency the more its slipstream becomes homogeneous.

6.6. Optimization of the sails of a yacht

As a second example of the application of optimization theory we consider the optimization of the sails of a yacht sailing close to wind. Under sailing close to wind we will understand the situation that the wind relative to the yacht (apparent wind) makes a small angle of $O(\varepsilon)$ with the course of the yacht. Then the thrust T developped by the sails will be $O(\varepsilon^2)$. We call in this case the sails optimum when T is as large as possible under the constraint of some conditions which we will formulate later on.

In figure 6.6.1 (X, Y, Z) represents a Cartesian coordinate system which is in rest with respect to the ship. The air is in the half space-region $Z > 0$, hence we neglect the existence of waves on the water surface. The relative velocity of the wind makes a small angle α with the X axis and has the magnitude U. For simplicity we omit the boundary layer of the wind at the water surface and take U independent of the Z coordinate. Because we assume $\alpha = O(\varepsilon)$ the vortex sheets of foresail and mainsail are separated by a distance of $O(\varepsilon)$ and, within the accuracy of our theory, can be assumed, to coincide. As we discussed already, in linearized optimization theory, lifting surfaces can be replaced without

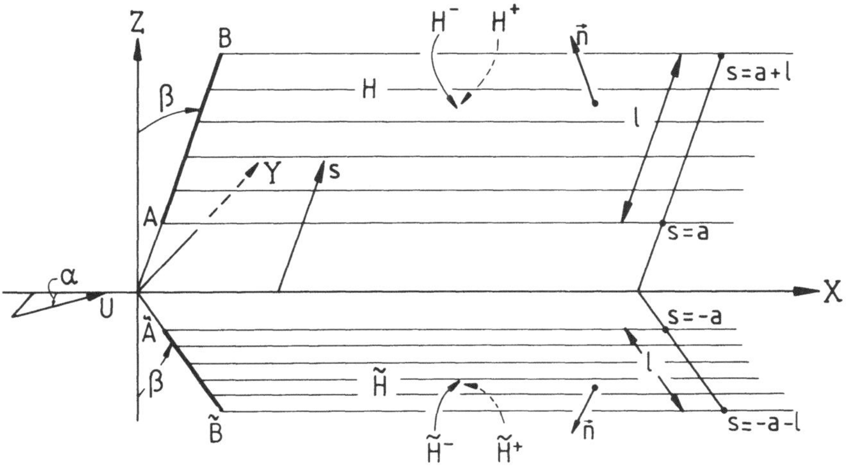

Fig. 6.6.1. Lifting lines $(A-B)$ and $(\tilde{A}-\tilde{B})$ and free vortex sheets.

loosing generality, by lifting lines chosen at an arbitrary place at the vortex sheets. Hence the lifting lines representing foresail and mainsail can be assumed to coincide. So we replace the sails of the ship by one lifting line $(A-B)$ through the origin O, in the (Y, Z) plane and with a heeling angle β, $0 \leq \beta \leq \frac{1}{2}\pi$. The free vortex sheet H stretches downstream from $(A-B)$. Because the undisturbed mainstream has only a component of $O(\varepsilon)$ perpendicular to the X axis, H can be chosen within the accuracy of the theory parallel to the X axis.

In order to simulate the boundary between air and water, we consider as usual, the image $(\tilde{A}-\tilde{B})$ of $(A-B)$ and the image $\tilde{H}$ of H, both reflections are with respect to the plane $Z=0$. We define the $+(-)$ side of H and $\tilde{H}$ to be oriented in the positive (negative) Y direction and the unit normal vector $\boldsymbol{n}$ on H and on $\tilde{H}$ from the $+$side toward the $-$side. On H and $\tilde{H}$ we introduce a Cartesian coordinate system (X, s) where on H the coordinate s is the distance from a point to the X axis, while on $\tilde{H}$ this coordinate is minus this distance. The boundaries of H are denoted by $s=a$ and $s=a+l$ with $l>0$, and of $\tilde{H}$ by $s=-a$ and $s=-a-l$, where l is the width of H and $\tilde{H}$ or the length of the lifting line. After this we assume the whole space to be filled with air of which the incoming velocity for $Z<0$ is the same as for $Z>0$. This air is assumed to be incompressible, inviscid and of density μ.

The driving force or thrust $T(O(\varepsilon^2))$ of the sails is defined as the component in the X direction of the force acting on the bound vorticity of the lifting line $(A-B)$ it is reckoned positive in the negative X direction. It is our intention to determine within the realm of the linear theory the maximum thrust T_m.

First we show that an upperbound exists for T. Suppose we have at

$(A-B)$ and $(\tilde{A}-\tilde{B})$ some spanwise circulation distribution $\nu\Gamma(s)$ where ν is some factor, $\nu \geq 0$. The bound vorticity $\Gamma(s)$ is reckoned positive with a right hand screw in the positive s direction. By symmetry we have

$$\Gamma(s) = \Gamma(-s). \tag{6.6.1}$$

We introduce the function $\nu w_n(s)$ which is the velocity induced in the direction $\boldsymbol{n}$ at $(A-B)$ by the free vortex sheets at H and $\tilde{H}$. The bound vortices at $(A-B)$ and $(\tilde{A}-\tilde{B})$ do not contribute. Then the thrust T with $T > 0$ in the negative X direction becomes

$$T = \mu \int_a^{a+l} \nu\Gamma(s)\{-\alpha U \cos\beta + \nu w_n(s)\}\, \mathrm{d}s. \tag{6.6.2}$$

For sufficiently small values of ν it follows that the term linear in ν dominates and that $\Gamma(s)$ has to be negative for positive values of T. It is easily seen that $w_n(s)$ is positive for negative values of $\Gamma(s)$. When ν increases, the linear term increases and hence the same holds for T. However, the induced velocity $\nu w_n(s)$ also increases and will more and more counteract the component $-U\alpha \cos\beta$ until for some larger value of ν the thrust T has reached its maximum value and then starts to decrease. Hence because T depends quadratically on ν, one maximum value exists for the chosen $\Gamma(s)$. When we assume that an optimum distribution $\Gamma_m(s)$ exists the same argument holds and the thrust will have for this distribution an optimum value denoted by T_m. Note that this discussion is valid because the two terms between brackets in (6.6.2) are of the same order of magnitude ($\alpha = O(\varepsilon)$ and $w_n = O(\varepsilon)$).

We next define the heeling force F acting on $(A-B)$ as the force perpendicular to H, reckoned positive in the negative $\boldsymbol{n}$ direction. The constraint that F has to have a prescribed valued F_h which in general will be chosen to be smaller than the value which belongs to T_m has the form

$$-\mu U \int_a^{a+l} \Gamma(s)\, \mathrm{d}s = F_h. \tag{6.6.3}$$

Analogously the constraint that the heeling moment has to have a prescribed value M_h, which is reckoned positive when it is connected with a right-hand screw to the negative X direction is

$$-\mu U \int_a^{a+l} (s_0 + s)\Gamma(s)\, \mathrm{d}s = M_h, \tag{6.6.4}$$

where s_0 is some constant depending on the line with respect to which the moment is calculated. Also here the value M_h will be chosen to be smaller than the value which belongs to T_m.

In this case of sailing close to wind, the heeling force on the sails is large with respect to the thrust. The first one is $O(\varepsilon)$ while the second one is only $O(\varepsilon^2)$. This is an analogous situation as the one we discussed in the first paragraph of section 4.9 with respect to the unsteady propulsion of regime ia and regime ib.

We now change the formulation in such a way that we obtain a problem of energy extraction out of a slightly disturbed fluid. Consider the Cartesian coordinate system (x, y, z) which is related to the former system by

$$x = X - Ut, \qquad y = Y, \qquad z = Z. \tag{6.6.5}$$

The air is in rest with respect to (x, y, z) except for a small homogeneous flow of magnitude $U\alpha = O(\varepsilon)$ in the y direction and for a negligible flow of $O(\varepsilon^2)$ in the x direction. Hence in our new coordinate system (x, y, z) we have a fluid disturbed by

$$\boldsymbol{v}_0^* = (0, \alpha U, 0). \tag{6.6.6}$$

The lifting lines $(A - B)$ and $(\tilde{A} - \tilde{B})$ translate in the negative x direction with velocity U, along the two strips H and $\tilde{H}$ which stretch from $x = -\infty$ towards $x = +\infty$ and on which from now on we use the coordinates (x, s). The circulation $\Gamma(s)$ of these lines has to be such that under the constraints (6.6.3) and (6.6.4) the resulting force in the negative x direction is maximum, hence the energy extraction out of the fluid has to be maximum.

As it is shown in section (5.2) we can replace $\boldsymbol{v}_0^*$ by $\boldsymbol{v}_0$ which has the same normal component at H and $\tilde{H}$ and of which the vorticity is denoted by $\boldsymbol{\gamma}_0(s)$. This vorticity is parallel to the x axis and reckoned positive with a right-hand screw in the positive x direction. Because $\boldsymbol{\gamma}_0(s)$ has only a component in the x direction we will denote it simply by $\gamma_0(s)$ which is the vorticity per unit of length in the s direction. By symmetry it follows $\gamma_0(s) = -\gamma_0(-s)$. As we discussed in section (5.2) we have to take

$$\int_a^{a+l} \gamma_0(s)\, \mathrm{d}s = 0. \tag{6.6.7}$$

The potential Φ_0 of the field $\boldsymbol{v}_0$ satisfies at H and $\tilde{H}$

$$\frac{\partial \Phi_0}{\partial n} = -\alpha U \cos\beta. \tag{6.6.8}$$

The bound vortices $\Gamma(s)$ shed free vorticity $\boldsymbol{\gamma}(s)$ which is also in the x

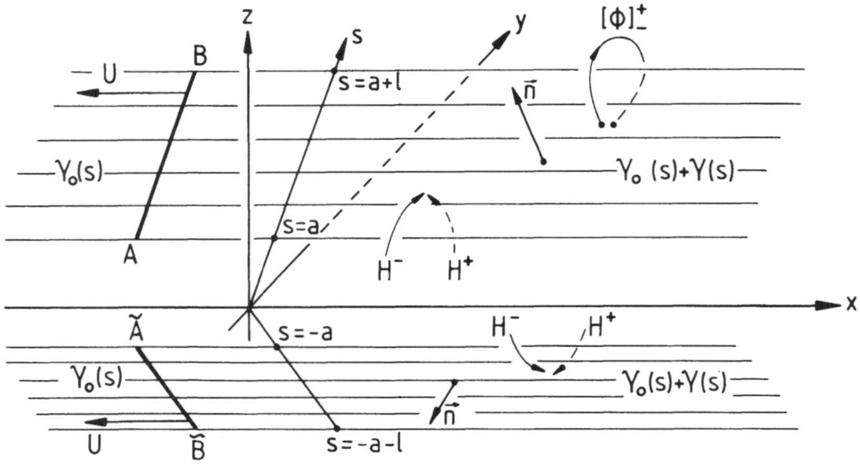

Fig. 6.6.2. The two-sided infinite free vortex sheets H and $\tilde{H}$.

direction. We make the same agreement with respect to notation and sign for $\boldsymbol{\gamma}(s)$ as we did for $\boldsymbol{\gamma}_0(s)$. We have the relation

$$\gamma(s) = -\frac{\mathrm{d}\Gamma(s)}{\mathrm{d}s}. \tag{6.6.9}$$

When the bound vortices representing the sails and the mirrored ones, are far away in the negative x direction, the potential Φ of the velocity field induced by $\gamma(s)$ has become independent of time and of the x coordinate hence $\Phi = \Phi(s)$. The jump in the potential Φ across the strip H for some value of s; equals minus the amount of $\gamma(s)$ in the interval $(s, a+l)$, hence

$$[\Phi]_-^+(s) = -\Gamma(s) \tag{6.6.10}$$

and analogously for $\tilde{H}$.

The constraints (6.6.3) and (6.6.4) can now be written as

$$\mu U \int_{-a-l}^{a+l} [\Phi]_-^+(s)\, \mathrm{d}s = 2F_h, \tag{6.6.11}$$

and

$$\mu U \int_{-a-l}^{a+l} [\Phi]_-^+(s)(s_0 + |s|)\, \mathrm{d}s = 2M_h, \tag{6.6.12}$$

which have the form (5.3.7). The interval $-a < s < a$ gives no contribution to the integral because there $[\Phi]_-^+$ vanishes. Hence we arrived at a variational problem as treated in section (5.4). We have in order to obtain the maximum thrust, to minimize the kinetic energy E (5.3.1) (where we take $b=1$) under the constraints (6.6.11) and (6.6.12). Note that in this problem we have no constraint on the thrust. Hence using (6.6.8) we obtain from (5.4.5)

$$\frac{\partial \Phi(s)}{\partial n} = \alpha U \cos\beta + \lambda_1 + \lambda_2(s_0 + |s|), \tag{6.6.13}$$

where λ_1 and λ_2 are two Lagrange multipliers.

We next introduce two potential functions $\varphi_1(y, z)$ and $\varphi_2(y, z)$ which satisfy at H and $\tilde{H}$

$$\frac{\partial \varphi_1}{\partial n} = 1, \qquad \frac{\partial \varphi_2}{\partial n} = |s|. \tag{6.6.14}$$

The potentials φ_1 and φ_2 can be determined by calculating their vorticity $\gamma_1(s)$ and $\gamma_2(s)$ reckoned positive with a right-hand screw in the x direction, on H and $\tilde{H}$. Using Biot and Savart's law it follows from (6.6.14) that $\gamma_1(s)$ and $\gamma_2(s)$ have to satisfy the following singular integral equations

$$-\frac{1}{2\pi}\oint_a^{a+l} \gamma_j(\sigma)\left\{\frac{1}{(\sigma - s)} - \frac{(s + \sigma\cos 2\beta)}{(\sigma^2 + 2\sigma s\cos 2\beta + s^2)}\right\} \mathrm{d}\sigma = g_j(s),$$

$$a \le s \le a + l, \qquad j = 1, 2, \tag{6.6.15}$$

where $g_1(s) = 1$ and $g_2(s) = s$. The second term in the kernel arises from the vorticity at $\tilde{H}$. Because the total circulation of each strip H or $\tilde{H}$ has to be zero we have

$$\int_a^{a+l} \gamma_j(s)\,\mathrm{d}s = 0, \qquad j = 1, 2, \tag{6.6.16}$$

by which the solutions for $j=1$ and $j=2$ of (6.6.15), become unique. These solutions have to be calculated by numerical means.

Integration of the vorticity yields

$$[\varphi_j]_-^+(s) = -\int_s^{a+l} \gamma_j(s)\,\mathrm{d}s, \qquad j = 1, 2, \tag{6.6.17}$$

hence these jumps $[\varphi_j]_-^+$ which are needed in the following, can be assumed to be known. From (6.6.13) and (6.6.14) it follows that the potential Φ belonging to the shed vorticity in the optimum case can be written as

$$\Phi(y, z) = (\alpha U\cos\beta + \lambda_1 + \lambda_2 s_0)\varphi_1(y, z) + \lambda_2\varphi_2(y, z). \tag{6.6.18}$$

The Lagrange multipliers λ_1 and λ_2 follow from (6.6.11) and (6.6.12).

Using (6.6.13) we find for the total velocity w_n in the direction of $\boldsymbol{n}$ at the lifting line $(A - B)$,

$$w_n(s) = -\alpha U \cos\beta + \frac{1}{2}\frac{\partial \Phi}{\partial n}(s) = -\tfrac{1}{2}\alpha U \cos\beta + \tfrac{1}{2}\{\lambda_1 + \lambda_2(s_0 + |s|)\}. \tag{6.6.19}$$

The factor $\frac{1}{2}$ of $\partial\Phi/\partial n$ arises because behind $(A - B)$ the shed vorticity $\gamma(s)$ is "only half infinitely" long. Hence the optimum thrust T has the value

$$T = \mu\int_a^{a+l} \Gamma(s) w_n(s)\,\mathrm{d}s = -\mu\int_a^{a+l} [\Phi]_-^+(s) w_n(s)\,\mathrm{d}s$$

$$= -\mu\int_a^{a+l} \{(\alpha U\cos\beta + \lambda_1 + \lambda_2 s_0)[\varphi_1]_-^+(s) + \lambda_2[\varphi_2]_-^+(s)\}$$

$$\times\{-\tfrac{1}{2}\alpha U\cos\beta + \tfrac{1}{2}(\lambda_1 + \lambda_2(s_0 + |s|))\}\,\mathrm{d}s. \tag{6.6.20}$$

In case we have no constraints, we can neglect equations (6.6.11) and (6.6.12) and put $\lambda_1 = \lambda_2 = 0$. Then it follows from (6.6.18) and (6.6.10) that the optimum circulation distribution is proportional to αU when the heeling angle β is fixed. It follows from (6.6.11) and (6.6.12) that then the heeling force and heeling moment are proportional to αU^2. We denote by μF_1 and μM_1 the heeling force and moment when $\alpha = 1$ and $U = 1$. Then the actual heeling force and moment, again in the case of no constraint, are $\alpha U^2 \mu F_1$ and $\alpha U^2 \mu M_1$. Using these notations we can slightly change the conditions (6.6.11) and (6.6.12). We introduce ν_1 which is the quotient of the admitted heeling force F_h and the heeling force $\alpha U^2 \mu F_1$ in the optimum situation when there is no constraint. In the same way ν_2 is the quotient of the admitted heeling moment M_h and the heeling moment $\alpha U^2 \mu M_1$ in the optimum situation without constraints. Hence

$$F_h = \nu_1 \alpha U^2 \mu F_1, \qquad M_h = \nu_2 \alpha U^2 \mu M_1. \tag{6.6.21}$$

The potential $\tilde{\Phi}(y, z)$ which belongs to the case of no constraints and $\alpha = U = 1$ satisfies the boundary conditions

$$\frac{\partial \tilde{\Phi}}{\partial n} = \cos\beta, \tag{6.6.22}$$

hence $\tilde{\Phi}(y, z) = \cos\beta\,\varphi_1(y, z)$. Substitution of $\tilde{\Phi}$ and $U = 1$ into the left hand sides of (6.6.11) and (6.6.12) yields the values of $2F_1$ and $2M_1$.

6.7. Numerical results

In this section we present a number of graphs which show numerical results of the previous theory. We assumed that the constant s_0 (6.6.4) is zero. The value of a represents to a certain extent the gap between "sail and deck". We did not put a constraint on the heeling force F but only on the heeling moment M. This means that we neglect (6.6.11) and take $\lambda_1 = 0$. The results have been made dimensionless in an obvious way. Figure 6.7.1a shows the maximum thrust for the heeling angle $\beta = 0°$ as a function of the relative width a/l of the gap between "sail and deck" and of the heeling moment constraint coefficient ν_2. For $\nu_2 = 1$ we have no constraint on this moment and for $\nu_2 = 0$, this moment has to be zero. On the horizontal axis we have plotted a/l on a logarithmic scale, which is appropriate because of the sensitivity of the maximum thrust to narrow gaps. At the right hand side of the lines the values of ν_2 are given. At the T axis the values for $a/l = 0$ are denoted. It is seen that the relative width a/l of the gap affects strongly the optimum thrust. In case of no constraint $\nu_2 = 1$ and for $a/l = 10^{-4}$ still about 20% of the maximum thrust is lacking. In figure 6.7.1b we show the same results but now for $\beta = 45°$. It is seen that a heeling angle of 45° nearly halves the maximum possible thrust.

Figure 6.7.2 shows analogously the values of the heeling moment. The

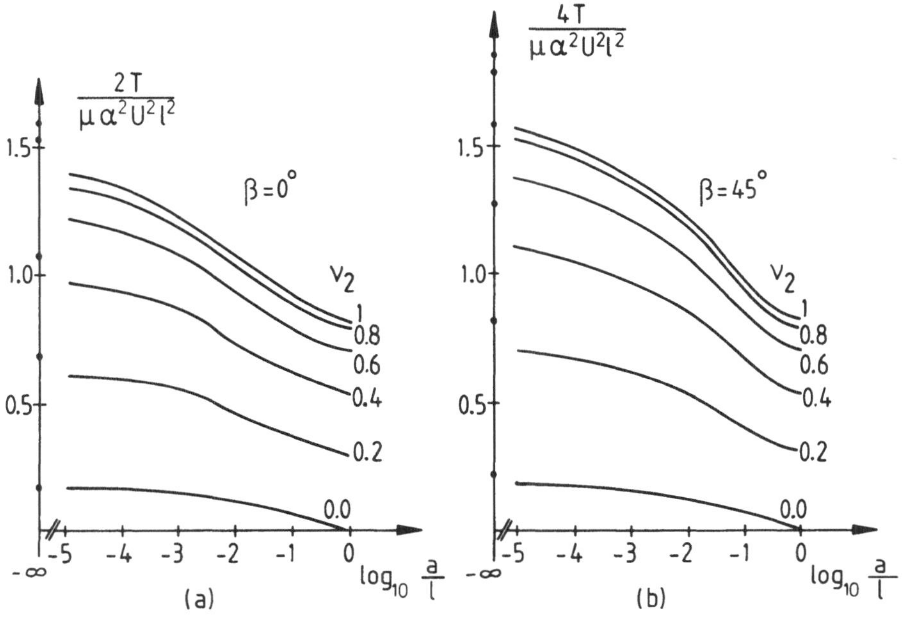

Fig. 6.7.1. Dependence of maximum thrust on a/l and on ν_2.

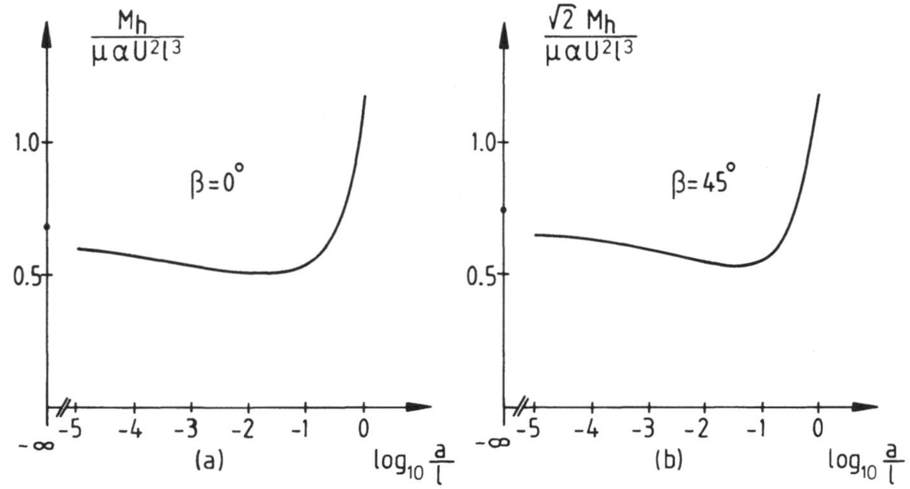

Fig. 6.7.2. Dependence of heeling moment on a/l for $\nu_2 = 1$.

lines for $\nu_2 \neq 1$ follow directly from the lines for $\nu_2 = 1$ by multiplication by ν_2.

It turned out from the computations that the values given in figures 6.7.1 and 6.7.2 for $\beta = 0°$, differ only by 6% at most from the corresponding values for $\beta = 30°$. This confirms a statement of Milgram [45] that mostly it will be sufficient to consider the case $\beta = 0°$.

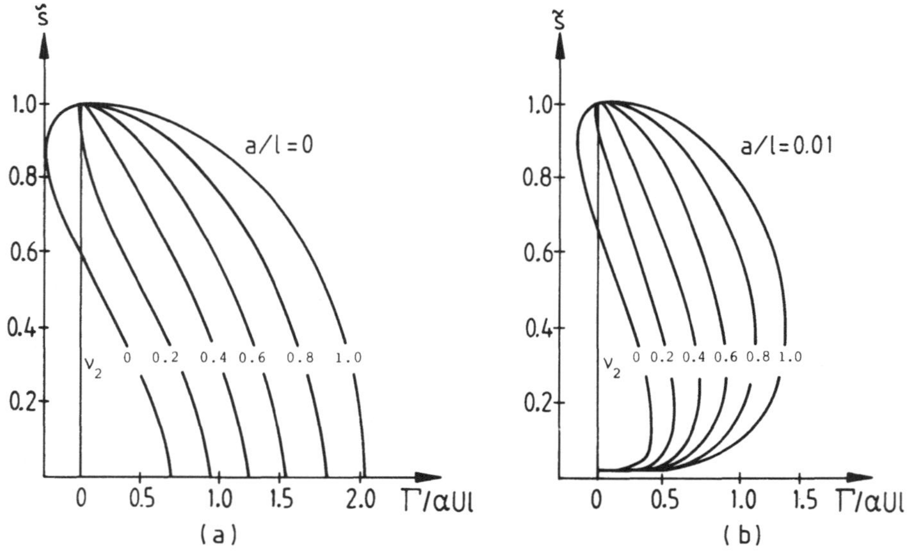

Fig. 6.7.3. Dependence of optimum circulation around the sail as a function of $\tilde{s}$ on ν_2; $\beta = 0°$.

Figure 6.7.3 shows the optimum circulation distribution $\Gamma(s)$ around the lifting line as a function of the dimensionless coordinate $\tilde{s}$, $a/l \leq \tilde{s} \leq (a/l)+1$, along the mast. Two cases are considered $a/l=0$ and $a/l=0.01$, both for $\beta=0°$. The numbers inside the graphs denote the values of ν_2. Again the very strong dependence of optimum quantities on a narrow gap is evident.

We next comment upon some of the given results. From figure 6.7.1 it is seen that with a moderate constraint on the heeling moment ($\nu_2 \approx 0.6$) not too much of the maximum possible thrust is lost. Hence it even is possible in connection with the transverse stability of the ship, that by such a constraint the maximum thrust increases because the heeling angle β will be smaller. This means that in fact we have to consider a more complicated problem which involves the combined action of sails and keel. The keel of a yacht considered as a lifting surface is needed to neutralize in combination with the hull the side force, which is the horizontal component of the heeling force, produced by the sails. The righting moment which occurs by the weight connected to the keel and the shape of the hull has to balance the heeling moment. It is clear that there is no fundamental difference between sail and keel. Both act as lifting surfaces one in the air, the other in the water. In such a problem we have to optimize the circulation distribution of the sails as well as the circulation of the keel, from the point of view of their combined action and taking into account the righting moment. For this we refer to [71].

When we consider a sail of span $l=20$ meters and a gap $a=1$ centimeters, we have $a/l=5.10^{-4}$, hence $\log_{10} a/l = -3.3$. Then for a not too large heeling angle it follows from figure 6.7.1a that a thrust reduction of about 25% would occur. The question arises if this happens in a real situation when the air is endowed with viscosity and turbulence. With respect to the viscosity, this seems not too important for a gap of that width. The turbulence of the air which has passed along the deck of the ship can induce the so-called eddy viscosity. However, the scale of the turbulence is possibly too large to have much influence. Hence it is not likely that the inviscid character of the flow has changed very much. It is possible, however that the air which has passed through the gap will not follow smoothly the lee side of the sail but forms a two dimensional jet which partly stretches along the deck. Then possibly the effect is overestimated in our theory. Finally, we emphasize that the decline caused by the gap, of the maximum possible thrust is calculated here. In general a sail will not be shaped in such a way that it is able to produce the maximum thrust. Then of course it is not known what exactly the influence of a narrow gap will be. What really happens must be investigated by experiments.

6.8. Optimum unsteady propulsion, large amplitude, two dimensional

We now discuss the two dimensional case of the two regimes of large amplitude unsteady propulsion of section 4.1. We have assumed in regime ii (flexible profile) that the shed free vorticity is at the reference surface H and in regime iii (rigid profile) that this vorticity is situated on the wake H^* of which the definition is given in section 4.8. The wake H^* in the neighbourhood of the rigid profile, is moving under influence of the motion of the profile, however infinitely far downstream it has come to rest and has become periodic, then it was denoted also by H. This means that in both cases we have to determine the optimum shed free vorticity on a periodic surface (figure 6.8.1).

We describe the shape of H by

$$H: y = g(x), \qquad g(x+b) = g(x), \tag{6.8.1}$$

with $\max g(x) = g_1$ and $\min g(x) = g_2$. We assume that there is no disturbance velocity field hence

$$\boldsymbol{v}_0^* \equiv 0. \tag{6.8.2}$$

The flow with potential Φ; induced by the optimum shed free vorticity far behind the propeller has to satisfy by (5.4.5) and (5.3.5) the boundary condition at H

$$\frac{\partial \Phi}{\partial n} = \lambda \cos_{(n,x)}(x) = -\frac{\lambda g'(x)}{\left(1 + g'(x)^2\right)^{1/2}}, \tag{6.8.3}$$

where λ is the Lagrange multiplier. Besides (6.8.3) the potential has to satisfy

$$\lim_{y\to\infty} \Phi(x,y) = c^+, \qquad \lim_{y\to-\infty} \Phi(x,y) = c^-, \tag{6.8.4}$$

c^+ and c^- being arbitrary constants, because the only demand is that the

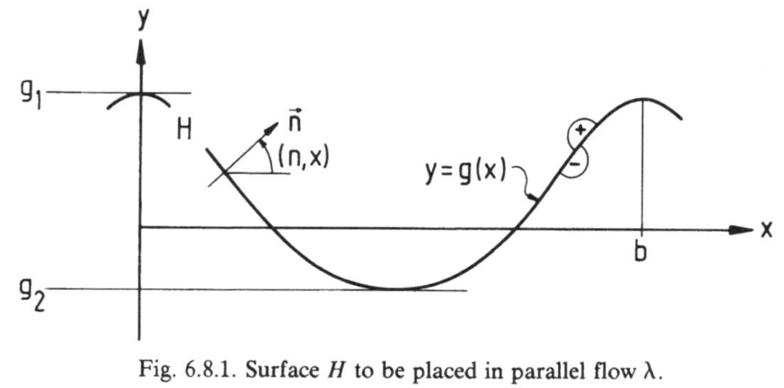

Fig. 6.8.1. Surface H to be placed in parallel flow λ.

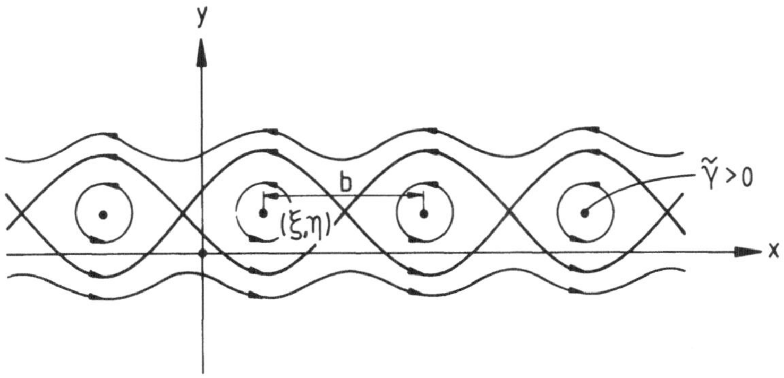

Fig. 6.8.2. Infinite row of concentrated vortices.

velocities induced by the shed vorticity tend to zero at infinity.

In order to reformulate this boundary value problem for $\Phi(x, y)$ as an integral equation we consider an infinite row of concentrated vortices of strength $\tilde{\gamma}$ spaced equally at the distance b. One of these vortices is placed at (ξ, η). Introducing the complex variable $z = x + iy$ we can write the complex velocity potential $\tilde{F}(z)$ of this row of vortices ([39] page 187), as

$$\tilde{F}(z) = \tilde{\Phi}(x, y) + i\tilde{\Psi}(x, y) = \frac{\tilde{\gamma}}{2\pi i} \ln\left(\sin\frac{\pi}{b}(z - \zeta)\right), \tag{6.8.5}$$

where $\zeta = \xi + i\eta$ and $d\tilde{F}(z)/dz = \tilde{v}_x - i\tilde{v}_y$. The velocity belonging to this potential has the real components

$$\left(\tilde{v}_x(x, y), \tilde{v}_y(x, y)\right) = \frac{\tilde{\gamma}}{2b} \frac{\left\{\sinh\frac{2\pi}{b}(y - \eta), \sin\frac{2\pi}{b}(x - \xi)\right\}}{\left\{\cos\frac{2\pi}{b}(x - \xi) - \cosh\frac{2\pi}{b}(y - \eta)\right\}}. \tag{6.8.6}$$

We next assume the points (x, y) and (ξ, η) to be placed at H, then the component $K(\xi, \eta)$ of the velocity at $(x, g(x))$, normal to $y = g(x)$ induced by the row of vortices of unit strength $(\tilde{\gamma} = 1)$, becomes

$$K(x, \xi) = -\frac{1}{2b} \frac{\left\{g'(x)\sinh\frac{2\pi}{b}(g(x) - g(\xi)) - \sin\frac{2\pi}{b}(x - \xi)\right\}}{\left(1 + g'(x)^2\right)^{1/2}\left\{\cos\frac{2\pi}{b}(x - \xi) - \cosh\frac{2\pi}{b}(g(x) - g(\xi))\right\}}. \tag{6.8.7}$$

Integration with respect to ξ yields the following singular integral equation for the function $\gamma(x)$, which is the unknown optimum shed vorticity apart from the factor λ,

$$\oint_0^b K(x,\xi)\gamma(\xi)\left(1+g'(\xi)^2\right)^{1/2} \mathrm{d}\xi = \cos_{(n,x)}(x), \tag{6.8.8}$$

where the integration is in the sense of the Cauchy principle value. A profile moving periodically along H leaves behind free vorticity of which the total strength per period is zero. This means that equation (6.8.8) has to be solved under the condition

$$\int_0^b \gamma(\xi)\left(1+g'(\xi)^2\right)^{1/2} \mathrm{d}\xi = 0, \tag{6.8.9}$$

then automatically condition (6.8.4) is satisfied.

Having solved numerically the integral equation (6.8.8) the function $\gamma(x)$ is known and we have to determine the scaling factor λ. This can be done for instance as follows. Assume that the wing W has a mean velocity U in the positive x direction and delivers a prescribed mean thrust $\overline{T}$ in this direction. Consider the surface H with the vorticity $\lambda\gamma(x)$ now placed in a fluid otherwise at rest and calculate the impulse I reckoned positive in the negative x direction induced by $\lambda\gamma(x)$, per period of length

$$I = -\mu\left\{\iint_{\substack{y<g(x)\\ 0\le x\le b}} \frac{\partial\Phi}{\partial x}(x,y)\,\mathrm{d}x\,\mathrm{d}y + \iint_{\substack{y>g(x)\\ 0\le x\le b}} \frac{\partial\Phi}{\partial x}(x,y)\,\mathrm{d}x\,\mathrm{d}y\right\}. \tag{6.8.10}$$

By an integration with respect to x and using the periodicity of $\Phi(x,y)$ which is a consequence of (6.8.9), we obtain

$$I = -\mu\int_0^b [\Phi]_-^+(x) g'(x)\,\mathrm{d}x. \tag{6.8.11}$$

It is easily seen that we can take

$$[\Phi]_-^+(x) = -\lambda\int_0^x \gamma(\xi)\left(1+g'(\xi)^2\right)^{1/2} \mathrm{d}\xi. \tag{6.8.12}$$

In this equation we made the choice $[\Phi]_-^+(0)=0$. This is allowed because $\Phi(x,y)$ above and below $y=g(x)$ are each determined upto a constant (6.8.4) and the values of these constants have no influence on I as given in (6.8.10) or (6.8.11).

Using (6.8.12) and (6.8.9) we can write (6.8.11) as

$$I=-\mu\lambda\int_0^b\gamma(x)g(x)\big(1+g'(x)^2\big)^{1/2}\,\mathrm{d}x. \tag{6.8.13}$$

Then we calculate λ from the relation $\overline{T}=IU/b$, hence

$$\lambda=-\frac{b\overline{T}}{U\mu}\left\{\int_0^b\gamma(x)g(x)\big(1+g'(x)^2\big)^{1/2}\,\mathrm{d}x\right\}^{-1}. \tag{6.8.14}$$

The Lagrange multiplier λ can also be calculated by considering the force exerted by a concentrated bound vortex of strength $\Gamma=[\Phi]_-^+$ moving along $y=g(x)$. The kinetic energy left behind per period b of the motion and per unit of span can, by the periodicity of the potential, be written as

$$E_p=-\tfrac{1}{2}\mu\int_0^b[\Phi]_-^+\frac{\partial\Phi}{\partial n}\big(1+g'(x)^2\big)^{1/2}\,\mathrm{d}x. \tag{6.8.15}$$

Using (6.8.3), (6.8.12) and (6.8.14) we obtain

$$E_p=\frac{b^2\overline{T}^2}{2U^2\mu}\left\{\int_0^b\gamma(x)g(x)\big(1+g'(x)^2\big)^{1/2}\,\mathrm{d}x\right\}^{-1}. \tag{6.8.16}$$

Next we give the kinetic energy E_i left behind by an ideal propeller over a length b in the x direction and per unit of span. By (5.7.6) and (5.7.8) we find

$$E_i=\frac{\overline{T}^2b}{2\mu U^2(g_1-g_2)}. \tag{6.8.17}$$

Then the quality number q follows from

$$q=\frac{E_i}{E_p}\le 1. \tag{6.8.18}$$

In section (5.8) among others, the values of q for two functions $y=g(x)$ (5.8.1) are given, namely for

$$y=0.15\cos 2\pi x \tag{6.8.19}$$

and for

$$y=0.15\cos 2\pi x+0.0375\cos 4\pi x, \tag{6.8.20}$$

hence $b=1$ in this case and for both functions holds $(g_1-g_2)=0.3$. The values of q follow approximately from figure 5.8.2 for "large" values of δ

because then the interaction between the two reference surfaces, which was the subject of that section, has disappeared. We find $q = 0.395$ and $q = 0.435$ for (6.8.19) and (6.8.20) respectively.

The optimum efficiency η_T is written as

$$\eta_T = \frac{\bar{T}b}{\bar{T}b + E_p} = \left(1 + \frac{E_i}{q\bar{T}b}\right)^{-1} = \left(1 + \frac{\bar{T}}{2\mu U^2 q(g_1 - g_2)}\right)^{-1}. \qquad (6.8.21)$$

This optimum efficiency can be realized theoretically still by all kinds of added motions of the profile. The only thing that matters is the free vorticity distribution along the reference surfaces $y = g(x)$.

6.9. Angles of incidence, large amplitude, two dimensional

We next indicate the calculation of an optimum motion of the profile. As we mentioned at the end of the preceding section, the only condition we have to satisfy in connection with optimization theory is that the free vorticity shed by the profile equals $\lambda\gamma(x)$. We assume $\lambda\gamma(x)$ to be known by the numerical solution of (6.8.8) and by (6.8.14).

First we consider regime ii. For a consistent theory the profile has to be flexible however when it is rigid and has a sufficiently small chord length $2l$ with respect to the radii of curvature of the reference surface H, the theory can be expected to hold. We direct our attention to this situation. Along the profile we have again the length parameter s which starts at the midpoint M of $(A - B)$, hence $s = -l$ at A and $s = l$ at B.

In order to fix the motion of the profile we assume that some chosen point R with $s = b$, moves along H when the profile carries out the base motion. In the same way as we did in section 4.8 we can determine the angle $\alpha_0(x_Q)$ between the profile and the x axis for which the circulation

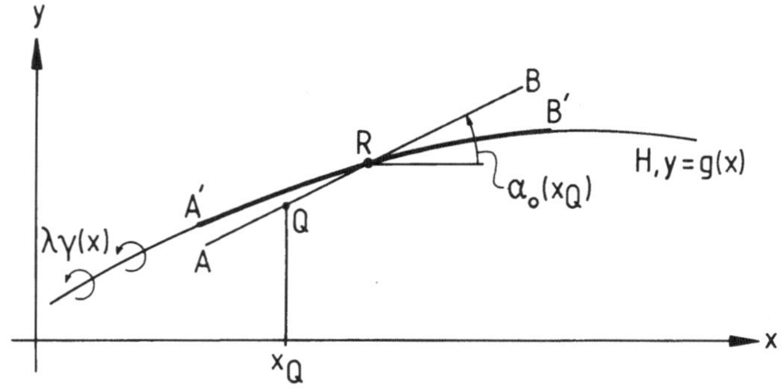

Fig. 6.9.1. Representation $(A' - B')$ of profile $(A - B)$, regime ii.

of the profile is zero, where x_Q is the x coordinate of the three quarter chord point Q. Then the vorticity $\Gamma_0(s, t)$ at the profile is given by (4.8.11). In the following we will use the notations of section 4.8.

Because the profile $(A - B)$ is assumed to be close to H we satisfy its boundary conditions at the part $(A' - B')$ of H, where the lengths of $(A - R)$ and $(A' - R)$ and of $(R - B)$ and $(R - B')$ are equal. Then we can introduce a length parameter along $(A' - B')$ with the same range as s and which we denote also by s. By using the representation $(A' - B')$ of the profile we have made a smooth junction of the known shed vorticity $\lambda\gamma(x)$ behind A' and the still unknown vorticity at $(A' - B')$.

Next we have to determine the added motion which yields the desired mean value of the thrust. This added motion consists for instance of an extra angle of rotation $\alpha_1(x_Q)$ around some chosen point S of the profile with $s = c$, which causes an extra total vorticity $\Gamma_1(s, t)$ situated at the profile. Both $\alpha_1(x_Q)$ and $\Gamma_1(s, t)$ are $O(\varepsilon)$, they have to satisfy equation (4.8.16) which reads

$$\frac{1}{2\pi}\oint_{-l}^{l}\frac{\Gamma_1(\sigma, t)}{(s-\sigma)}\,d\sigma = -\alpha_1 V_Q + (s - c)\frac{d}{dt}\alpha_1 - \tilde{v}_n, \tag{6.9.1}$$

where $\tilde{v}_n$ is the normal component of the velocity induced at the profile by the free vorticity $\lambda\gamma(x)$ at H behind the profile. The condition that the desired free vorticity $\lambda\gamma(x)$ is shed by the profile has the form

$$\frac{d}{dt}\left(\int_{-l}^{l}\Gamma_1(\sigma, t)\,d\sigma\right) = \lambda\gamma(\tilde{x}), \tag{6.9.2}$$

where $\tilde{x}$ is the x coordinate of the trailing edge. In this case equations (6.9.1) and (6.9.2) are assumed to hold at H. It is easily seen that Γ_1 and α_1 are proportional to λ and hence to $\bar{T}$. So if for one value of $\bar{T}$ the function $\alpha_1(x_Q)$ has been calculated it is known for all values of $\bar{T}$, while also the efficiency for these cases is known by (6.8.21).

The case of regime iii is somewhat more complicated. Again referring to section 4.8 we have to satisfy (4.8.16) or (6.9.1). Because now we admit profiles with a larger chord length we satisfy the boundary conditions, as has been discussed in section 4.8, at the position $(A - B)$ of the base motion of the profile (figure 4.8.4) and not at the given H as we proposed for regime ii. In equation (6.9.1) $\tilde{v}_n$ is the velocity component normal to $(A - B)$, induced by the shed free vorticity at H^*. Now the shape of H^* is time dependent and different from the ultimate periodic wake H far behind the profile. Hence we have to calculate the free vorticity density at H^*, in the neighbourhood of the profile up to the trailing edge, from the known free vorticity density at H, hence far behind the profile. This can be done by considering two particles of fluid q_1 and q_2 spaced rather closely which move by the base motion along the profile and leave it, one after the other at the trailing edge. These particles come to rest ultimately at H, suppose their distance has become Δ. At that place the optimum free vorticity density is known and equals $\lambda\gamma(x)$. It follows that the amount of free vorticity in between the two particles q_1 and q_2 is $\lambda\gamma\Delta$. This amount however is the same as the vorticity between q_1 and q_2, when they have not yet reached their rest position. Suppose at any place at H^* their distance is Δ^*, then the vorticity density $(\lambda\gamma)^*$ at that small interval is

$$(\lambda\gamma)^* = \lambda\gamma \frac{\Delta}{\Delta^*}. \tag{6.9.3}$$

Hence along H^* the shed vorticity can be determined and we can calculate $\tilde{v}_n$ in equation (6.9.1). Also we know the strength of the free vorticity $(\lambda\gamma)^*(\tilde{x})$ where $\tilde{x}$ is the place of the trailing edge, hence we have instead of (6.9.2) the relation

$$\frac{d}{dt}\int_{-l}^{l} \Gamma_1(\sigma, t)\, d\sigma = (\lambda\gamma)^*(\tilde{x}). \tag{6.9.4}$$

6.10. Optimum large amplitude unsteady propulsion, finite span

In section 5.8 we discussed for the two dimensional case the values of the quality number q for a number of base motions of pairs of possibly flexible profiles. It turned out that in the optimum case the interaction of the profiles was favourable and could be even very favourable. In this section we discuss some results with respect to possibly flexible optimum wings of finite span which carry out a large amplitude motion. In this case the optimization problem is three dimensional. The propulsion systems deliver a mean value of thrust $\overline{T}$ and have a mean velocity of advance U.

The types of reference surfaces are indicated in figure 6.10.1. First, one wing W of span h moving in the neighbourhood of the reference surface H (figure 6.10.1a). The projection of H on the (x, z) plane has the form

$$z = a \cos 2\pi x, \tag{6.10.1}$$

hence the width of the working region is $2a$ and the period of the motion in the x direction equals one. This chosen value of the period is not a restriction of generality because we can always take this length as the unit of length. This holds also for the other two cases.

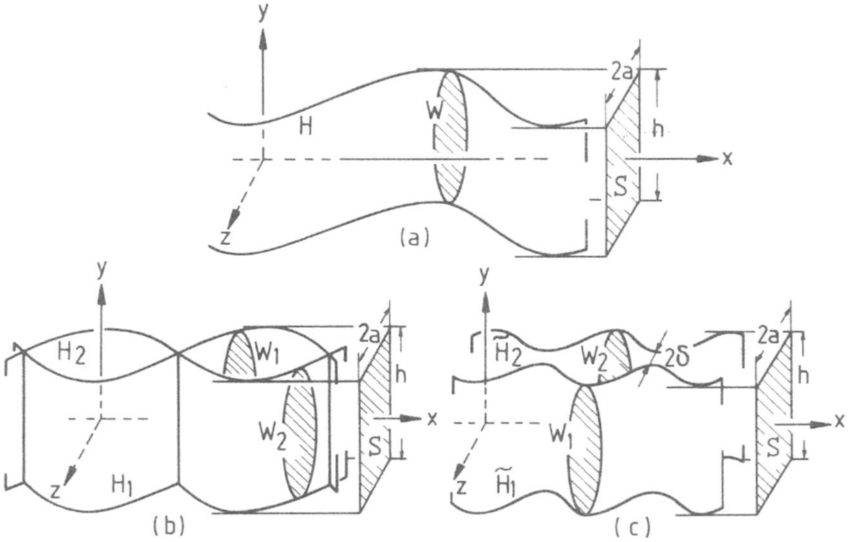

Fig. 6.10.1. Three types of reference surfaces.

Second, two wings W_1 and W_2 each of span h moving in the neighbourhood of two reference surfaces H_1 and H_2 respectively (figure 6.10.1b). The projections of the intersecting H_1 and H_2 on the (x, z) plane are given by

$$z = \pm a \cos 2\pi x, \tag{6.10.2}$$

where the $+(-)$ sign refers to $H_1(H_2)$. As has been mentioned already at the end of section 4.9 the wings W_1 and W_2 have to move in such a way that they do not collide with each other.

Third, two wings $\tilde{W}_1$ and $\tilde{W}_2$ moving "side by side" each in the neighbourhood of the non-intersecting reference surfaces $\tilde{H}_1$ and $\tilde{H}_2$ respectively (figure 6.10.1c). The projections of $\tilde{H}_1$ and $\tilde{H}_2$ on the (x, z) plane are given by

$$z = \pm \tfrac{1}{2}\{(a - \delta) \cos 4\pi x + (a + \delta)\}, \tag{6.10.3}$$

where the $+(-)$ sign refers to $\tilde{H}_1(\tilde{H}_2)$. Because the amplitude of each wing in this case is about the half (when δ is small) of the amplitude in cases (a) and (b) we took the length of the period of the motion in the x direction equal to a half. Then the maximum slope of the paths of the wings in all three cases is about the same.

In all three cases we have, as we discussed in chapter 5, to place the reference surfaces as impermeable rigid surfaces in a homogeneous flow of still unknown velocity, parallel to the x axis in order to find the optimum shed free vorticity. The resulting boundary value problems are of the Neumann type which can be solved approximately by the vortex lattice method. Two checks on the method have been made. For large values of h the results calculated by means of the vortex lattice method can be compared with those of the two dimensional case [18] a part of which are given in section 5.8. For small values of h the results obtained by slender body theory can be used for comparison. It turned out that in both checks the agreement was satisfactory. For a more detailed discussion of the method of computation we refer to Potze and Sparenberg [49].

For the calculation of the quality factor q we took in all cases, hence also in case (c) for $\delta \neq 0$, the working area S of the propulsion system equal to $2ah$. This differs from the choice made in section 5.8 with respect to figure 5.8.1 where we took, in the notation of that section, $S = S_1 + S_2$. There the slit inbetween the two reference surfaces did not belong to S. The reason is that here we want to compare the three different types of propellers (a), (b) and (c), which have to propel the same ship and which are allowed to use the full breadth of the hull. However when the wings $\tilde{W}_1$ and $\tilde{W}_2$ move side by side as we assumed in case (c), it will be necessary to keep some clearance 2δ between them when they approach each other. Hence this clearance can be looked upon as an unfavourable property of the device which has to be taken into account in the comparison.

First we given in table 6.10.1 the quality numbers q for case (a), hence for one wing. Here and in the following the column $h \ll 1$ follows from slender body theory. It is seen that for $h = 0.05$ slender body theory gives a reasonable approximation for q. The last column gives q for the two dimensional case.

Next in table 6.10.2 we give the values of q for case (b) hence for two wings of which the reference surfaces intersect each other.

In case (c) we have the additional parameter δ which determines the smallest value of the distance between the two wings. Table 6.10.3 shows besides the dependence of q on a and on h also the dependence of q on δ.

Table 6.10.1
Values of q, case (a).

a	h							
	$h \ll 1$	0.05	0.2	0.4	0.8	1.6	3.2	∞
0.12	$0.782h$	0.038	0.132	0.208	0.268	0.302	0.318	0.334
0.25	$1.175h$	0.058	0.202	0.327	0.429	0.487	0.516	0.545
0.38	$1.339h$	0.066	0.232	0.380	0.508	0.582	0.621	0.648

Table 6.10.2
Values of q, case(b).

a	h							
	$h \ll 1$	0.05	0.2	0.4	0.8	1.6	3.2	∞
0.12	$1.564h$	0.079	0.295	0.481	0.617	0.690	0.725	0.755
0.25	$2.35h$	0.112	0.363	0.550	0.684	0.753	0.789	0.823
0.38	$2.677h$	0.126	0.396	0.583	0.719	0.791	0.827	0.861

From the numerical values of the tables we can draw the following conclusions. When we enlarge the span h of the wings or the amplitude of the motions or both at the same time, the optimum quality number q increases. When we increase the distance δ in between the two wings which move side by side the value of q decreases.

The maximum efficiency of a propulsion device mentioned above can be calculated by means of (5.7.9) which reads in this case

$$\eta_T = \left(1 + \frac{\bar{T}}{4\mu ahqU^2}\right)^{-1}. \tag{6.10.4}$$

We emphasize that this is the highest possible efficiency of this type of propellers in an unbounded inviscid fluid according to a linearized theory.

In practice a number of uncertainties occur. Of course analogous difficulties arise in this case as are discussed in the introduction to

Table 6.10.3
Values of q, case (c).

δ	a	h							
		$h \ll 1$	0.05	0.2	0.4	0.8	1.6	3.2	∞
0	0.12	$1.564h$	0.081	0.299	0.466	0.574	0.627	0.653	0.670
	0.25	$2.351h$	0.116	0.379	0.554	0.668	0.723	0.753	0.774
	0.38	$2.677h$	0.130	0.416	0.599	0.721	0.778	0.812	0.830
0.02	0.12	$1.137h$	0.055	0.170	0.237	0.276	0.295	0.305	0.314
	0.25	$2.090h$	0.100	0.300	0.410	0.476	0.507	0.523	0.537
	0.38	$2.503h$	0.120	0.362	0.495	0.578	0.616	0.638	0.653
0.04	0.12	$0.759h$	0.036	0.107	0.145	0.167	0.178	0.183	0.188
	0.25	$1.832h$	0.087	0.256	0.344	0.396	0.420	0.433	0.444
	0.38	$2.328h$	0.111	0.331	0.449	0.520	0.553	0.572	0.585

chapter 3 with respect to the screw propeller. However here we have, even for a previously undisturbed fluid without boundaries, some other difficulties which do not arise in relation with the screw propeller or the unsteady two dimensional propulsion of regime iii.

First the question can be raised, does there exist a base motion for a rigid wing of finite span, which does not shed any free vorticity in the fluid. This does not seem to be very likely because of the tip vortices which can be expected to appear. Then a consistent linearized theory is no longer possible because these vortices when the chord length of the wing is $O(\varepsilon^0)$, are of $O(\varepsilon^0)$ while the desired free vorticity at the reference surfaces is $O(\varepsilon)$. However this need not to be the reason that the theory will be of no use, because in applications we need some finite vorticity on the reference surfaces. Hence the problem is; can we find a suitable total motion of the rigid wing or wings of finite span such that the desired vorticity on the reference surfaces is reasonably well approximated by the shed free vorticity. This is a complicated problem because most probably also the shape of the planform of the wings will be important. Further research will be needed to clear up this question.

6.11. On the optimum Voith-Schneider propeller

In this section we discuss shortly the optimization of a Voith-Schneider propeller (section 4.7) with many blades. In order to simulate the bottom of the ship which is above the propeller, we mirror the blades (figure 4.7.1) with respect to this bottom and consider a vertical axis propeller of which the blades are twice as long as the real ones but now working in an unbounded fluid. The working region of the Voith-Schneider propeller is the cylinder with rectangular cross section denoted by G in figure 6.11.1. Because we will assume that we have "infinitely many very slender"

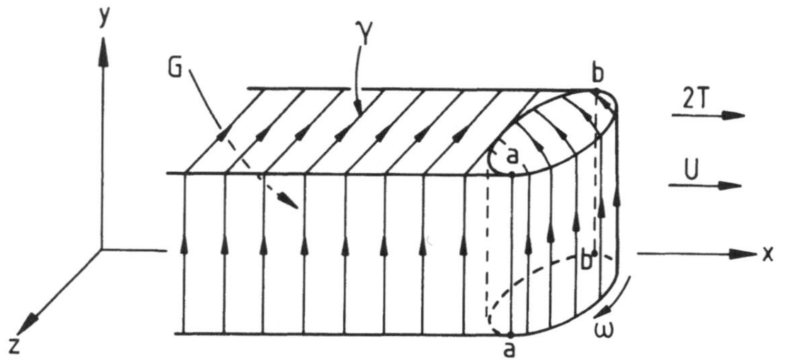

Fig. 6.11.1. The vorticity of an optimum Voith-Schneider propeller with "many" blades.

blades this working region will be crowded with reference surfaces H_l behind these blades.

We apply the reasoning as given in sections 5.6 and 5.7 with respect to ideal lifting surface systems or more specificly to propellers. At each of the reference surfaces we have condition (5.7.3) but now for $l = 1, 2, 3, \ldots$. Hence we have to place as usual the impermeable H_l in a homogeneous flow of unknown velocity λ parallel to the x axis in order to find the free vorticity shed by the propeller blades on the H_l. The propeller is assumed to be already at infinity, hence the region G and the reference surfaces are two-sided infinitely long. It follows that the shed free vorticity γ in the optimum case is entirely at the boundary ∂G of G. This vorticity density is independent of the x coordinate and consists of rectangular "vortices" perpendicular to the x axis, the direction of which is with right hand screw denoted in figure 6.11.1. It follows, as we discussed in section 5.7, that this model of the Voith-Schneider propeller yields an ideal propeller with a rectangular cross section of which the quality number $q = 1$.

Suppose that the bound vorticity of the blades per unit of length along the propeller circle is Γ in front and $-\Gamma$ at the aft position of the blades. When this bound vorticity passes the line $a - a$, it changes from Γ to $-\Gamma$, hence leaves behind free vorticity at the vertical boundary of G of strength

$$\gamma = \frac{2\Gamma\omega R}{U}, \tag{6.11.1}$$

per unit of length in the x direction, where R is the radius of the propeller circle and ω the rotational velocity of the blades. The inverse happens at the line $b - b$.

The vorticity shed by the continuously distributed tips of the blades is drawn also in figure 6.11.1. The free vorticity shed by the front tips lies along cycloidal lines. It is a matter of simple algebra to find that the tip vorticity of the front and aft tips combine behind the latter ones to vorticity of strength (6.11.1) at the lower and upper boundary surfaces of G and is perpendicular to the x axis. Here with the optimum bound vorticity Γ of the blades is found, which creates the optimum free vorticity γ demanded by the theory. In this case the thrust is equally distributed over the front and aft position of the blades.

When h is the length of the blades and T the prescribed thrust of the propeller, the strength γ follows from

$$2\int_0^{\pi} \mu\omega R \sin\varphi h \Gamma R \, d\varphi = T, \tag{6.11.2}$$

or

$$\Gamma = \frac{T}{4\mu\omega R^2 h}, \tag{6.11.3}$$

where in (6.11.2) φ is a suitably chosen angle which determines points at the propeller circle. It is not difficult to show that the momentum of the fluid in the wake caused by the free vorticity γ (6.11.1) with Γ from (6.11.3) is in agreement with the thrust T.

In [43] van Manen describes some results from experiments with Voith-Schneider propellers. Three cases where tested, 1) the thrust of the propeller was delivered by the blades in the front position, 2) the thrust was delivered by the blades in the aft position and 3) the thrust was equally divided over the front and the aft position. It turned out that the latter case had the highest efficiency. This corresponds with our result: bound vorticity $+\Gamma$ for the front position and $-\Gamma$ for the aft position of the blades.

We still mention two points of interest. First, when the propeller is underneath the aft of a hull, it is working in the boundary layer of the bottom. By this the incoming flow is inhomogeneous its velocity in the x direction depends on the y coordinate. This means that the propeller works in a disturbance field $\mathbf{v}_0^*$. Then the optimum circulation distribution Γ of the blades has to become an appropriate function of y.

Second, the blades of the propeller are rigid. This leads to the question; to which extent can we approximate by a suitable choice of the chord length of the blades and the profiles, both as a function of y, and of the oscillatory motion of the blades as a whole, the optimum spanwise circulation distribution. This is the same complicated problem as we discussed at the end of the previous section.

6.12. On the optimum large hub propeller

At last we discuss the optimization of lifting surfaces which act in a fluid with disturbance velocities of $O(\varepsilon^0)$, under the assumption that the free vorticity shed into the fluid is $O(\varepsilon)$. These considerations have some points in common with the discussion given in section 4.1 with respect to the unsteady propulsion of regime iii. Although it is possible to give a more general discussion of these problems we will confine ourselves to the propeller with a large hub of finite length. By this the defectiveness of the model of an infinitely long cylindrical hub, especially in optimization theory becomes clear. This type of propeller is also discussed by Andrews and Cummings in [2], from a more practical point of view.

We assume that the hub C induces disturbance velocities of $O(\varepsilon^0)$ and that the prescribed thrust T is of $O(\varepsilon)$, to be delivered by the propeller as one propulsion unit, hence by blades and hub together.

For simplicity we first consider a one bladed propeller of which the blade is assumed to be without thickness. The reference surface for the blade can be found as follows. Choose some, not necessarily straight, line L which is connected to the hub and which rotates and translates with it. Along L we have a parameter σ which is zero at the hub and has the value $\sigma = \sigma_1$ at the end, $0 \leq \sigma \leq \sigma_1$. Now we take some point A_1 with parameter value $\sigma(A_1)$ at L and consider the fluid particles which pass "through" this point while L moves with the hub. These particles lie at some line denoted by the dashed line A_1, A_2, A_3, A_4, A_5 of figure 6.12.1. Doing this for all points of L these lines form a more or less helicoidal surface H^*. This surface when considered as a rigid and impermeable surface does not disturb the flow around the hub c when it rotates and translates with C, hence it can be used as a reference surface for the blade. The projection of the blade on this reference surface is called its planform W.

The reference line of a profile of the blade is the line A_2, A_3 along which the fluid particles, within an accuracy of $O(\varepsilon)$, pass along the blade. The points A_2 and A_3 lie at the boundary of W. We can also consider a surface formed by straight lines, cutting the reference line A_2A_3 and perpendicular to H^*. Then a profile is the section of the blade by this surface.

The surface H^* behind the planform W is, in the neighbourhood of the hub C, influenced by the flow around C. However far downstream, the surface will have assumed a periodic limit shape H. The vorticity of the blade will be in our linearized theory at the planform W and the shed

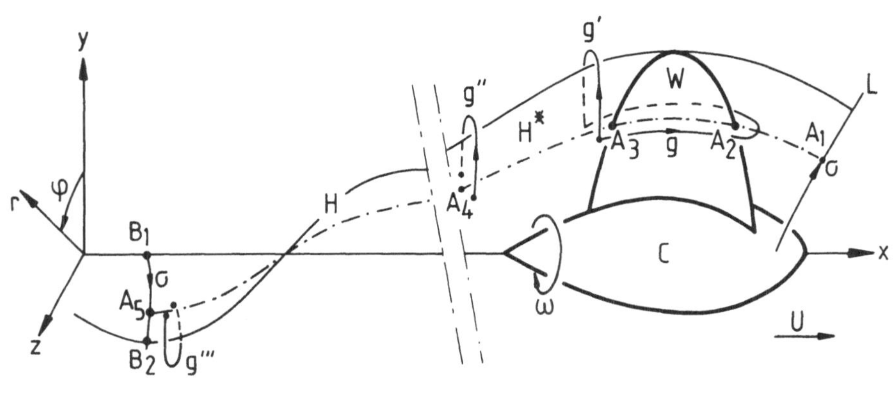

Fig. 6.12.1. One blade with vortex sheet of a large hub propeller.

free vorticity at H^* and ultimately at H. This free vorticity will lie at H along helicoidal lines of the type

$$\varphi - ax = b, \qquad r = c, \qquad a = \frac{\omega}{U}, \tag{6.12.1}$$

for different values of the constants b and c.

Now we consider a lifting line (B_1, B_2) far behind the propeller at the limit surface H. This line has to be curved in one way or another. Along (B_1, B_2) we have also the parameter σ by assigning to the point A_5 the same parameter value as the parameter value σ of A_1 and L. The lifting line (B_1, B_2) moves along H with a velocity U in the positive x direction. The bound vorticity $\Gamma(\sigma)$ of (B_1, B_2) is chosen in such a way that behind (B_1, B_2) the free vorticity of the propeller is annihilated. It follows that within the accuracy of the theory the resultant thrust (of $O(\varepsilon)$) of the one bladed propeller and the thrust of (B_1, B_2) are equal and opposite. Hence the large hub propeller and the lifting line (B_1, B_2) provided with bound vorticity $\tilde{\Gamma}(\sigma) = -\Gamma(\sigma)$ are equivalent in the sense that both yield the same thrust at the cost of the same energy losses.

An exactly analogous reasoning can be given for the case of a propeller with more blades. Note, as we mentioned already, that with respect to the thrust of the propeller we mean the resultant thrust of blades and hub together, because the induced pressures on the hub will have a non zero resultant in general.

From the foregoing it follows that instead of the original large hub propeller we can also optimize, when we have n blades, a system $(B_1, B_2)_k$ $(k = 1, \ldots, n)$ of lifting lines moving far behind the propeller along the periodic surfaces H_k under the constraint of a prescribed thrust T. This optimization resembles much the problem of sections (6.1)–(6.5) for homogeneous inflow and no endplates, be it that the reference surfaces now have to be calculated from the flow along the hub C in the way as we described. This however does not introduce essential difficulties. Hence the optimum free vorticity behind the lifting lines $(B_1, B_2)_k$ can be considered to be known. From this follows by a simple integration the optimum bound vorticity $\tilde{\Gamma}_k(\sigma_k)$ of the lifting lines $(B_1, B_2)_k$ where σ_k is the parameter along that line.

The optimum free vorticity at the H_k cannot have a concentrated vortex at the x axis, which is the line in common to the H_k, otherwise the efficiency would be zero. It follows that the optimum bound vorticity $\tilde{\Gamma}_k(\sigma_k)$ is zero at the x axis hence for $\sigma_k = 0$.

Next we consider the optimum circulation around the profiles of the blades W_k of the original propeller for a certain value of σ_k. This circulation has to be calculated along a contour around the blade and close to the line along which fluid particles are moving or with other words the profile reference line (A_2, A_3). It is not difficult to show that this circulation equals the circulation $\tilde{\Gamma}_k(\sigma_k)$ of the optimum lifting line $(B_1, B_2)_k$. This is most easily done by deforming the contour $g \to g' \to g'' \to g'''$ (figure 6.12.1) without cutting any vorticity.

Because $\tilde{\Gamma}_k(0) = 0$ it follows that the optimum circulation of the blades at their roots has to be zero, in contradistinction to the results for an infinitely long cylindrical hub (figure 6.5.1, $\rho = 0$). The reason is that in the latter case the shed free concentrated vorticity flowing from the roots of the blades, never enters freely the fluid. It always lies along the infinite hub and because it is "mirrored" in its surface it cannot induce any velocity in the fluid. Hence from this point of view an infinitely long hub is a poor approximation of reality.

It is not impossible to make a more complicated model of a propeller with a hub of finite length which allows in the optimum case the circulation of the blades to be non zero at their roots, while still the efficiency is not zero. In that case we have to admit a cavity with swirl around it behind the hub, see for instance Schmidt and Sparenberg [51].

When we make an assumption for the loading of the blade or pressure jump along a profile, hence in chordwise direction, we know the vorticity at the blades as well as the free vorticity behind them. Then we can calculate analogously as we have done in chapter 3, the shape of the profiles, however here we have to take into account more specificly the hub. This can be done in two different ways.

The first one seems to be the more common way of approach. We calculate the velocities induced by the optimum vorticity and demand that the hub remains axisymmetric. Then we have to add unknown vorticity on the hub in order to compensate there the normal component of the induced velocity. Hence we have to solve an integral equation at the surface of the hub.

The second method is, we calculate also the normal component of the induced velocities at the hub but then change its shape so that a nonaxisymmetric hub comes out of which the surface follows the induced velocities. This new hub in our theory only deviates from the original one by an amount of $O(\varepsilon)$. Although this approach is possibly more expensive in the process of manufacturing, a more streamlined shape of the propeller as a whole arises. Then the blades and hub form one integrated unit.

7. On the existence of optimum propellers

For most problems in hydrodynamics we have by experience sufficient information about the existence of a solution. In fact hydrodynamic problems usually reduce to the solution of a boundary value problem or of an integral equation of some well known type. In section 5.6 however we discussed already the nonexistence of an optimum propeller of a class of rotating screw propellers for which the rotational velocity was left free. This phenomenon becomes more complicated in the case of unsteady propulsion where the base motion or the added motion or both can have an infinite number of degrees of freedom. As we mentioned in the preface the nonexistence of an optimum motion of a class of propellers, does not mean that there are no propulsive systems with a high efficiency in that class. It only states that it is not possible by using some algorithm to find a propulsive system of which the efficiency is better than or equal to the efficiency of any other system of the class.

We will discuss four special cases, chosen from the unsteady propellers considered in section 4.1, in order to show some of the difficulties which can be met. In all cases the problem is two dimensional and it is demanded that the propeller delivers a prescribed mean value of thrust, at the cost of lowest possible energy losses. In these two dimensional problems relevant quantities are per unit of span, which will not always be mentioned.

We start with two cases of nonexistence. First, the small amplitude propulsion by means of a flexible profile (section 4.1, regime ia). In this case the base motion, which is a uniform translational motion along the x axis, is given. The added motion, which yields the thrust $\overline{T} = \mathrm{O}(\varepsilon^2)$, consists of small periodic deviations of $\mathrm{O}(\varepsilon)$ from the base motion. It will be shown that when we only put a constraint on the amplitude of the added motion, no optimum added motion need to exist.

Second, the semilinear finite amplitude propulsion (section 4.1, regime iii). Here the profile is assumed to be rigid. For each fixed base motion we can find an infinite number of added motions each of which yields the desired thrust $\overline{T} = \mathrm{O}(\varepsilon)$. The existence of the optimum added motions is determined by the existence of the solution of a Neumann problem by which the optimum shed free vorticity is calculated (6.8.3). This does not yield any essential difficulty from the point of view of applied mathematics. However, when we go one step further and want to optimize these optimum propellers with respect to their base motion a difficulty arises. It will be shown that when we only put a constraint on the amplitude of the base motion no optimum base motion need to exist.

Next we will consider two existence proofs. The first one discusses small amplitude propulsion (section 4.1, regime ia) of which also the amplitude is bounded. But now the added motion of the profile is much more restricted than in the just mentioned corresponding case of non-existence. The profile is assumed to be rigid and only heaving motions are allowed.

The second one is a finite amplitude motion of a flexible profile (section 4.1, regime ii). For each chosen sufficiently smooth reference surface H, there exist optimum motions of the profile in an ε neighbourhood of H. This is again a question of the possibility of solving the Neumann problem (6.8.3). Next we allow the reference surfaces, hence the base motions, to be varied. We discuss, under what conditions does there exist an optimum base motion or in other words an optimum reference surface H. This problem is treated by Urbach [63] in full detail, because its mathematics is very complicated, we will discuss the results.

7.1. Small amplitude flexible profile

We consider the two dimensional version of the small amplitude propulsion of section 4.1, regime ia (figure 4.1.1), hence the span of W in the z direction is infinite. In order to simplify the formulation we assume the profile to be at rest and the fluid coming in with a velocity U as denoted in figure 7.1.1. The profile has no thickness, its shape is given by

$$y = h(x, t), \qquad -l \leq x \leq l, \tag{7.1.1}$$

where $h(x, t) = O(\varepsilon)$ and $0 \leq l \leq \tilde{l}$, for some $\tilde{l} > 0$. The prescribed mean value of the thrust is given by $\overline{T} = O(\varepsilon^2)$. We assume the period of time $\tilde{\tau}$ of the motion to be given

$$h(x, t) = h(x, t + \tilde{\tau}), \qquad -\infty < t < \infty. \tag{7.1.2}$$

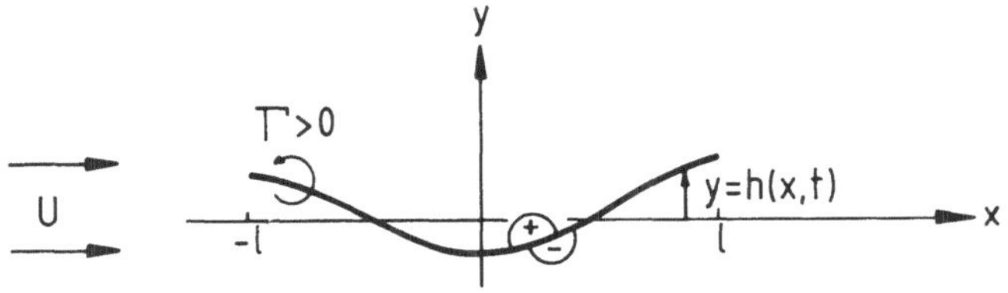

Fig. 7.1.1. The flexible profile $y = h(x, t)$, $-l \leq x \leq l$.

Also we assume that the amplitude of the motion is bounded

$$|h(x, t)| \leq B. \tag{7.1.3}$$

The problem we consider is the following [55]. Does an added motion $y = h(x, t)$ exist for prescribed values of U, $\overline{T}$, $\tilde{l}$ and $\tilde{\tau}$ which satisfies (7.1.3) and which leaves behind the least amount of kinetic energy per unit of time?. Note that because $h(x, t) = O(\varepsilon)$, this problem has no connection with the theory of chapter 5.

When we prescribe the period of time to be $\tilde{\tau}$, motions with period $\tilde{\tau}/j$, $j = 1, 2, 3 \ldots$ satisfy equation (7.1.2) and hence also can be considered to have time period $\tilde{\tau}$. For given $\tilde{\tau}$, we can always find an integer $\tilde{j}$ and a length l such that

$$\frac{\tilde{\tau}}{\tilde{j}} = \frac{2l}{U}, \qquad l \leq \tilde{l}. \tag{7.1.4}$$

We choose $2l$ as the length of our profile and consider motions with time period $\tau = \tilde{\tau}/\tilde{j}$, which by the foregoing also have the desired period $\tilde{\tau}$. Instead of (7.1.2) we have

$$h(x, t) = h(x, t + \tau), \qquad \tau = 2l/U. \tag{7.1.5}$$

Our first aim is to construct a series of profile motions $y = h_n(x, t)$ which satisfy our conditions and of which the shed kinetic energy E_n per unit of length in the x direction, tends to zero for $n \to \infty$.

The force exerted at the fluid by a two-sided infinite bound vortex of strength $\Gamma^*(t)$ placed in a homogeneous flow with velocity U, is per unit of span

$$\mu U \Gamma^*(t), \tag{7.1.6}$$

in the positive y direction. When this vortex is placed at the point $x = \xi$ of the x axis, it sheds a free vortex layer which at the place x with $x > \xi$, has the strength

$$-\frac{1}{U}\frac{\partial}{\partial t}\Gamma^*\left(t - \frac{(x-\xi)}{U}\right) \tag{7.1.7}$$

per unit of length in the x-direction.

We now consider a profile $y = h_n(x, t)$ which creates a pressure jump

$$\begin{aligned}[p(x, t)]_-^+ &= p(x, +0, t) - p(x, -0, t) \\ &= (-1)^n \mu U a \sin \pi n \frac{x}{l} \sin \pi n \frac{Ut}{l}, \qquad -l \leq x \leq l, \end{aligned} \tag{7.1.8}$$

where n can assume any one of the values 1, 2, 3,.... This profile can be generated by elementary external forces acting at the fluid, of strength

$$[p(x,t)]_-^+ \, dx, \qquad -l \le x \le l, \tag{7.1.9}$$

in the positive y direction. In this way we find by (7.1.6) and (7.1.7), for the total density of the vorticity $\Gamma(x, t)$ at the profile

$$\Gamma(x,t) = \frac{1}{\mu U}[p(x,t)]_-^+ - \frac{1}{\mu U^2}\int_{-l}^{x} \frac{\partial}{\partial t}\left[p\left(\xi, t - \frac{(x-\xi)}{U}\right)\right]_-^+ d\xi,$$

$$-l \le x \le l. \tag{7.1.10}$$

For $x > l$, that is behind the profile, the vorticity density is denoted by $\gamma(x, t)$ and is calculated as

$$\gamma(x,t) = (-1)^n \pi a n \sin\left(\pi n \frac{(Ut - x)}{l}\right), \qquad l < x. \tag{7.1.11}$$

It is easily checked that at the trailing edge

$$\gamma(l,t) = -\frac{1}{U}\frac{d}{dt}\int_{-l}^{l} \Gamma(x,t)\, dx$$

$$= -\frac{1}{\mu U^2}\int_{-l}^{l} \frac{\partial}{\partial t}\left[p\left(\xi, t - \frac{(l-\xi)}{U}\right)\right]_-^+ d\xi = \Gamma(l,t). \tag{7.1.12}$$

Next we estimate the velocity component in the y direction $v_y(x, 0, t)$, $-l \le x \le l$, induced by the vorticity (7.1.10) and (7.1.11)

$$v_y(x,0,t) = \frac{1}{2\pi}\oint_{-l}^{l} \frac{\Gamma(\xi,t)}{(x-\xi)} d\xi + \frac{1}{2\pi}\int_{l}^{\infty} \frac{\gamma(\xi,t)}{(x-\xi)} d\xi, \tag{7.1.13}$$

where the first integral is in the sense of Cauchy's principle value. An elementary but rather complicated calculation, which we do not reproduce here, shows that there exist constants C and N such that

$$|v_y(x,0,t)| \le Can, \qquad -l \le x \le l, \qquad n > N, \tag{7.1.14}$$

where C is independent of a and n. In the following we will use C also for other such constants.

The shape $h_n(x, t)$ of a profile which creates the pressure difference (7.1.8) satisfies the equation

$$\frac{\partial}{\partial t}h_n(x,t) + U\frac{\partial}{\partial x}h_n(x,t) = v_y(x,0,t), \tag{7.1.15}$$

where $v_y(x, 0, t)$ is given by (7.1.13) and estimated by (7.1.14). This partial differential equation for $h_n(x, t)$ can easily be solved, we find

$$h_n(x,t) = \frac{1}{U}\int_{-l}^{x} v_y\left(\xi, 0, t + \frac{(\xi - x)}{U}\right) \mathrm{d}\xi + g(x - Ut)$$

$$\stackrel{\text{def}}{=} f_n(x,t) + g(x - Ut). \qquad (7.1.16)$$

The function $g(x - Ut)$ is an "arbitrary" function of $(x - Ut)$ and represents the general solution of the homogeneous part of (7.1.15). By (7.1.14) we have

$$|h_n(x,t)| \leq \frac{Can}{U}(l + x) + |g(x - Ut)|, \quad -l \leq x \leq l, \ n \geq N. \qquad (7.1.17)$$

Now we make a choice $g_n(x)$ for $g(x)$ which is essential for the non-existence proof. It will be a step function as shown in figure 7.1.2, of which the horizontal parts have the length l/n and the steps have the height $b = O(\varepsilon)$. Although we have chosen in this way a "strange function" with vertical parts (at the x values of these parts the function is not uniquely defined), it will be clear that this is not essential. We could have taken instead, functions with steep parts, so that everywhere the slope would be finite. Then this steepness could be increased so fast with n, that for $n \to \infty$ the new functions give results, which differ from those belonging to the corresponding original ones with vertical parts, by an amount which tends to zero. We also could have rounded off the sharp edges so that the functions would have an infinite number of derivatives (C^∞ functions). These refinements however, are not essential and would obscure the analysis considerably. Hence we take $g_n(x)$ as shown in figure 7.1.2.

The profile $g_n(x - Ut)$ is pushed to the right with velocity U, annihilated for $x > l$, and completed at the left-hand side such that it preserves its character and stretches over the range $-l \leq x \leq l$. Note that each moment the pressure difference across the profile given in equation (7.1.8) exerts a nonzero force to the left at the profile $g_n(x - Ut)$ except

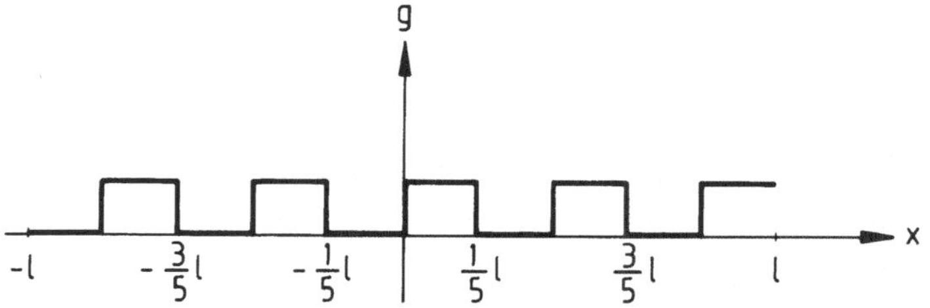

Fig. 7.1.2. The function $g_n(x - Ut)$ for $n = 5$, $t = 0$.

for those times where $[p]_-^+ = 0$. In this way the profile $g_n(x - Ut)$ is used to generate thrust without shedding additional vorticity. By this the efficiency increases. Perhaps this increase in efficiency of the propulsive action of a flexible profile by means of the function $g_n(x - Ut)$ has some relation with the fact that a swimming fish often propagates waves along its body from heat to tail [74]. A difference with the fish motion is that there the amplitude of the waves at the thicker forepart of the body is smaller than at the tail.

First, we estimate the mean value $\bar{T}_1$ of the thrust, delivered by the part $f_n(x, t)$ of the profile motion (7.1.16), reckoned positive in the negative x direction. Using equations (7.1.15) and (7.1.16) we can write

$$\begin{aligned} \bar{T}_1 &= -\frac{1}{\Delta t_1}\int_0^{\Delta t_1}\int_{-l}^{l}[p(x,t)]_-^+ \frac{\partial}{\partial x} f_n(x,t)\,\mathrm{d}x\,\mathrm{d}t \\ &= -\frac{1}{\Delta t_1}\int_0^{\Delta t_1}\int_{-l}^{l}[p(x,t)]_-^+ \left\{\frac{v_y(x,0,t)}{U}\right. \\ &\quad \left. -\frac{1}{U^2}\int_{-l}^{x}\frac{\partial}{\partial t}v_y\left(\xi, 0, t+\frac{(\xi - x)}{U}\right)\mathrm{d}\xi\right\}\mathrm{d}x\,\mathrm{d}t, \end{aligned} \tag{7.1.18}$$

where $\Delta t_1 = 2l/nU$ is one period of time of the motion. It can be shown by (7.1.8) and (7.1.14) that

$$|\bar{T}_1| \le Ca^2n^2, \tag{7.1.19}$$

where C is some constant independent of a and n. The factor n^2 in (7.1.19) arises because the velocity component v_y in (7.1.18) is differentiated with respect to time.

Second, we consider the contribution $\bar{T}_2$ of the part $g_n(x - Ut)$ of the profile motion (7.1.16) to the mean value of the thrust. At $t = 0$ this profile is in the position as denoted in figure 7.1.2, it is pushed to the right with velocity U, annihilated for $x > l$ and completed at the leading edge of the profile, as we mentioned already. This profile does not disturb the fluid however it perceives the pressures (7.1.8) induced by $f_n(x, t)$. Hence

$$\begin{aligned} \bar{T}_2 &= -\frac{1}{\Delta t_2}\int_0^{\Delta t_2}\int_{-l}^{l}[p(x,t)]_-^+ \cdot \frac{\partial}{\partial x} g_n(x - Ut)\,\mathrm{d}x\,\mathrm{d}t \\ &= -\frac{1}{\Delta t_2}\int_0^{\Delta t_2}\int_{-l}^{l}[p(x,t)]_-^+ \sum_{j=0}^{2n-1}(-1)^{j+1} \\ &\quad \times b\delta\left(x - Ut + \frac{(n-j)l}{n}\right)\mathrm{d}x\,\mathrm{d}t, \end{aligned} \tag{7.1.20}$$

where we take $\Delta t_2 = l/nU$ and $\delta(x)$ is the delta function of Dirac. During the time interval Δt_2, the profile $g_n(x - Ut)$ moves to the right over a distance l/n. Substitution of the pressure jump (7.1.8) into (7.1.20) yields

$$\bar{T}_2 = \mu Uanb. \tag{7.1.21}$$

We now take

$$an = A = \text{const.}, \tag{7.1.22}$$

where A is some constant which can be chosen at will, but once chosen remains fixed. Then it is possible to find for each n a value $b = b_n$ for the height of the steps of the profile $g_n(x - Ut)$, such that

$$\bar{T}_1 + \bar{T}_2 = \bar{T}, \tag{7.1.23}$$

where $\bar{T}$ is the prescribed mean value of the thrust. It is easily seen that we have an upperbound b^* for the values $|b_n|$, for instance we can choose

$$b^* = \frac{CA^2 + \bar{T}}{\mu UA}. \tag{7.1.24}$$

From (7.1.17) and (7.1.24) it follows that when B (7.1.3) is sufficiently large the foregoing procedure is possible and the thrust $\bar{T}$ can be generated.

The wasted kinetic energy per unit of length in the x direction far behind the profile can be calculated easily. We consider the free vorticity $\gamma(x, t)$ (7.1.11), with respect to a coordinate system which moves with the free stream. Then we have a vortex sheet of strength

$$\pi an \sin \pi n \frac{x}{l}, \tag{7.1.25}$$

which can be assumed to extend from $x = -\infty$ towards $x = +\infty$. The kinetic energy E_n of the fluid per unit of length in the x direction, induced by this sheet becomes

$$E_n = \tfrac{1}{8}\pi\mu la^2 n = \tfrac{1}{8}\pi\mu l\frac{A^2}{n}. \tag{7.1.26}$$

Hence we have the result that for $n \to \infty$ this kinetic energy E_n tends to zero. This means that we have constructed a minimizing sequence of profile motions, each of which delivers a mean value of thrust $\bar{T}$ and for which the efficiency tends to one. We note that also for flexible wings of finite span minimizing sequences in the above mentioned sense can be constructed.

7.2. Nonexistence of optimum added motion

We now show that the constraint on the amplitude of the hydrofoil motion need not to be sufficient for the existence of an optimum motion. We first prove that when no free vorticity is shed behind a moving profile, no mean value of the thrust can occur. This has been proved in section 1.2 with respect to nonlinear potential theory, however, it is not quite obvious in the approximate linearized small amplitude theory. It is even not true as we discussed in section 4.1 (figure 4.1.2) and in section 4.6 for a slotted profile.

Consider some profile $h(x, t)$ (figure 7.1.1) with total density of vorticity

$$\Gamma(x, t), \qquad -l \leq x \leq l. \tag{7.2.1}$$

We use Bernoulli's theorem for unsteady flow (1.1.6) which reads in its linearized version

$$p = -\mu\left(U\frac{\partial \Phi}{\partial x} + \frac{\partial \Phi}{\partial t}\right) + \text{const.}, \tag{7.2.2}$$

where $\Phi(x, y, t)$ is the potential of the disturbance velocity field. Hence we can write

$$\begin{aligned}[p(x, t)]_-^+ &= -\mu\left[U\frac{\partial \Phi}{\partial x} + \frac{\partial \Phi}{\partial t}\right]_-^+ \\ &= \mu\left\{U\Gamma(x, t) + \frac{\partial}{\partial t}\int_{-l}^{x} \Gamma(\xi, t)\, \mathrm{d}\xi\right\}. \end{aligned} \tag{7.2.3}$$

Then the mean value of the thrust becomes

$$\begin{aligned}\bar{T} &= -\frac{1}{\tau}\int_0^\tau \int_{-l}^{l} [p(x, t)]_-^+ \frac{\partial h}{\partial x}(x, t)\, \mathrm{d}x\, \mathrm{d}t \\ &= -\frac{\mu}{\tau}\int_0^\tau \int_{-l}^{l}\left\{U\Gamma(x, t) + \frac{\partial}{\partial t}\int_{-l}^{x} \Gamma(\xi, t)\, \mathrm{d}\xi\right\}\frac{\partial h}{\partial x}(x, t)\, \mathrm{d}x\, \mathrm{d}t, \end{aligned} \tag{7.2.4}$$

where τ is the period of motion. We suppose that no free vorticity is shed, hence

$$\frac{\mathrm{d}}{\mathrm{d}t}\int_{-l}^{l} \Gamma(x, t)\, \mathrm{d}x = 0. \tag{7.2.5}$$

Integrating the second term of the integrand at the righthand side of

(7.2.4) first partially with respect to x, then with respect to t and using (7.2.5) and the periodicity of the motion we find

$$\bar{T} = -\frac{\mu}{\tau}\int_0^\tau \int_{-l}^{l} \Gamma(x,t)\left\{U\frac{\partial h}{\partial x}(x,t) + \frac{\partial h}{\partial t}(x,t)\right\} \mathrm{d}x\,\mathrm{d}t$$

$$= -\frac{\mu}{\tau}\int_0^\tau \int_{-l}^{l} \Gamma(x,t) v_y(x,0,t)\,\mathrm{d}x\,\mathrm{d}t. \tag{7.2.6}$$

Again using the fact that there is no vorticity for $x > l$, we can replace (7.2.6) by

$$\bar{T} = -\frac{\mu}{\tau}\int_0^\tau \int_{-l}^{l} \Gamma(x,t) \oint_{-l}^{l} \frac{\Gamma(\xi,t)}{(x-\xi)}\,\mathrm{d}\xi\,\mathrm{d}x\,\mathrm{d}t. \tag{7.2.7}$$

Assume first that we only admit profile motions $h(x, t)$ such that $\Gamma(x, t)$ satisfies

$$\lim_{x\to -l} \Gamma(x,t) \le \frac{A(t)}{(l+x)^\alpha}, \qquad 0 \le \alpha < \tfrac{1}{2}. \tag{7.2.8}$$

Then we have no serious singularity in the integrand of (7.2.7) for $x \to -l$ and $\xi \to -l$ and this formula can be written as

$$\bar{T} = -\frac{\mu}{\tau}\int_0^\tau \left\{\int_{-l}^{l}\int_{-l}^{l} \frac{\Gamma(x,t)\Gamma(\xi,t)}{(x-\xi)}\,\mathrm{d}\xi\,\mathrm{d}x\right\}\mathrm{d}t, \tag{7.2.9}$$

where the integration with respect to ξ and x is as follows. Exclude from the region of integration a strip of width δ at each side of the diagonal $x = \xi$, perform the integration and take the limit $\delta \to 0$. By the antisymmetry of the integrand with respect to the diagonal, it follows that

$$\bar{T} = 0. \tag{7.2.10}$$

The case that we admit shapes $h(x, t)$ of the profile such that

$$\lim_{x\to -l} \Gamma(x,t) \approx \frac{A(t)}{(l+x)^{1/2}}, \tag{7.2.11}$$

is slightly more complicated. In this case a suction force (1.11.14) arises at the leading edge of the profile which contributes to the thrust. We discuss this case by a limit procedure, which is analogous to the one we mentioned at the end of section (1.11). Consider instead of the correct $\Gamma(x, t)$, the following mutilated one

$$\Gamma^*(x, t) = 0, \qquad -l \le x \le -l + \tilde{\varepsilon}, \tag{7.2.12}$$
$$\Gamma^*(x, t) = \Gamma(x, t), \qquad -l + \tilde{\varepsilon} \le x \le l. \tag{7.2.13}$$

Now we can apply (7.2.9) to $\Gamma^*(x, t)$ and of course find again (7.2.10) for each $\tilde{\varepsilon}$ with $\tilde{\varepsilon} \to 0$. That by this procedure we have correctly taken into account the suction force follows from the consideration of

$$\bar{T}_\delta = -\frac{\mu}{\tau}\int_0^\tau \int_{-l}^{-l+\delta} \Gamma^*(x, t) \oint_{-l}^{l} \frac{\Gamma^*(\xi, t)}{(x-\xi)}\,\mathrm{d}\xi\,\mathrm{d}x\,\mathrm{d}t, \qquad 0 < \delta, \tag{7.2.14}$$

which is the thrust exerted by the interval $(-l \le x \le -l + \delta)$ in the neighbourhood of the leading edge of the profile. When we consider the following limit, first, for fixed $\delta > 0$ we let $\tilde{\varepsilon} \to 0$ and second, $\delta \to 0$, we find that the integral (7.2.14) does not tend to zero but to the suction force at the leading edge. Hence it follows that also in the case of (7.2.11) we have $\bar{T} = 0$.

Note that such a limit procedure is the only correct way to incorporate the suction force in (7.2.7). When for instance we take $\Gamma(x, t) = (l + x)^{-1/2}$, the successive integrals in (7.2.7) can be calculated analytically in closed form. It is found that the result equals minus the suction force, hence the integral is not zero. Addition of the suction force then gives the result $\bar{T} = 0$, which is the correct value. The reason is that the suction force is not represented in the integrand in the case of (7.2.11). When we consider the case of (7.2.12) and (7.2.13) the suction force is, by the strongly curved shape of the profile in the neighbourhood of the leading edge, perceived by the integral.

The case for which the exponent in (7.2.8) satisfies $\frac{1}{2} < \alpha$ is of no interest. Then both the suction force, as derived in section (1.11) and the kinetic energy of the fluid become infinite. So we can restrict ourselves to vortex distributions $\Gamma(x, t)$ which satisfy (7.2.5) and

$$\lim_{x \to 0} \Gamma(x, t) \approx \frac{A(t)}{(l + x)^\alpha}, \qquad 0 \le \alpha \le \tfrac{1}{2}, \tag{7.2.15}$$

then it follows that $\bar{T} = 0$ when no free vorticity is shed.

Inversely when we know of some vortex distribution $\Gamma^*(x, t)$ for which (7.2.15) holds, that it yields a thrust $\bar{T}^* \neq 0$, then (7.2.5) cannot be true. Hence free vorticity density of $O(\varepsilon)$ has to be shed and kinetic energy E^* of $O(\varepsilon^2)$ per unit of time is left behind, with

$$E^* > 0. \tag{7.2.16}$$

This also applies to an optimum vortex distribution it it exists.

In the previous section however we have constructed a minimizing series of profile motions $h_n(x, t)$ for which the lost kinetic energy E_n per unit of time satisfies

$$E_n \to 0; \qquad n \to \infty. \tag{7.2.17}$$

Hence no optimum profile motion can exist because its lost kinetic energy would be larger than the lost energy of the $h_n(x, t)$ for n larger than some suitable number N.

This result is valid when for given U, $\tilde{\tau}$, $\tilde{l}$, $\overline{T}$ the bound B (7.1.3) is sufficiently large. This does not mean that for small values of B an optimum motion will exist, only that our construction of the minimizing sequence of profile motions does not work.

7.3. Large amplitude rigid profile

The following is concerned with the existence or better the nonexistence of an optimum base motion in the case of the semilinear theory for a rigid flat profile as considered in section 4.8. The profile is drawn in figure 7.3.1, it moves with its three quarter chord point Q along a periodic line L, while it remains tangent to L. Its rotational velocity is $\omega = \dot{\alpha}(t)$, hence its vorticity $\Gamma(s, t)$ (4.8.4) with $a = -\frac{1}{2}l$, can be written as

$$\Gamma(s, t) = \dot{\alpha}(t)(l - 2s)\left(\frac{l+s}{l-s}\right)^{1/2}, \qquad -l \leq s \leq l, \tag{7.3.1}$$

of which the total circulation is zero.

We suppose the motion of the point $Q = (x_Q, y_Q)$ to be described by the functions

$$x_Q(t) = x_Q(t+\tau) + U\tau, \qquad y_Q(t) = y_Q(t+\tau), \tag{7.3.2}$$

where U is the mean velocity of advance and τ is the time period. Then $\alpha(t)$ follows from

$$\operatorname{tg} \alpha(t) = \frac{\dot{y}_Q(t)}{\dot{x}_Q(t)}, \tag{7.3.3}$$

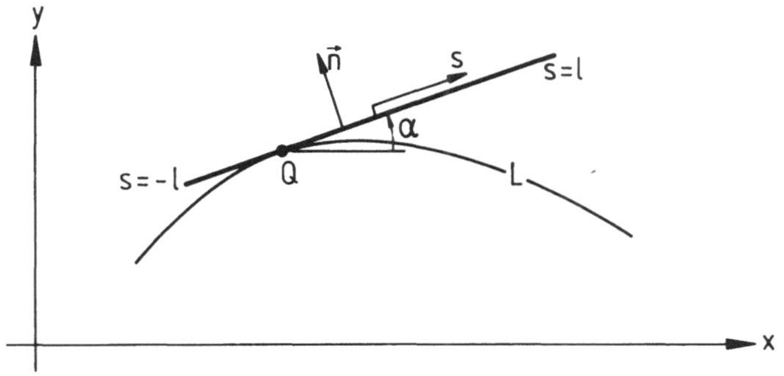

Fig. 7.3.1. Flat profile moving along L.

with

$$\alpha(t)=\alpha(t+\tau). \tag{7.3.4}$$

In this way we have given the base motion of the profile because, as is discussed in section 4.8 no vorticity is shed into the fluid.

Next we put constraints on the base motion of the profile. Its translational and its rotational velocity are assumed to be continuous functions of time and

$$|x_Q(t)-Ut|\leq A_1, \qquad |y_Q(t)|\leq A_2, \tag{7.3.5}$$

where A_1 and A_2 are prescribed quantities. Inequalities (7.3.5) mean that the point Q of the profile has to stay within a rectangle with sides $2A_1$ and $2A_2$, which translates in the direction of the positive x axis with velocity U. Hence the profile has a bounded amplitude and cannot move backwards and forwards over arbitrary large distances.

We now consider the set of all base motions with continuous $x_Q(t)$, $y_Q(t)$ and $\alpha(t)$, which satisfy (7.3.5). To each base motion we assign an optimum added motion such that a prescribed mean value of thrust $\bar{T}$ is delivered by the profile. This means that to each element of the set of admitted base motions is attached an amount of kinetic energy loss per unit of time. As follows from optimization theory this loss of kinetic energy does not depend on the optimum added motion but only on the thrust $\bar{T}$. The question we will consider is, does there exist an optimum base motion for which this energy loss is a minimum (Sparenberg and Takens [56]).

7.4. The wagging motion

Because it is our intention to prove the nonexistence of an optimum base motion we will as we did in the previous nonexistence problem, try to find again a minimizing sequence but now of base motions instead of added motions. This sequence will be such that the lost kinetic energy of the corresponding optimum propellers which deliver a prescribed thrust, tends to zero. Then the assumption that an optimum base motion exists yields a contradiction.

We start with an important part of the base motion we want to construct. We say that the profile carries out one wagging motion with amplitude 1, when $x_Q(t)$ and $y_Q(t)$ are as follows

$$\begin{aligned}
&t \in \left[0, \tfrac{1}{4}\right] : x_Q(t) = 0; \qquad && y_Q(t),\ -1 \to +1;\ \alpha(t) = \tfrac{1}{2}\pi,\\
&t \in \left[\tfrac{1}{4}, \tfrac{1}{2}\right] : x_Q(t) = 0; && y_Q(t) = 1;\ \alpha(t),\ \tfrac{1}{2}\pi \to -\tfrac{1}{2}\pi,\\
&t \in \left[\tfrac{1}{2}, \tfrac{3}{4}\right] : x_Q(t) = 0; && y_Q(t),\ +1 \to -1;\ \alpha(t) = -\tfrac{1}{2}\pi,\\
&t \in \left[\tfrac{3}{4}, 1\right] : x_Q(t) = 0; && y_Q(t) = -1;\ \alpha(t),\ -\tfrac{1}{2}\pi \to \tfrac{1}{2}\pi.
\end{aligned} \tag{7.4.1}$$

The line L along which the profile slides during the wagging motion is drawn in figure 7.4.1., where we have to take the limit of zero deviation from the y axis. Note that this wagging motion is only part of the total periodic motion of $x_Q(t)$ and $y_Q(t)$, which still has to be specified.

We want to investigate how the fluid particles have moved after one such a motion. That is, what can be said about the transformation M which is defined as follows; if $x(t)$, $y(t)$ is a solution of (4.8.12) and (4.8.13) when the profile carries out one wagging motion $t = 0 \to t = 1$, then

$$M(x(0), y(0)) = (x(1), y(1)), \tag{7.4.2}$$

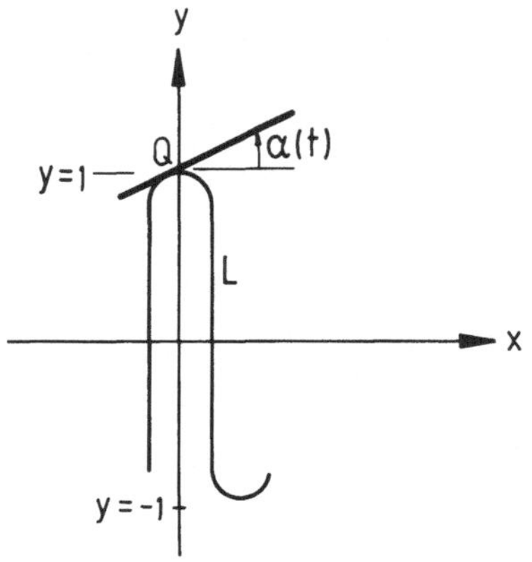

Fig. 7.4.1. Wagging motion.

where $(x(0), y(0))$ is any point of the (x, y) plane, hence M maps the (x, y) plane into itself. It is clear that with respect to M the interval of time $0 \leq t \leq 1$, used in definition (7.4.1) is not essential and can be changed into any other interval of time.

We define the displacement vector $\boldsymbol{D}$ belonging to the transformation M (7.4.2) by

$$M(x(0), y(0)) = (x(1), y(1)) = (x(0), y(0)) + \boldsymbol{D}(x(0), y(0)). \quad (7.4.3)$$

We shall prove that

$$\boldsymbol{D}(x, y) = \int_0^1 \{\dot{x}(x, y, t), \dot{y}(x, y, t)\}\, \mathrm{d}t + \mathrm{O}\left((x^2+y^2)^{-5/2}\right), \quad (7.4.4)$$

where we kept in the integrand the values of x and y unchanged. Here and in the following, we use $f(x, y) = \mathrm{O}((x^2+y^2)^{\beta})$ to denote that there exist C_1 and C_2, such that $|f(x, y)| \leq C_1(x^2+y^2)^{\beta}$ for $(x^2+y^2) \geq C_2$. From (4.8.12), (4.8.13) and from the fact that the total circulation of the profile is zero (4.8.4) ($a = -\frac{1}{2}l$) it follows that

$$\{\dot{x}(x, y, t), \dot{y}(x, y, t)\} = \mathrm{O}\left((x^2+y^2)^{-1}\right), \quad (7.4.5)$$

and

$$\left\{\frac{\partial \dot{x}}{\partial x}(x, y, t), \frac{\partial \dot{x}}{\partial y}(x, y, t), \frac{\partial \dot{y}}{\partial x}(x, y, t), \frac{\partial \dot{y}}{\partial y}(x, y, t)\right\}$$
$$= \mathrm{O}\left((x^2+y^2)^{-3/2}\right). \quad (7.4.6)$$

By this we have

$$|\{x(0), y(0)\} - \{x(t), y(t)\}| = \mathrm{O}\left((x^2(0)+y^2(0))^{-1}\right), \qquad 0 \leq t \leq 1.$$
$$(7.4.7)$$

Hence

$$\boldsymbol{D}(x(0), y(0)) = \int_0^1 \{\dot{x}(x(t), y(t)\ t), \dot{y}(x(t), y(t), t)\}\, \mathrm{d}t$$
$$= \int_0^1 \{\dot{x}(x(0), y(0), t), \dot{y}(x(0), y(0), t)\}\, \mathrm{d}t$$
$$+ \mathrm{O}\left((x(0)^2+y(0)^2)^{-3/2}\right) \cdot \mathrm{O}\left((x(0)^2+y(0)^2)^{-1}\right),$$
$$(7.4.8)$$

this proves (7.4.4).

Next we show that the displacement vector $\boldsymbol{D}$ satisfies

$$|\boldsymbol{D}(x,y)-2C_3\left(\frac{x^3-3xy^2}{(x^2+y^2)^3},\frac{3x^2y-y^3}{(x^2+y^2)^3}\right)|=O\left((x^2+y^2)^{-2}\right). \tag{7.4.9}$$

To this end we introduce the following functions of x and y

$$X(x,y)=-\frac{1}{2\pi}\int_{-\frac{1}{2}\pi}^{\frac{1}{2}\pi}\int_{-l}^{l}\frac{(l-2s)\left(\dfrac{l+s}{l-s}\right)^{1/2}\{y-(s+\frac{1}{2}l)\sin\alpha\}\,ds\,d\alpha}{\{x-(s+\frac{1}{2}l)\cos\alpha\}^2+\{y-(s+\frac{1}{2}l)\sin\alpha\}^2}, \tag{7.4.10}$$

$$Y(x,y)=\frac{1}{2\pi}\int_{-\frac{1}{2}\pi}^{\frac{1}{2}\pi}\int_{-l}^{l}\frac{(l-2s)\left(\dfrac{l+s}{l-s}\right)^{1/2}\{x-(s+\frac{1}{2}l)\cos\alpha\}\,ds\,d\alpha}{\{x-(s+\frac{1}{2}l)\cos\alpha\}^2+\{y-(s+\frac{1}{2}l)\sin\alpha\}^2}. \tag{7.4.11}$$

These functions arise from (4.8.12) and (4.8.13) by an integration with respect to time. The point (x, y) is assumed to be far away from the origin, or $|x|$ and $|y|$ large with respect to l. The three quarter point x_Q, y_Q is chosen at the origin. Under the integral signs the small changes of x and y have been neglected. Hence the functions $X(x, y)$ and $Y(x, y)$ represent an approximation of the displacement of far away points when the profile carries out a rotation with its three quarter point at the origin, from $\alpha = -\frac{1}{2}\pi$ towards $\alpha = \frac{1}{2}\pi$.

Expanding the integrands in (7.4.10) and (7.4.11) for large values of $|x|$ and $|y|$ and using again the property that the total circulation of the profile is zero it can be shown that there exists a positive constant C_4, such that

$$\left|\frac{\partial X}{\partial y}(x,y)-C_4\left\{\frac{x^3-3xy^2}{(x^2+y^2)^3}\right\}\right|=O\left((x^2+y^2)^{-2}\right), \tag{7.4.12}$$

$$\left|\frac{\partial Y}{\partial y}(x,y)-C_4\left\{\frac{3x^2y-y^3}{(x^2+y^2)^3}\right\}\right|=O\left((x^2+y^2)^{-2}\right). \tag{7.4.13}$$

From (7.4.1) and (7.4.4), it follows

$$D(x,y)=\left(\int_{\frac{1}{4}}^{\frac{1}{2}}+\int_{\frac{3}{4}}^{1}\right)\{\dot{x}(x,y,t),\dot{y}(x,y,t)\}\,\mathrm{d}t$$

$$+\mathrm{O}\left((x^2+y^2)^{-5/2}\right). \tag{7.4.14}$$

By (7.4.10) and (7.4.11)

$$\int_{\frac{1}{4}}^{\frac{1}{2}}\{\dot{x}(x,y,t),\dot{y}(x,y,t)\}\,\mathrm{d}t=-\{X(x,y-1),\,Y(x,y-1)\}, \tag{7.4.15}$$

$$\int_{3/4}^{1}\{\dot{x}(x,y,t),\dot{y}(x,y,t)\}\,\mathrm{d}t=\{X(x,y+1),\,Y(x,y+1)\}, \tag{7.4.16}$$

where the arguments $y-1$ and $y+1$ show that the rotations in (7.4.1) and hence in (7.4.4) take place at (0, 1) and (0, −1). Hence

$$D(x,y)=\{X(x,y+1),\,Y(x,y+1)\}-\{X(x,y-1),\,Y(x,y-1)\}$$

$$+\mathrm{O}\left((x^2+y^2)^{-5/2}\right)$$

$$=2\left\{\frac{\partial X}{\partial y}(x,y+\theta_1),\,\frac{\partial Y}{\partial y}(x,y+\theta_2)\right\}+\mathrm{O}\left((x^2+y^2)^{-5/2}\right), \tag{7.4.17}$$

where θ_1 and θ_2 depend on (x,y) and are in $[-1,+1]$. By (7.4.12) and (7.4.13) we find that (7.4.9) is correct.

In figure 7.4.2 we have drawn a sketch of the direction of motion of the fluid particles at a large distance from the origin, induced by the wagging of the profile. We now introduce the set l_a, which consists of points (x,y) with $x\leq -4a$, $|y|=a$ (figure 7.4.2). A curve σ will be called of type a and denoted by $\sigma(a)$ if it contains a curve $\sigma^*(a)$ such that the points (x,y) of $\sigma^*(a)$ satisfy

$$x\leq -4a,\qquad |y|\leq a, \tag{7.4.18}$$

while $\sigma^*(a)$ connects the two components of l_a.

We will prove the following, for some $a_0>0$ holds, that for every curve σ of the type $a\geq a_0$ and any $b>a$ there is an N such that for any $n\geq N$, $M^N(\sigma(a))$ is of type b. By $M^N(\sigma(a))$ we mean the new curve that arises by applying N times the transformation M to all points of $\sigma(a)$. This statement shows that it is possible to "blow up" the curve σ of type

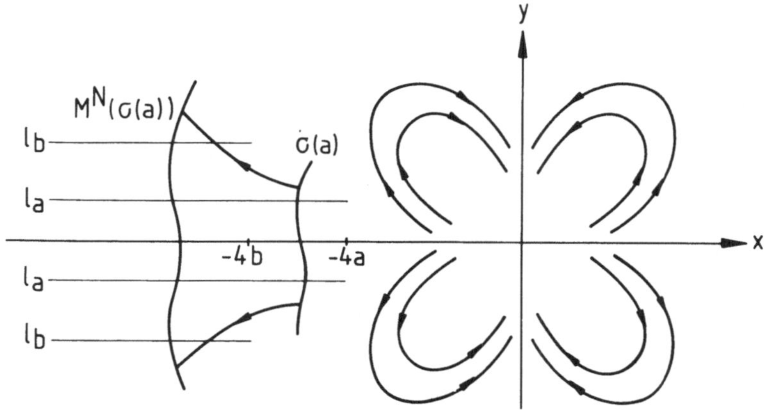

Fig. 7.4.2. Directions of displacement vector $\boldsymbol{D}(x, y)$.

a into a "larger" curve of type b by a suitable number N of wagging motions of the profile (figure 7.4.2).

The proof is as follows. For each point (x, y) of $\sigma^*(a)$ belonging to a $\sigma(a)$ with a larger than some sufficiently large a_0, there holds by (7.4.9) for the x-component $D_1(x, y)$ of $\boldsymbol{D}(x, y)$

$$\frac{Q_1}{\left(x^2+y^2\right)^{3/2}} \leq D_1(x, y) \leq \frac{Q_2}{\left(x^2+y^2\right)^{3/2}} < 0, \tag{7.4.19}$$

for some constants $Q_1 < 0$ and $Q_2 < 0$. The inequalities (7.4.19) follow from the fact that at $\sigma^*(a)$ there holds $x \leq -4|y|$.

For any point $(x, y) \in l_a$ we have for the y-component $D_2(x, y)$ of $\boldsymbol{D}(x, y)$

$$\frac{Q_3 y}{\left(x^2+y^2\right)^2} \leq (\geq) D_2(x, y), \qquad y > 0, \qquad (y < 0), \tag{7.4.20}$$

for some constant $Q_3 > 0$. The number a_0 being such that when we take greater values of a_0, the values of Q_1, Q_2 and Q_3 do not change.

Now repeat the wagging motion a sufficient number N_1 of times such that the x coordinates of

$$\left(M^{N_1}(\sigma(a))\right)^*(a), \tag{7.4.21}$$

are smaller than $-4b$. This is possible because of (7.4.19). Then repeat the wagging motion a sufficient number N_2 of times so that the points of (7.4.21) with $y = a$ and $y = -a$, pass the two components of l_b. This is possible by (7.4.20) and the first inequality of (7.4.19). Hence the curve σ

of type a is changed into a curve of type b by a number of $N = N_1 + N_2$ wagging motions.

7.5. Nonexistence of optimum base motion

In figure 7.5.1 we have schematically drawn the line $L_N = (x_Q(t), y_Q(t))$ periodic in the x direction with period $U\tau$, along which the three quarter chord point Q of the profile moves, while the profile remains tangent to L_N. First the profile carries out one wagging motion with a "large" amplitude A_2 at $x = -U\tau$ then it slides along the x axis and in the neighbourhood of $x = 0$, carries out N wagging motions with amplitude 1. Then a new period starts. Contrary to the drawing, which is for stimulating the imagination, the wagging motions are thought to happen at $x = nU\tau$, $n = -1, 0, 1, 2, \ldots$.

The reason for choosing the path L_N in this way is as follows. In the first part of the period for instance at $x = -U\tau$, we carry out the large wagging, its wake (see definition below (4.8.14)) is a line which reaches large distances from the x axis. In the second part of the period at $x = 0$, by the N wagging motions of amplitude 1 this wake is "blown up" to still farther distances at both sides of the x axis. Then the process is repeated.

We now determine the constants τ (7.3.2), A_1 and A_2 (7.3.5) and N so that we achieve our aim. A_1 can be chosen as any number larger than zero, in order to let Q follow the vertical parts of L with finite velocity. Next, we choose A_2 and $U\tau$ such that the wake produced in the first part of the period is a line of type a_0, used in the last part of the previous section, with respect to the wagging motions of the second half of the period. It is easy to see that such A_2 and $U\tau$ exist. Then we consider the effect of the mth period, which starts at the position of the profile with $x_Q = (m-2)U\tau$, $y_Q = 0$. From (4.8.12) and (4.8.13) it follows that $\dot{x}$ and $\dot{y}$

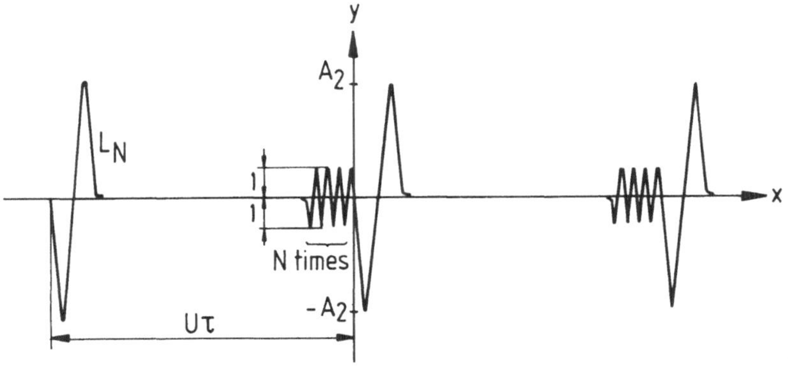

Fig. 7.5.1. Path L_N of three-quarter chord point Q.

are $O(\{(x-x_Q)^2+(y-y_Q)^2\}^{-1})$. Hence there exist constants $q>0$ and $b_0>0$ such that, if σ is a curve of type a with $a>b_0$, due to the first part of the mth period σ is moved to a curve which is anyhow of type

$$a-q\{(m-2)U\tau+4a\}^{-2}, \qquad a>b_0. \tag{7.5.1}$$

From the previous section it follows that if σ is a curve of type $a>a_0$ (a_0 as before), then σ is moved due to the second part of the mth period, to a curve which is again of type a.

Now we have to sum up the effects. Consider the curve σ_0, which is the wake produced in the first part of the first period and which is of type a_0. By the N wagging motions of the second part of the first period, this curve is changed into σ_1, which will be of the type a_1, with $a_1>a_0$. Let σ_n be this curve after n periods, then σ_n is a curve of type a_n with

$$a_n=a_{n-1}-q\{(n-2)U\tau+4a_{n-1}\}^{-2}, \tag{7.5.2}$$

provided that $a_{n-1}>b_0$. When σ_n is of type a_n it is also of type $\tilde{a}_n$ with $\tilde{a}_n<a_n$; then the following calculation will be clear

$$a_2=a_1-q(4a_1)^{-2}, \tag{7.5.3}$$

$$a_3=a_2-q(U\tau+4a_2)^{-2},\ a_3>a_2-q(U\tau)^{-2}, \tag{7.5.4}$$

$$a_4=a_3-q(2U\tau+4a_3)^{-2},\ a_4>a_2-q\{(U\tau)^{-2}+(2U\tau)^{-2}\}, \tag{7.5.5}$$

$$a_n>a_2-q\sum_{j=1}^{n}(jU\tau)^{-2}, \tag{7.5.6}$$

as long as $a_n>b_0$ and $a_n>a_0$.

From (7.5.3) and (7.5.6), we find by estimating the summation

$$a_n>a_1-q\left\{\frac{1}{(4a_1)^2}+\frac{2}{(U\tau)^2}\right\}\overset{\text{def}}{=}\beta. \tag{7.5.7}$$

Hence if we choose the number N of wagging motions of amplitude 1, in the second half of each period, so large that a_1 becomes so large that $\beta>a_0$ and $\beta>b_0$, the foregoing analysis is valid. Next if we want the wake to go out of the region $|y|\leq B$, hence to contain a curve of type B, we have to choose N so large that also $\beta>B$.

It follows from these considerations that for l, U, τ, A_1 and A_2 suitably chosen we can construct a periodic base motion of which the wake

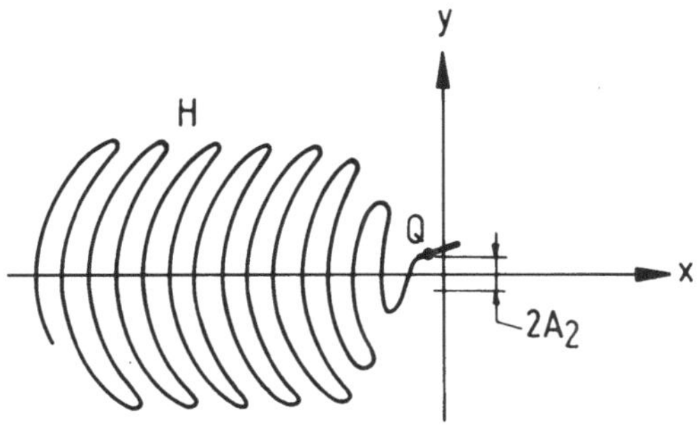

Fig. 7.5.2. Impression of a profile with an ultimate wide wake H.

becomes as wide as we want. Also it follows from (7.4.5) that when the profile moves on, the wake far behind it cannot become wide beyond all bounds, hence its width will tend to a finite value. Because the motion of the profile is periodic for $t > 0$, it is clear that also the wake becomes periodic with period $U\tau$ in the x direction. Its shape will have the appearance of the wake drawn in figure 7.5.2.

We now consider an optimum added motion of $O(\varepsilon)$, of the profile which has to be such that for a fixed line L_N (figure 7.5.1) the mean value of the thrust becomes $\bar{T} = 0(\varepsilon)$ at the cost of least energy losses. The shed free vorticity, which in this semilinear theory is assumed to lie at the wake, follows from the shape of the ultimate wake, hence far behind the profile, in an analogous way as we discussed in case of the large hub propeller of section 6.12. This means that here also we have to translate the ultimate periodic wake H as an impermeable rigid surface with an unknown velocity λ parallel to the x axis in an inviscid and incompressible fluid. Then the vorticity needed on H is the free vorticity shed by "an" optimum added motion of the profile. The unknown constant λ follows again from the demand that the mean value of the thrust has to be equal to $\bar{T}$.

We now consider a series of base motions determined by lines $L_N = (x_Q(t), y_Q(t))$ with an increasing number N of wagging motions in the second half of the period. Then the periodic ultimate wakes become wider and wider. It is clear that then the optimum free vorticity is concentrated relatively more and more at the outer edges of the wake. This means that over an arbitrary wide neighbourhood of the x axis a parallel flow by the free vorticity is induced of which the backwards directed impulse per length U in the x direction equals the thrust of the profile. Hence by increasing N the kinetic energy in the wake per unit of length in the x direction tends to zero. It means that we have found a minimizing sequence of base motions for which the efficiency tends to one. Note that in section 7.1. we constructed a minimizing sequence of added motions, the base motion in that section is purely translational.

In this semilinear theory we have to start from a base motion of which the profile has zero circulation, hence the flat profile has to move with its three-quarter chordpoint along a line L_N and has to be tangent to it. Consider for instance a line $L \in C_2^{(2)}$ (continuous functions $x_Q(t)$ and $y_Q(t)$ with continuous derivatives up to an including the second order). Then its derivatives upto the second order are uniformly bounded and it is not difficult to show that the width of the wake of the profile is also bounded. Suppose there exists an optimum base motion determined by $L^* \in C_2^{(2)}$ for given l, U, τ, A_1 and A_2. In order to deliver a mean thrust $\overline{T} = O(\varepsilon)$ it has in connection with the impulse theorem to shed free vorticity of $O(\varepsilon)$ and hence it leaves behind kinetic energy E^* per unit of length with $E^* > 0$, $E^* = O(\varepsilon^2)$. However when l, U, τ, A_1 and A_2 are such that our previous theory about the creation of an arbitrary wide wake applies, we can always find a base motion L_n of our minimizing sequence which for the same $\overline{T}$ leaves behind less kinetic energy E_n

$$E_n < E^*. \tag{7.5.8}$$

By a suitable parametrization with respect to time $L_n \in C_2^{(2)}$, hence no optimum base motion exists in the class $C_2^{(2)}$ of function pairs $x_Q(t)$, $y_Q(t)$. Ofcourse this class can be enlarged while still no optimum motion exists.

Hence we have shown that there are constraints for which no optimum base motion can be found.

7.6. Small amplitude heaving motion

We consider the two dimensional problem of section 4.1, regime ia for a purely heaving motion of a flat and infinitely thin rigid profile. The profile lies in the neighbourhood of the x axis and stretches from $x = -l$ to $x = l$, it is placed in a parallel flow of velocity U. Its periodic heaving motion is given by

$$y = h(t) = h\left(t + \frac{2\pi}{\omega}\right), \qquad |h(t)| \leq B = O(\varepsilon). \tag{7.6.1}$$

In this case, thrust is generated only by means of suction forces at the leading edge of the profile. Ofcourse this type of thrust production is not very practical, because it is liable to disturbances caused by flow separation at the leading edge. However, the analysis shows that when the wriggling type of motion of figure 7.1.2 is prevented, the profile being rigid, we can prove that an optimum motion exists (Sparenberg and

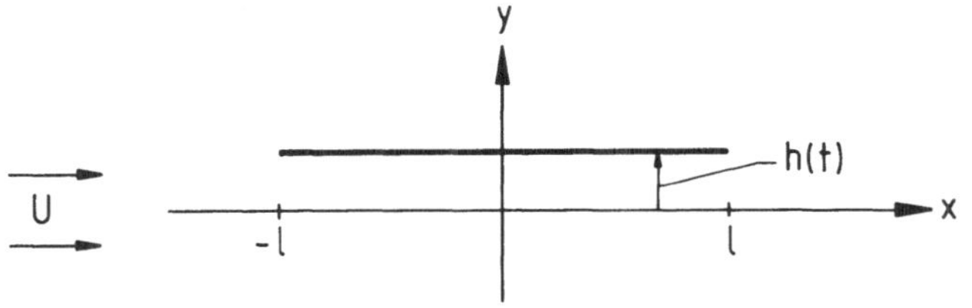

Fig. 7.6.1. Heaving motion of the profile.

Thomas [57]). The same holds possibly for more realistic motions of thrust producing rigid profiles.

We use the results of sections 4.2 and 4.3. First we specialize to

$$h(t) = ae^{j\omega t}, \tag{7.6.2}$$

hence in formula (4.3.16) we have $h(x) = a$. Then by (4.2.11), (4.2.21) and (4.3.22)

$$V(x) = j\omega a, \qquad \psi_1(x) = a\{-j\omega U + \omega^2(x+l)\}. \tag{7.6.3}$$

In the complex acceleration potential $f(z, t)$ (4.3.10) we have to substitute for A (4.3.29)

$$A = \frac{1}{\pi}\int_{-l}^{l} \frac{a\{-j\omega U + \omega^2(x+l)\}}{(l^2-\xi^2)^{1/2}}\,d\xi$$

$$+\frac{aj\omega U}{\pi l}\int_{-l}^{l} \frac{1}{(l^2-\xi^2)^{1/2}}\left\{\frac{jlH_1^{(2)}(\alpha) + \xi H_0^{(2)}(\alpha)}{H_0^{(2)}(\alpha) - jH_1^{(2)}(\alpha)}\right\} d\xi, \tag{7.6.4}$$

where as in section 4.3, $\alpha = \omega l/U$. Using the relations (4.3.30) we can split A into its real and imaginary part and write

$$(\omega aU)^{-1}A = A_1(\alpha) + iA_2(\alpha), \tag{7.6.5}$$

with

$$A_1(\alpha) = -\left[\alpha + \frac{\{J_0(\alpha)J_1(\alpha) + Y_0(\alpha)Y_1(\alpha)\}}{\{J_0(\alpha) - Y_1(\alpha)\}^2 + \{Y_0(\alpha) + J_1(\alpha)\}^2}\right], \tag{7.6.6}$$

$$A_2(\alpha) = \frac{\left\{J_0^2(\alpha) + Y_0^2(\alpha) + \dfrac{2}{\pi\alpha}\right\}}{\{J_0(\alpha) - Y_1(\alpha)\}^2 + \{Y_0(\alpha) + J_1(\alpha)\}^2}. \tag{7.6.7}$$

For the calculation of the power, which is a quadratic quantity, needed to maintain the motion we go one step further and assume $h(t)$ to be real

$$h(t) = a \cos \omega t. \tag{7.6.8}$$

This means that in (4.4.1) we take $h_1(x) = a$, $h_2(x) = 0$. The mean value P with respect to time of the power becomes

$$P = \frac{\omega}{2\pi}\int_0^{2\pi/\omega}\left\{\int_{-l}^{l}[p]_-^+ \frac{\partial h}{\partial t}\,\mathrm{d}x\right\}\mathrm{d}t = -a\omega\mu\int_{-l}^{l} \operatorname{Im}_j \varphi^+(x)\,\mathrm{d}x. \tag{7.6.9}$$

After some reductions we find

$$P = \frac{l}{\pi\mu U^3 a^2}\alpha^2\{1 - A_2(\alpha)\} \overset{\text{def}}{=} \frac{l}{\pi\mu U^3 a^2}P^*(\alpha), \tag{7.6.10}$$

where $P^*(\alpha)$ is dimensionless. The mean value $\bar{T}$ of the suction force at the leading edge is by (1.11.14)

$$\bar{T} = \frac{\omega}{2\pi}\int_0^{2\pi/\omega} \tfrac{1}{4}\mu\pi\left\{\lim_{x\to -l}\varphi^+(t)(l+x)^{1/2}\right\}^2 \mathrm{d}t, \tag{7.6.11}$$

where $\varphi^+(x, t) = \varphi(x, +0, t)$ is given in (4.4.4) for $h_1(x) = a$, $h_2(x) = 0$. We obtain

$$\bar{T} = \frac{l}{\pi\mu U^2 a^2}\alpha^2\left\{(1 - A_2(\alpha))^2 + (\alpha + A_1(\alpha))^2\right\}$$

$$\overset{\text{def}}{=} \frac{l}{\pi\mu U^2 a^2}T^*(\alpha), \tag{7.6.12}$$

where $T^*(\alpha)$ is dimensionless.

Note that when we had started in (7.6.8) from $h(t) = a \sin \omega t$ we had found the same mean values for the power and the thrust.

Next we consider some properties of $P^*(\alpha)$ and $T^*(\alpha)$. We will show that

$$P^*(\alpha),\ T^*(\alpha),\ P^*(\alpha)/T^*(\alpha), \tag{7.6.13}$$

are positive and increasing functions of α.

First we show that

$$\alpha\{1 - A_2(\alpha)\} = \frac{\alpha^2\{J_1^2(\alpha) + Y_1^2(\alpha)\} + \dfrac{2\alpha}{\pi}}{\alpha\left[\{J_0(\alpha) - Y_1(\alpha)\}^2 + \{Y_0(\alpha) + J_1(\alpha)\}^2\right]}$$

$$\overset{\text{def}}{=} \frac{I(\alpha) + \dfrac{2\alpha}{\pi}}{II(\alpha)}, \tag{7.6.14}$$

where we introduced $I(\alpha)$ and $II(\alpha)$, is an increasing function of α. For the proof we use Nicholson's integral [67]

$$\{J_\nu^2(\alpha) + Y_\nu^2(\alpha)\} = \frac{8}{\pi^2}\int_0^\infty K_0(2\alpha \sinh t) \cosh 2\nu t \, dt, \tag{7.6.15}$$

where $K_0(\alpha)$ is a modified Bessel function. Then with $\nu = 1$, the derivative of $I(\alpha)$ can be written as

$$\frac{d}{d\alpha} I(\alpha) = \frac{8}{\pi^2} \frac{d}{d\alpha} \alpha^2 \int_0^\infty K_0(2\alpha \sinh t) \cosh 2t \, dt$$

$$= \frac{8\alpha}{\pi^2} \int_0^\infty K_0(2\alpha \sinh t) \cosh 2t$$

$$\times \left\{2 - \frac{1}{\cosh^2 t} - 2 \operatorname{tgh} t \operatorname{tgh} 2t\right\} dt. \tag{7.6.16}$$

It is easily seen that the expression between brackets in the integral of (7.6.16) is positive, hence $I(\alpha)$ and by this the numerator of the right handside of (7.6.14) is an increasing function of α. Next we write

$$II(\alpha) = \alpha\{J_0^2(\alpha) + Y_0^2(\alpha) + J_1^2(\alpha) + Y_1^2(\alpha)\} + 4\pi, \tag{7.6.17}$$

$$\frac{d}{d\alpha} II(\alpha) = \frac{8}{\pi^2} \int_0^\infty K_0(2\alpha \sinh t)\left\{1 + \cosh 2t - \frac{(1 + \cosh t)}{\cosh^2 t}\right.$$

$$\left. - 2 \operatorname{tgh} t \sinh 2t\right\} dt. \tag{7.6.18}$$

Now the expression between brackets in (7.6.18) is negative, hence II(α) is a decreasing function of α. Then it follows that $\alpha\{1 - A_2(\alpha)\}$ is an increasing function of α. Hence $P^*(\alpha)$ (7.6.10) is a positive and increasing function of α.

Next we show that

$$-\alpha\{\alpha+A_1(\alpha)\}=\frac{\alpha^2\{J_0(\alpha)J_1(\alpha)+Y_0(\alpha)Y_1(\alpha)\}}{\alpha\left[\{J_0(\alpha)-Y_1(\alpha)\}^2+\{Y_0(\alpha)+J_1(\alpha)\}^2\right]}$$

$$=\frac{III(\alpha)}{II(\alpha)}, \tag{7.6.19}$$

where we introduced $III(\alpha)$, is positive and an increasing function of α. That (7.6.19) is positive follows from

$$\{J_0(\alpha)J_1(\alpha)+Y_0(\alpha)Y_1(\alpha)\}=-\frac{\mathrm{d}}{\mathrm{d}\alpha}\{J_0^2(\alpha)+Y_0^2(\alpha)\}, \tag{7.6.20}$$

which by (7.6.15) turns out to be positive. Consider

$$\frac{\mathrm{d}}{\mathrm{d}\alpha}III(\alpha)=-\frac{\mathrm{d}}{\mathrm{d}\alpha}\alpha^2\frac{\mathrm{d}}{\mathrm{d}\alpha}\{J_0^2(\alpha)+Y_0^2(\alpha)\}$$

$$=\frac{8}{\pi^2}\int_0^\infty K_0(2\alpha\sinh t)\left\{\frac{3\sinh^2 t}{\cosh^4 t}\right\}\mathrm{d}t. \tag{7.6.21}$$

Because the integrand in (7.6.21) is positive we have the desired result.

Using this result and the fact that (7.6.14) is an increasing function of α we find that $T^*(\alpha)$ (7.6.12) is an increasing function of α.

Next we consider $P^*(\alpha)/T^*(\alpha)$ which is clearly positive. Instead of showing this function to be increasing with α, it is more easy to show that

$$\frac{T^*(\alpha)}{P^*(\alpha)}-1=\frac{T^*(\alpha)-P^*(\alpha)}{P^*(\alpha)}. \tag{7.6.22}$$

is a decreasing function of α. Using the well known relation

$$J_0(\alpha)Y_1(\alpha)-J_1(\alpha)Y_0(\alpha)=-\frac{2}{\pi\alpha}, \tag{7.6.23}$$

we find after a straight forward calculation that (7.6.22) can be written as

$$-\frac{2}{\pi}\left[\alpha\{J_1^2(\alpha)+Y_1^2(\alpha)\}+\frac{2}{\pi}\right]^{-1}. \tag{7.6.24}$$

From (7.6.15) it follows in an analogous way as before that (7.6.24) is a decreasing function of α. Hence $P^*(\alpha)/T^*(\alpha)$ is increasing with α.

At last we mention the asymptotic behaviour of $P^*(\alpha)$, $T^*(\alpha)$ and

$P^*(\alpha)/T^*(\alpha)$ for large values of α. From the well known asymptotic representations of Bessel functions we find

$$\lim_{\alpha\to\infty} \frac{P^*(\alpha)}{\alpha^2} = \frac{1}{2}, \quad \lim_{\alpha\to\infty} \frac{T^*(\alpha)}{\alpha^2} = \frac{1}{4}, \quad \lim_{\alpha\to\infty} \frac{P^*(\alpha)}{T^*(\alpha)} = 2. \tag{7.6.25}$$

7.7. The optimization problem

We assume the time to be scaled such that the heaving motion of the profile $y = h(t)$, has period 2π and is given by

$$h(t) = \sum_{n=1}^{\infty} (a_n \sin nt + b_n \cos nt); \; a_{2n} = b_{2n} = 0, \; n = 1, 2, 3, \ldots. \tag{7.7.1}$$

From this representation it follows that $h(t+\pi) = -h(t)$, hence the profile carries out the same motion above and below its neutral position. We introduce the mean value P_n of the power and the mean value T_n of the thrust, which belong to a purely sinusoidal motion $h(t) = \sin(nt + \gamma)$ where γ is any phase angle. Then we can write the mean values of power and thrust belonging to a general motion as

$$P(h) = \sum_{n=1}^{\infty} P_n \cdot (a_n^2 + b_N^2), \qquad T(h) = \sum_{n=1}^{\infty} T_n \cdot (a_n^2 + b_n^2). \tag{7.7.2}$$

From the previous section it follows

$$P_n > P_m, \quad T_n > T_m, \quad \frac{P_n}{T_n} > \frac{P_m}{T_m}; \qquad n > m, \tag{7.7.3}$$

$$\lim_{n\to\infty} \frac{P_n}{n^2} = \frac{\pi\mu U^3}{2l} > 0, \qquad \lim_{n\to\infty} \frac{T_n}{n^2} = \frac{\pi\mu U^2}{4l} > 0. \tag{7.7.4}$$

The optimization problem is the following; minimize $P(h)$ under the condition that the prescribed mean value $\bar{T}$ of the thrust is generated while the amplitude of the motion is smaller than or equal to a given number B. In formula

$$\min P(h); \qquad T(h) = \bar{T}, \quad |h(t)| \leq B. \tag{7.7.5}$$

It follows from (7.7.4) that for each prescribed $\bar{T} > 0$ and $B > 0$ there exist motions, for instance

$$h(t) = \left(\frac{\bar{T}}{T_n}\right)^{1/2} \sin nt, \tag{7.7.6}$$

by choosing n sufficiently large for which the thrust $T(h)=\overline{T}$ and $|h(t)|\leq B$. We emphasize again that the motion (7.7.6) has also time period 2π.

We now establish a simple property of the power P. Consider two motions (7.7.1) $h^{(j)}(t)$, $(j=1, 2)$ with coefficients $a_n^{(j)}$, $b_n^{(j)}$, power $P^{(j)}$ and equal thrust $T^{(j)}=\overline{T}$. Suppose that the coefficients of $h^{(1)}(t)$ and $h^{(2)}(t)$ satisfy

$$\left(a_n^{(2)}\right)^2+\left(b_n^{(2)}\right)^2=\left(a_n^{(1)}\right)^2+\left(b_n^{(1)}\right)^2+\Delta_n, \tag{7.7.7}$$

$$\Delta_n\leq 0,\quad 1\leq n\leq N_1;\qquad \Delta_n\geq 0,\quad N_1<n\leq N_2, \tag{7.7.8}$$

for some numbers N_1 and N_2, while other coefficients are equal. Then because the thrust is the same

$$\sum_{n=1}^{N_1} T_n\Delta_n+\sum_{n=N_1+1}^{N_2} T_n\Delta_n=0. \tag{7.7.9}$$

For the difference in power we have

$$P^{(2)}-P^{(1)}=\sum_{n=1}^{N_1} P_n\Delta_n+\sum_{n=N_1+1}^{N_2} P_n\Delta_n. \tag{7.7.10}$$

Using (7.7.3) and (7.7.9) we find

$$\sum_{n=N_1+1}^{N_2} P_n\Delta_n\geq\frac{P_{N_1+1}}{T_{N_1+1}}\sum_{n=N_1+1}^{N_2} T_n\Delta_n=-\frac{P_{N_1+1}}{T_{N_1+1}}\sum_{n=1}^{N_1} T_n\Delta_n. \tag{7.7.11}$$

Hence we have shown under the assumptions (7.7.7) and (7.7.8)

$$P^{(2)}-P^{(1)}\geq\sum_{n=1}^{N_1} T_n\left(\frac{P_n}{T_n}-\frac{P_{N_1+1}}{T_{N_1+1}}\right)\Delta_n\geq 0. \tag{7.7.12}$$

In (7.7.12) we have strict inequalities if there are $\Delta_n\neq 0$.

Stated somewhat loosely, admitted propulsive motions $y=h(t)$ yielding a prescribed thrust $\overline{T}$, need less power according as their dominant frequencies are lower.

Now consider the simple case that the bound B (7.7.5) on the amplitude of the motion is sufficiently large so that we can generate the prescribed thrust $T(h)=\overline{T}$ by means of the lowest frequency $(n=1)$

$$h(t)=(a_1\sin t+b_1\cos t),\qquad \left(a_1^2+b_1^2\right)^{1/2}=\left(\frac{\overline{T}}{T_1}\right)^{1/2}<B. \tag{7.7.13}$$

Then by our previous property of P, this is an optimum motion.

Next we reduce B until

$$B < \left(\frac{\bar{T}}{T_1} \right)^{1/2}, \tag{7.7.14}$$

then in order to generate the desired thrust $\bar{T}$, higher frequencies have to occur in the motion. It is clear from the property of P, that in this case the optimum motion of the profile, if it exists, has to touch its boundaries at least one time per period at $y = +B$ and at $y = -B$. Otherwise it is possible to enlarge amplitudes of lower frequencies and to diminish amplitudes of higher frequencies without changing $T(h)$. Then as we have shown, P decreases and hence our motion was not optimum.

In the next section, we prove that also in the case (7.7.14) an optimum motion exists.

7.8. Existence of optimum added motion

In the two previous cases of this chapter, nonexistence proofs could be carried out by the construction of a minimizing sequence of propellers of which the efficiency tends to one. From the point of view of applied mathematics these can be found by classical analysis. However when we try to prove the existence of an optimum propeller it seems not well possible to develop a convincing theory without the use of the more abstract and general methods of functional analysis. For general reference on this subject we refer to [42] and [61].

We introduce the Hilbert space S of real valued and absolutely continuous and periodic functions on the interval $0 \leq t \leq 2\pi$ by

$$S = \left\{ h:[0, 2\pi] \to \mathbb{R};\ \frac{\mathrm{d}h}{\mathrm{d}t} \in L^2;\ h(t) = -h(t+\pi), 0 \leq t \leq \pi \right\}, \tag{7.8.1}$$

where $\mathbb{R}$ are the real numbers and L^2 are the quadratic integrable functions in the sense of Lebesque. The inner product of two elements $h^{(1)}$ and $h^{(2)}$ is defined by

$$(h^{(1)}, h^{(2)}) = \int_0^{2\pi} \frac{\mathrm{d}}{\mathrm{d}t} h^{(1)}(t) \frac{\mathrm{d}}{\mathrm{d}t} h^{(2)}(t) \mathrm{d}t. \tag{7.8.2}$$

Hence the norm in this space is defined by

$$\|h\|^2 = \int_0^{2\pi} \left\{ \frac{\mathrm{d}}{\mathrm{d}t} h(t) \right\}^2 \mathrm{d}t. \tag{7.8.3}$$

When $\|h\| = 0$ it follows that $h(t) = \text{const.}$, however because of the demand $h(t) = -h(t+\pi)$ we find $h(t) \equiv 0$. Substitution of the representation (7.7.1) in (7.8.2) and in (7.8.3) yields

$$(h^{(1)}, h^{(2)}) = \pi \sum_{n=1}^{\infty} n^2 \left(a_n^{(1)} a_n^{(2)} + b_n^{(1)} b_n^{(2)} \right), \tag{7.8.4}$$

$$\|h\|^2 = \pi \sum_{n=1}^{\infty} n^2 \left(a_n^2 + b_n^2 \right). \tag{7.8.5}$$

Because $h(t)$ is a continuous function and $h(t) = -h(t+\pi)$, there is at least one point $t = t_0$ such that $h(t_0) = 0$, hence

$$h(t) = \int_{t_0}^{t} \frac{\mathrm{d}}{\mathrm{d}\xi} h(\xi)\, \mathrm{d}\xi. \tag{7.8.6}$$

Using Schwarz' inequality we find

$$|h(t)|^2 \le |t - t_0|^2 \int_{t_0}^{t} \left\{ \frac{\mathrm{d}}{\mathrm{d}\xi} h(\xi) \right\}^2 \mathrm{d}\xi \le 4\pi^2 \|h\|^2. \tag{7.8.7}$$

It follows that the maximum norm

$$\|h\|_\infty = \max_t |h(t)|, \qquad 0 \le t \le 2\pi, \tag{7.8.8}$$

is a continuous functional on S. This means that for each $\varepsilon > 0$ there exists a $\delta > 0$ such that from $\|h^{(1)} - h^{(2)}\| \le \delta$ it follows that $\|h^{(1)} - h^{(2)}\|_\infty \le \varepsilon$.

On S we have two functionals (7.7.2), one representing the mean value $P(h)$ of the power and the other the mean value $T(h)$ of the thrust. From (7.7.4) it follows that there exist suitable constants C_1 and C_2 such that

$$C_1 \|h\|^2 \le P(h) \le C_2 \|h\|^2, \tag{7.8.9}$$

hence the functional $P(h)$ is continuous on S and its square root can be considered as an equivalent norm, related to the inner product

$$(h^{(1)}, h^{(2)})_P = \sum_{n=1}^{\infty} P_n \cdot \left(a_n^{(1)} a_n^{(2)} + b_n^{(1)} b_n^{(2)} \right). \tag{7.8.10}$$

The same holds for $T(h)$.

We introduce the set $G \subset S$ with

$$G = \left\{ h \in S : T(h) \le \bar{T};\ \|h\|_\infty \le B \right\}. \tag{7.8.11}$$

This set is bounded because $T(h)$ is an equivalent norm. Because both

$T(h)$ and $\|h\|_\infty$ are continuous on S, the set G is closed. It is also easily seen that for $h^{(1)}) \in G$ and $h^{(2)} \in G$

$$\left(\nu h^{(1)} + (1-\nu)h^{(2)}\right) \in G, \qquad 0 \le \nu \le 1, \tag{7.8.12}$$

hence G is convex.

We now prove that G is weakly compact. First we note that the functional

$$h \to h(t_1), \qquad h \in S, \qquad 0 \le t_1 \le 2\pi, \tag{7.8.13}$$

is obviously linear and by (7.8.7) it is continuous. Hence this functional is also continuous with respect to the weak topology. This means that the set

$$h \in S, \qquad |h(t_1)| \le A, \qquad 0 \le t_1 \le 2\pi, \tag{7.8.14}$$

is closed with respect to the weak topology. By this the set $\tilde{G}$ defined by

$$\tilde{G} = \bigcap_{t_1 \in [0,\, 2\pi]} \{|h(t_1)| \le A\} \tag{7.8.15}$$

is weakly closed. The set $\tilde{\tilde{G}}$ defined by

$$\tilde{\tilde{G}} = \{h \in S\colon T(h) \le \bar{T}\}, \tag{7.8.16}$$

is also weakly closed because it is closed and convex. Hence

$$G = \tilde{G} \cap \tilde{\tilde{G}} \tag{7.8.17}$$

is weakly closed. Because $\{T(h)\}^{1/2}$ is an equivalent norm on S the set G can be considered as the intersection of G with a sufficiently large sphere in S. A sphere in the Hilbert space S is weakly compact, hence also G is weakly compact.

In the following we need the notion of an extreme element of a set which will be discussed first. An extreme element $h^{(1)}$ of a set G is defined by the following property. When $h^{(2)}$ and $h^{(3)} \in G$ are such that

$$h^{(1)} = \tfrac{1}{2}\left(h^{(2)} + h^{(3)}\right), \tag{7.8.18}$$

then $h^{(1)} = h^{(2)} = h^{(3)}$. The set of extreme elements of G is denoted by $E(G)$.

We now consider the set $G_1 \subset G$

$$G_1 = \{h \in S: T(h) = \bar{T}; \|h\|_\infty \le B\}, \tag{7.8.19}$$

and show that each extreme element of G belongs to G_1. Suppose $h \in E(G)$ and $T(h) < \bar{T}$, otherwise $h \in G_1$ already. Because there exists a t_0 with $h(t_0) = 0$, there exists an interval with $|h(t)| < B$. Hence we can find a for instance smooth function $\tilde{h} \in S$ with $\|\tilde{h}\| \neq 0$, such that

$$\|h + \tilde{h}\|_\infty \le B, \qquad T(h \pm \tilde{h}) < \bar{T}. \tag{7.8.20}$$

Then both $h + \tilde{h}$ and $h - \tilde{h} \in G$ and $h + \tilde{h} \neq h - \tilde{h}$. However

$$h = \tfrac{1}{2}\{(h + \tilde{h}) + (h - \tilde{h})\}, \tag{7.8.21}$$

hence $h \notin E(G)$ and we arrive at a contradiction.

Inversely each element of G_1, is extreme in G, because each element $h \in G$ with $T(h) = \bar{T}$, is even an extreme element of the "sphere" S_T with

$$S_T = \{h \in S: T(h) \le \bar{T}\} \tag{7.8.22}$$

of which G is a subset. That such an h is an extreme element of S_T follows directly by the application of the parallelogram law for the Hilbert space with norm $\{T(h)\}^{1/2}$. Summing up we have found

$$G_1 = E(G). \tag{7.8.23}$$

We now write

$$\lambda_n = \left(\lim_{m \to \infty} \frac{P_m}{T_m}\right) - \frac{P_n}{T_n} = C - \frac{P_n}{T_n}, \tag{7.8.24}$$

where by (7.6.10), (7.6.12) and (7.6.25), $C = 2U$. It follows from the monotony properties that

$$\lambda_n > 0, \qquad \lim_{n \to \infty} \lambda_n = 0. \tag{7.8.25}$$

Next we introduce the operator K and the functional F by

$$\begin{aligned} P(h) &= CT(h) - \sum_{n=1}^{\infty} T_n \lambda_n (a_n^2 + b_n^2) \\ &= C\bar{T} - (Kh, h) = C\bar{T} - F(h), \end{aligned} \tag{7.8.26}$$

where

$$K(h) = \sum_{n=1}^{\infty} \mu_n (a_n \sin nt + b_n \cos nt), \qquad \mu_n = \frac{T_n \lambda_n}{\pi n^2} > 0, \tag{7.8.27}$$

and $\lim_{n\to\infty} \mu_n = 0$.

The operator K is compact because it is the limit in norm of a sequence of compact operators K_N which follow from (7.8.27) by replacing the upper summation limit ∞ by N.

Because the operator K is compact it follows that the functional F is continuous with respect to the weak topology. It can also be seen easily that F is a convex functional on the convex set G. Now it is known by Bauer's maximum principle ([9] page 102), that a continuous convex functional defined on a compact and convex set assumes its maximum in at least one of the extreme points of the set. This holds for any topology hence also for the weak topology used above. Hence by (7.8.26)

$$\min_{h\in G_1} P(h) = C\bar{T} - \max_{h\in G_1} F(h). \tag{7.8.28}$$

This proves that the set of motions $h(t)$ (7.8.1) with amplitude smaller than or equal to B, which deliver a mean value of the thrust $T(h) = \bar{T}$, contains at least one absolutely continuous motion for which the mean value of the power, needed to maintain the motion, is a minimum. The calculation of such a motion yields an interesting numerical problem.

7.9. Large amplitude propulsion, flexible profile

We now go on with sections 6.8 and 6.9 and direct our attention to the case of a flexible profile (regime ii, section 4.1). It is allowed that a periodic disturbance velocity field $\boldsymbol{v}_0^*$ is present. Our problem being two dimensional we give the shape of the reference surface H (figure 6.8.1) by

$$H: y = g(x), \qquad g(x+b) = g(x), \qquad |g(x)| \le A_0. \tag{7.9.1}$$

In an ε neighbourhood of H moves the flexible profile W which has a mean velocity of advance U and which delivers a prescribed mean value of thrust $\bar{T} = O(\varepsilon)$.

The potential $\Phi = \Phi_1 + \Phi_2$ (5.4.6) far behind the wing W has to satisfy by (5.4.5) and (5.3.5) the boundary condition at H

$$\frac{\partial \Phi_1}{\partial n} = -\boldsymbol{v}_0^* \cdot \boldsymbol{n}, \qquad \frac{\partial \Phi_2}{\partial n} = \lambda \cos_{(n,x)}(x), \tag{7.9.2}$$

where λ is the Lagrange multiplier which is determined by $\bar{T}$. Further we demand

$$\lim_{y\to\infty} \Phi(x,y)=c^+, \qquad \lim_{y\to-\infty} \Phi(x,y)=c^-. \tag{7.9.3}$$

The kinetic energy E_p (5.4.10) put into the fluid by the wing W per period of length in the x direction and per unit of span, has the value

$$E_p=\tfrac{1}{2}\mu\int_{-\infty}^{\infty}\int_0^b(\text{grad }\Phi_2)^2\,\mathrm{d}x\,\mathrm{d}y-\tfrac{1}{2}\mu\int_{-\infty}^{\infty}\int_0^b(\boldsymbol{v}_0)^2\,\mathrm{d}x\,\mathrm{d}y, \tag{7.9.4}$$

where $\boldsymbol{v}_0$ follows from $\boldsymbol{v}_0^*$ in the way as discussed in chapter 5. This minimum energy loss E_p is a functional on the space of admitted shapes $y=g(x)$ of the reference surface H.

The question we will consider is: what are sufficient constraints needed on H ($y=g(x)$), so that an optimum H exists. For the extensive treatment of this problem we refer to [63]. We now discuss the main result of that paper.

First some more accurate definitions are necessary. The disturbance velocity field $\boldsymbol{v}_0^*$ has to satisfy the following conditions. Its two components are quadratically Lebesque integrable functions

$$\int_{\infty}^{\infty}\int_0^b\left\{v^{*2}_{0x}+v^{*2}_{0y}\right\}\mathrm{d}x\,\mathrm{d}y<\infty, \tag{7.9.5}$$

hence the kinetic energy of $\boldsymbol{v}_0^*$ per period b in the x direction and per unit of span is finite. Further div $\boldsymbol{v}_0^*=0$ in the sense of distributions and for almost all (x, y) holds

$$\boldsymbol{v}_0^*(x+b,y)=\boldsymbol{v}_0^*(x,y). \tag{7.9.6}$$

We consider the space $W_b^{1,\infty}(\mathbb{R})$, which consists of continuous real functions defined on the real axis $\mathbb{R}$, with period b and of which the distributional derivative is essentially bounded. The norm on this space is

$$\|g\|_{1,\infty}=\max\left(\|g\|_\infty,\|g'\|_\infty\right). \tag{7.9.7}$$

Also we introduce the space K of functions $g(x)=\text{const.}$. For these functions we define $E_p=\infty$, because for each sufficiently smooth series of functions $g_n(x)$, $n=1, 2,\ldots,$ with

$$\max_{0\le x\le b}|g'(x)|\to 0, \qquad E_p\to\infty, \tag{7.9.8}$$

when the prescribed thrust $\bar{T}\neq 0$ is $\mathrm{O}(\varepsilon)$.

Then it is proved in [63]: the functional $g \to E_p(g)$ is continuous on $W_b^{1,\infty}(\mathbb{R}) \backslash K$.

From this it follows that E_p assumes a minimum on each compact subset of $W_b^{1,\infty}(\mathbb{R}) \backslash K$. Because we assigned the value $E_p = \infty$ to functions $g \in K$, we can state that E_p assumes a minimum on each compact subset of $W_b^{1,\infty}(\mathbb{R})$.

We now consider a special case of the general result. The space $C_b^k(\mathbb{R})$ consists of all those periodic functions $g(x)$ (period b) which are real, continuous and have continuous derivatives up to and including the order k. This space is equipped with the norm

$$\|g\|_{k,\infty} = \max\left(\|g\|_{\infty}, \|g'\|_{\infty}, \ldots, \|g^{(k)}\|_{\infty}\right). \tag{7.9.9}$$

Hence

$$C_b^k(\mathbb{R}) \subset W_b^{1,\infty}(\mathbb{R}), \qquad k = 1, 2, \ldots. \tag{7.9.10}$$

The main result implies that E_p also assumes its minimum on each compact subset of $C_b^1(\mathbb{R})$. Consider the set

$$G: \left\{ g \in C_b^2(\mathbb{R}), |g(x)| \leq A_0, |g'(x)| \leq A_1, |g''(x)| \leq A_2, 0 \leq x \leq b \right\}, \tag{7.9.11}$$

where A_0, A_1 and A_2 are prescribed numbers. When we take the closure of this set with respect to the norm of $C_b^1(\mathbb{R})$ we have a compact subset in $C_b^1(\mathbb{R})$.

When we have a suitable algorithm, we can calculate an optimum reference surface $H(y = g(x))$ with $g(x) \in C_b^1(\mathbb{R})$, hence which is a continuous function with a continuous first order derivative. About the second order derivative however nothing can be said. In this way the existence of an optimum base motion is proved.

Appendices

A. The Hilbert problem

The formulas of Plemelj, the Hilbert problem and the singular integral equation are thoroughly discussed in [47]. For direct reference however, we state some results. We do not enter into details but sketch the theory to an extent necessary for the understanding of the applications.

A.1. The formulas of Plemelj

Let L be a smooth arc in the complex plane, defined by

$$z = z(s) = x(s) + iy(s), \quad s_a \leq s \leq s_b, \quad z(s_a) = a, \quad z(s_b) = b, \qquad \text{(A.1.1)}$$

where s is a parameter and $x(s)$ and $y(s)$ have continuous first order derivatives which do not vanish simultaneously. Also we assume L to be simple, this means that $(x(s_1), y(s_1)) \neq (x(s_2), y(s_2))$ for $s_1 \neq s_2$.

When t denotes a point of L we consider a function $\varphi = \varphi(t)$ which satisfies the Hölder condition

$$|\varphi(t_2) - \varphi(t_1)| < A|t_2 - t_1|^{\mu}, \qquad \text{(A.1.2)}$$

A and μ are positive constants. Then we form the Cauchy integral

$$\Phi(z) = \frac{1}{2\pi i}\int_L \frac{\varphi(t)}{(t-z)}\mathrm{d}t, \qquad z \notin L. \qquad \text{(A.1.3.)}$$

The resulting function $\Phi(z)$ is analytic in the entire complex domain minus the arc L. At L the values of $\Phi(z)$ exhibit a jump by passing from one side of L to the other. We call the left hand side of L, with respect to the positive direction of s, the "+" side and the right hand side the "−" side. The limiting values of $\Phi(z)$ are denoted by $\Phi^+(t)$ and $\Phi^-(t)$ respectively (figure A.1.1). The values of $\Phi(z)$ are continuous up to L,

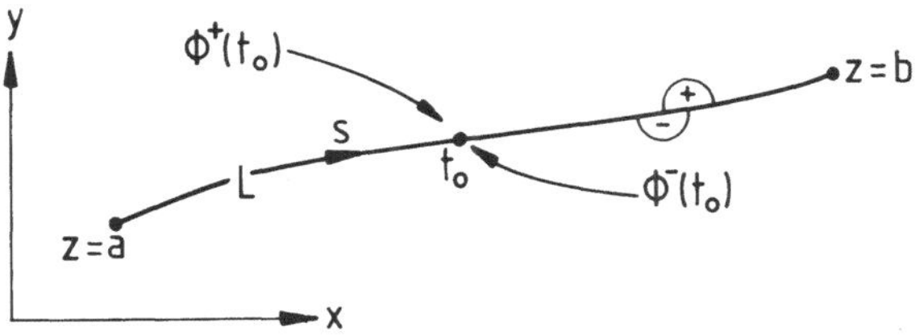

Fig. A.1.1. Arc of discontinuity for a sectionally holomorfic function.

possibly with the exception of the endpoints. In general we consider limiting values of Φ at the open arc hence at L without endpoints. The limiting values satisfy the following relations of Plemelj

$$\Phi^+(t_0)-\Phi^-(t_0)=\varphi(t_0) \tag{A.1.4}$$

and

$$\Phi^+(t_0)+\Phi^-(t_0)=\frac{1}{\pi i}\oint_L\frac{\varphi(t)}{(t-t_0)}\mathrm{d}t, \tag{A.1.5}$$

where the integration is in the sense of the Cauchy principal value.

Formulas (A.1.4) and (A.1.5) can be verified directly in the case that $\varphi(t)$ represents the values on L of a function $\varphi(z)$ analytic in a neighbourhood of L. In this case we may deform L slightly (figure A.1.2.) in order to calculate for instance the limit value

$$\Phi^+(t_0)=\frac{1}{2\pi i}\oint_L\frac{\varphi(t)}{(t-t_0)}\mathrm{d}t+\tfrac{1}{2}\varphi(t_0). \tag{A.1.6}$$

Analogously we find for the other side

$$\Phi^-(t_0)=\frac{1}{2\pi i}\oint_L\frac{\varphi(t)}{(t-t_0)}\mathrm{d}t-\tfrac{1}{2}\varphi(t_0). \tag{A.1.7}$$

Subtraction of (A.1.6) and (A.1.7) yields (A.1.4), addition yields (A.1.5).

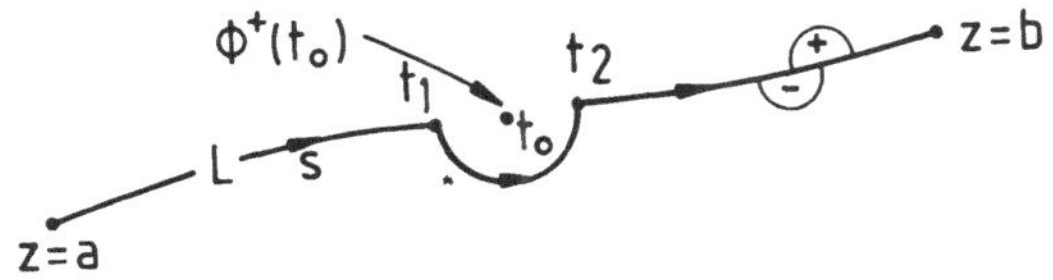

Fig. A.1.2. Deformation of L when $\varphi(z)$ is analytic.

A.2. The Hilbert problem for an arc

The problem is to find a sectionally holomorphic function $\Phi(z)$ which satisfies the relation

$$\Phi^{-}(t)-G(t)\Phi^{+}(t)=g(t), \qquad t\in L, \tag{A.2.1}$$

where $G(t)$ and $g(t)$ are given functions, which satisfy the Hölder condition and $G(t)\neq 0$.

First we consider the homogeneous equation

$$\Psi^{-}(t)=G(t)\Psi^{+}(t), \qquad t\in L. \tag{A.2.2}$$

We try to find a sectionally holomorphic function $\Psi(z)$ without zero's in the whole complex domain. Then ln $\Psi(z)$ is also sectionally holomorphic and satisfies

$$\{\ln \Psi(t)\}^{+}-\{\ln \Psi(t)\}^{-}=-\ln G(t), \qquad t\in L. \tag{A.2.3}$$

Comparing (A.2.3) and (A.1.4) we find by (A.1.3) as a solution of (A.2.2)

$$\Psi(z)=\exp\left(-\frac{1}{2\pi i}\int_L \frac{\ln G(t)}{(t-z)}\mathrm{d}t\right). \tag{A.2.4}$$

Multiplying this function by an arbitrary rational function $R_1(z)$, which is allowed to possess poles only at the ends of L, we do not disturb relation (A.2.2). Hence a more general solution of (A.2.2) is

$$\Psi(z)=\left\{\exp\left(-\frac{1}{2\pi i}\int_L \frac{\ln G(t)}{(t-z)}\mathrm{d}t\right)\right\}R_1(z). \tag{A.2.5}$$

In order to deal with the inhomogeneous equation we write (A.2.1) by using (A.2.2), in the form

$$\frac{\Phi^{-}(t)}{\Psi^{-}(t)}-\frac{\Phi^{+}(t)}{\Psi^{+}(t)}=\frac{g(t)}{\Psi^{-}(t)}. \tag{A.2.6}$$

Hence again by (A.1.4) and (A.1.3) we obtain

$$\Phi(z)=-\frac{\Psi(z)}{2\pi i}\left\{\int_L \frac{g(t)}{\Psi^{-}(t)(t-z)}\mathrm{d}t+R_2(z)\right\}, \tag{A.2.7}$$

where it is assumed that $\Psi(z)$ is chosen in such a way that the integral converges and $R_2(z)$ is also an arbitrary rational function with the same restriction as $R_1(z)$. In [47] it is shown that (A.2.7) and (A.2.5) are the general solutions of the equations (A.2.1) and (A.2.2) respectively, when the behaviour at infinity is prescribed to be algebraic.

A.3. Singular integral equations

Next we discuss the relation between a simple type of singular integral equation and the Hilbert problem. Consider the equation

$$\frac{1}{\pi i}\oint_{-1}^{1}\frac{\varphi(t)}{(t-t_0)}\mathrm{d}t = g(t_0), \tag{A.3.1}$$

where t and t_0 are real variables, $\varphi(t)$ is the unknown function and $g(t)$ is given. Introducing

$$\Phi(z) = \frac{1}{2\pi i}\int_{-1}^{1}\frac{\varphi(t)}{(t-z)}\mathrm{d}t, \tag{A3.2}$$

we find from (A.1.5)

$$\Phi^+(t) + \Phi^-(t) = g(t), \quad -1 \le t \le 1. \tag{A.3.3}$$

The arc L of section A.1 is here the interval $(-1, 1)$ of the real axis, with "+" side at the upper half plane and "−" side at the lower half plane (figure A.3.1.). Substitution of $G(t) = -1$ in (A.2.5) yields for the solution $\Psi(z)$ of the homogeneous part of (A.3.3)

$$\Psi(z) = \left\{\exp -\tfrac{1}{2}\int_{-1}^{1}\frac{\mathrm{d}t}{(t-z)}\right\}R_1(z) = \left(\frac{z+1}{z-1}\right)^{1/2}R_1(z), \tag{A.3.4}$$

where we took $\ln -1 = \pi \mathrm{i}$. Choosing $R_1(z) = 1$; $R_1(z) = (z+1)^{-1}$ or $R_1(z) = (z-1)(z+1)^{-1}$ yields a number of functions $\Psi(z)$ namely

$$\Psi(z) = \left(\frac{z+1}{z-1}\right)^{1/2}, \ \Psi(z) = (z^2-1)^{-1/2}, \ \Psi(z) = \left(\frac{z-1}{z+1}\right)^{1/2}, \tag{A.3.5}$$

respectively. The choice we make is not essential, however in connection

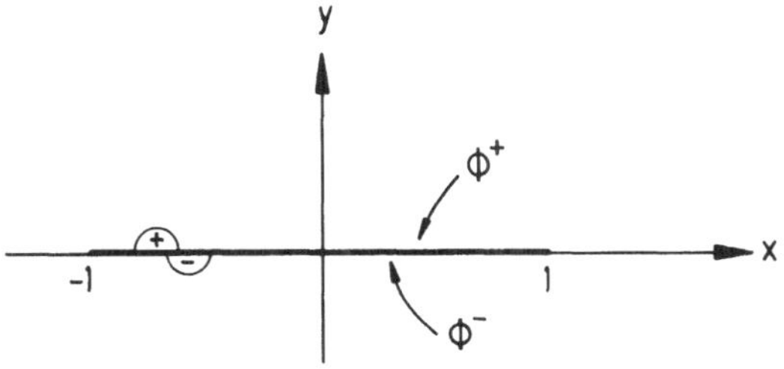

Fig. A.3.1. The real interval $[-1, 1]$.

with the problem of the second part of section 5.10 where we have to satisfy the Kutta condition at $x=1$ and where we tolerate only a square root singularity at $x=-1$, the third expression in (A.3.5) yields a more simple analysis. Of course we have to define the meaning of the square roots, we will take them real and positive for z real and $z>1$.

In this way we find for the solution (A.3.3) by using (A.2.7)

$$\Phi(z)=-\frac{1}{2\pi i}\left(\frac{z-1}{z+1}\right)^{1/2}\left\{\int_{-1}^{1}\frac{g(t)}{-i\left(\frac{1-t}{1+t}\right)^{1/2}(t-z)}\mathrm{d}t+R_2(z)\right\}. \tag{A.3.6}$$

From (A.3.2) it follows that $\Phi(z)$ has to tend to zero for $|z|\to\infty$ and because of the mentioned square root singularity at $z=-1$, the rational function $R_2(z)$ has to be identically zero, $R_2(z)\equiv 0$. By (A.1.6) and (A.1.7) for $-1<t_0<1$, we have

$$\Phi^+(t_0)=-\frac{i}{2\pi}\left(\frac{1-t_0}{1+t_0}\right)^{1/2}\left\{\oint_{-1}^{1}\left(\frac{1+t}{1-t}\right)^{1/2}\frac{g(t)}{(t-t_0)}\mathrm{d}t\right.$$
$$\left.+\pi i\left(\frac{1+t_0}{1-t_0}\right)^{1/2}g(t_0)\right\}, \tag{A.3.7}$$

$$\Phi^-(t_0)=\frac{i}{2\pi}\left(\frac{1-t_0}{1+t_0}\right)^{1/2}\left\{\oint_{-1}^{1}\left(\frac{1+t}{1-t}\right)^{1/2}\frac{g(t)}{(t-t_0)}\mathrm{d}t\right.$$
$$\left.-\pi i\left(\frac{1+t_0}{1-t_0}\right)^{1/2}g(t_0)\right\}. \tag{A.3.8}$$

Then by (A.1.4) we find the desired solution of (A.3.1)

$$\varphi(t_0)=-\frac{i}{\pi}\left(\frac{1-t_0}{1+t_0}\right)^{1/2}\oint_{-1}^{1}\left(\frac{1+t}{1-t}\right)^{1/2}\frac{g(t)}{(t-t_0)}\mathrm{d}t. \tag{A.3.9}$$

When we tolerate at each end of the interval, hence at $x=-1$ and at $x=1$ a square root singularity, we can use in (A.3.4) $R_1(z)=(z+1)^{-1}$ hence $\Psi(z)=(z^2-1)^{-1/2}$. Then with (A.2.7)

$$\Phi(z)=-\frac{1}{2\pi i(z^2-1)^{1/2}}\left\{\int_{-1}^{1}\frac{-i(1-t^2)^{1/2}g(t)}{(t-z)}\mathrm{d}t+R_2(z)\right\}. \tag{A.3.10}$$

When we demand again that $\Phi(z)$ tends to zero for $|z| \to \infty$, we find that the rational function $R_2(z)$ can only be a constant for which we take $-ic_2$. From (A.3.10) it follows

$$\Phi^+(t_0) = \frac{-i}{2\pi(1-t_0^2)^{1/2}} \left\{ \oint_{-1}^{1} \frac{(1-t^2)^{1/2} g(t)}{(t-t_0)} \mathrm{d}t + \pi i (1-t_0^2)^{1/2} g(t_0) + c_2 \right\}, \tag{A.3.11}$$

$$\Phi^-(t_0) = \frac{i}{2\pi(1-t_0^2)^{1/2}} \left\{ \oint_{-1}^{1} \frac{(1-t^2)^{1/2} g(t)}{(t-t_0)} \mathrm{d}t - \pi i (1-t_0^2)^{1/2} g(t_0) + c_2 \right\}. \tag{A.3.12}$$

Hence by (A.1.4)

$$\varphi(t_0) = \frac{-i}{\pi(1-t_0^2)^{1/2}} \left\{ \oint_{-1}^{1} \frac{(1-t^2)^{1/2} g(t)}{(t-t_0)} \mathrm{d}t + c_2 \right\}. \tag{A.3.13}$$

We now suppose $g(t) = c_1 =$ const., using (5.10.2) we obtain

$$\varphi(t_0) = \frac{i}{\pi(1-t_0^2)^{1/2}} \{ \pi c_1 t_0 - c_2 \}. \tag{A.3.14}$$

B. Curvilinear coordinates

In this appendix we give without argumentation some formulas needed for the calculation of operations such as grad, the Laplacian Δ, div and rot in curvilinear coordinates. These coordinates do not need to be orthogonal which is important with respect to the helicoidal coordinates used in section 6.1. We do not enter into a discussion about the concept of a tensor. For the incorporation of the introduced subjects in a systematic theory we refer to [53] or [6].

B.1. Tensor representation of some operators

Suppose (x^1, x^2, x^3) is a Cartesian coordinate system and (ξ^1, ξ^2, ξ^3) an arbitrarily curvilinear one. Their relation is defined by

$$x^l = x^l(\xi^k), \quad \xi^l = \xi^l(x^k); \qquad k, l = 1, 2, 3. \tag{B.1.1}$$

In the Cartesian system we have a point P with vector of position $\boldsymbol{R}$ (figure B.1.1.), the base vectors $\boldsymbol{e}_j$ are of unit length. With respect to the curvilinear system we have the base vectors

$$\boldsymbol{g}_i = \frac{\partial \boldsymbol{R}}{\partial \xi^i} = \frac{\partial}{\partial \xi^i}\left(x^1(\xi^k), x^2(\xi^k), x^3(\xi^k)\right), \qquad \text{(B.1.2)}$$

here as well as in the following the indices assume the values 1, 2 or 3. Equation (B.1.2) yields the components of the $\boldsymbol{g}_i$ with respect to the Cartesian system. The metric tensor of the curvilinear system is given by

$$g_{ij} = \boldsymbol{g}_i \cdot \boldsymbol{g}_j = \sum_{l=1}^{3} \frac{\partial x^l}{\partial \xi^i}\frac{\partial x^l}{\partial \xi^j} = g_{ji}. \qquad \text{(B.1.3)}$$

The length of a base vector $\boldsymbol{g}_i$ can be written as

$$|\boldsymbol{g}_i| = (\boldsymbol{g}_i \cdot \boldsymbol{g}_i)^{1/2} = (g_{ii})^{1/2}. \qquad \text{(B.1.4)}$$

In order that (B.1.1) is an acceptable transformation we demand that the Jacobian determinant

$$g = |g_{ij}| \neq 0. \qquad \text{(B.1.5)}$$

In the following we assume the summation convention which states that; if in an expression an index occurs twice, once in a lower position and once in an upper position, then this expression is summed with respect to this index over its values 1, 2 and 3. Because in (B.1.4) we have two lower indices i, we do not sum. When a summation sign Σ is used, this convention does not apply.

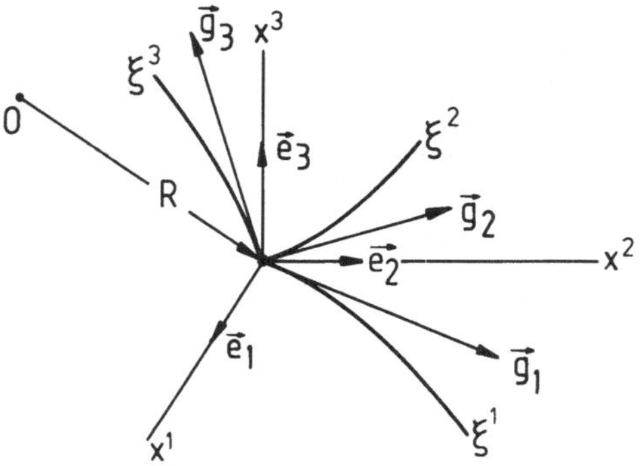

Fig. B.1.1. Base vectors $\boldsymbol{e}_j$ and $\boldsymbol{g}_j$.

By means of the metric tensor (B.1.3) we introduce an other fundamental one g^{ij} by

$$g^{ij}g_{jk} = \delta^i_k, \tag{B.1.6}$$

where δ^i_k is the Kronecker symbol, $\delta^i_k = 1$ for $i = k$, $\delta^i_k = 0$ for $i \neq k$. It follows that

$$g^{ij} = \frac{G(i,j)}{g}, \tag{B.1.7}$$

$G(i, j)$ being the cofactor (including sign) of g_{ij} in the expansion of $g = |g_{kl}|$.

We can now write any vector $\boldsymbol{v}$ at P as

$$\boldsymbol{v} = v^k \boldsymbol{e}_k, \tag{B.1.8}$$

or as

$$\boldsymbol{v} = \tilde{v}^k \boldsymbol{g}_k, \qquad \tilde{v}^k = \frac{\partial \xi^k}{\partial x^l} v^l. \tag{B.1.9}$$

We can by using (B.1.4), also decompose $\boldsymbol{v}$ with respect to the curvilinear system by using "base vectors" $\boldsymbol{c}_k$ of unit length along the coordinate directions

$$\boldsymbol{v} = \sum_{k=1}^{3} \tilde{v}^k (g_{kk})^{1/2} \left(\frac{\boldsymbol{g}_k}{(g_{kk})^{1/2}} \right) \stackrel{\text{def}}{=} \sum_{k=1}^{3} \overset{+}{v}{}^k \boldsymbol{c}_k. \tag{B.1.10}$$

We will call the components $\overset{+}{v}{}^k$ the "physical components" of a vector, which are useful in orthogonal curvilinear coordinate systems.

Next we introduce the ε^{klm} system by

$$\varepsilon^{klm} = \frac{1}{(g)^{1/2}}; \qquad \varepsilon^{klm} = -\frac{1}{(g)^{1/2}}; \qquad \varepsilon^{klm} = 0, \tag{B.1.11}$$

if (k, l, m) is an even permutation of (1, 2, 3); if (k, l, m) is an odd permutation of (1, 2, 3); if otherwise, respectively.

With the above introduced quantities we can write down the operations grad, Δ, div and rot with respect to curvilinear coordinates. We use the physical components $\overset{+}{v}{}^k$ of a vector $\boldsymbol{v}$ (B.1.10) and express grad and rot also in their physical components.

$$\operatorname{grad}\Phi = \sum_{i,j=1}^{3}\left\{(g_{jj})^{1/2} g^{ij}\frac{\partial\Phi}{\partial\xi^i}\right\}\boldsymbol{c}_j, \tag{B.1.12}$$

$$\Delta\Phi = g^{-1/2}\frac{\partial}{\partial\xi^i}\left\{g^{1/2}g^{ij}\frac{\partial\Phi}{\partial\xi^j}\right\}, \tag{B.1.13}$$

$$\operatorname{div}\boldsymbol{v} = \sum_{k=1}^{3} g^{-1/2}\frac{\partial}{\partial\xi^k}\left\{g^{1/2}(g_{kk})^{-1/2}\overset{+}{v}{}^k\right\}, \tag{B.1.14}$$

$$\operatorname{rot}\boldsymbol{v} = \sum_{k,l,m,\alpha,\beta,\gamma=1}^{3} \varepsilon^{klm} g_{l\alpha}\left\{\frac{\partial}{\partial\xi^k}\left(\frac{\overset{+}{v}{}^\alpha}{(g_{\alpha\alpha})^{1/2}}\right)\right.$$

$$\left. + \tfrac{1}{2}g^{\alpha\gamma}\left(\frac{\partial g_{\beta\gamma}}{\partial\xi^k} + \frac{\partial g_{kj}}{\partial\xi^\beta} - \frac{\partial g_{\beta k}}{\partial\xi^\gamma}\right)\frac{\overset{+}{v}{}^\beta}{(g_{\beta\beta})^{1/2}}\right\}(g_{mm})^{1/2}\boldsymbol{c}_m. \tag{B.1.15}$$

It is easily seen that when the transformation (B.1.1) is orthogonal, hence when (ξ^1, ξ^2, ξ^3) is again Cartesian, then (B.1.12) – (B.1.15) assume their common form for a Cartesian coordinate system.

B.2. Two applications

We give two examples of which some results are needed in section 2.4 and in section 6.1. First we pass from the Cartesian system (x, y, z) to the cylindrical coordinates (x, r, φ) (figure 2.2.3), which form an orthogonal curvilinear system. We introduce the index notation by

$$x^1 = x, \quad x^2 = y, \quad x^3 = z; \qquad \xi^1 = x, \quad \xi^2 = r, \quad \xi^3 = \varphi;$$

$$x^1 = \xi^1, \quad x^2 = \xi^2\cos\xi^3, \quad x^3 = \xi^2\sin\xi^3. \tag{B.2.1}$$

Then we obtain

$$\begin{aligned} &\boldsymbol{g}_1 = (1, 0, 0), \qquad \boldsymbol{g}_2 = (0, \cos\xi^3, \sin\xi^3),\\ &\boldsymbol{g}_3 = (0, -\xi^2\sin\xi^3, \xi^2\cos\xi^3), \end{aligned} \tag{B.2.2}$$

$$\begin{aligned} &g_{11} = 1, \quad g_{22} = 1, \quad g_{33} = (\xi^2)^2; \qquad g_{ij} = 0, i \neq j; \qquad g = (\xi^2)^2,\\ &g^{11} = 1, \quad g^{22} = 1, \quad g^{33} = (\xi^2)^{-2}; \qquad g^{ij} = 0, i \neq j. \end{aligned} \tag{B.2.3}$$

We denote in the cylindrical system, the physical components of a vector $\boldsymbol{v}$ by v_x, v_r and v_φ. Then we find from (B.1.12)–(B.1.15),

$$\operatorname{grad}\Phi=\left(\frac{\partial\Phi}{\partial x},\frac{\partial\Phi}{\partial r},\frac{1}{r}\frac{\partial\Phi}{\partial\varphi}\right), \tag{B.2.4}$$

$$\Delta\Phi=\frac{\partial^2\Phi}{\partial x^2}+\frac{1}{r}\frac{\partial}{\partial r}r\frac{\partial\Phi}{\partial r}+\frac{1}{r^2}\frac{\partial^2\Phi}{\partial\varphi^2}, \tag{B.2.5}$$

$$\operatorname{div}\boldsymbol{v}=\frac{\partial v_x}{\partial x}+\frac{1}{r}\frac{\partial}{\partial r}rv_r+\frac{1}{r}\frac{\partial v_\varphi}{\partial\varphi}, \tag{B.2.6}$$

$$\operatorname{rot}\boldsymbol{v}=\left(\frac{1}{r}\frac{\partial}{\partial r}rv_\varphi-\frac{1}{r}\frac{\partial}{\partial\varphi}v_r,\ \frac{1}{r}\frac{\partial}{\partial\varphi}v_x-\frac{\partial}{\partial x}v_\varphi,\ \frac{\partial}{\partial x}v_r-\frac{\partial}{\partial r}v_x\right), \tag{B.2.7}$$

where grad Φ and rot $\boldsymbol{v}$ are given by their physical components in the x, r and φ directions, hence with respect to the "base vectors" of unit length.

Next we pass from the Cartesian system (x, y, z) to the helicoidal coordinate system (ρ, ζ, σ) (6.1.6),

$$x^1=x,\quad x^2=y,\quad x^3=z;\qquad \xi^1=\rho,\quad \xi^2=\zeta,\quad \xi^3=\sigma; \tag{B.2.8}$$

$$x^1=\frac{1}{a}(\xi^2-\xi^3),\qquad x^2=\frac{\xi^1}{a}\cos\xi^3,\qquad x^3=\frac{\xi^1}{a}\sin\xi^3. \tag{B.2.9}$$

Then we obtain

$$\begin{aligned}&\boldsymbol{g}_1=\left(0,\frac{1}{a}\cos\xi^3,\frac{1}{a}\sin\xi^3\right),\qquad \boldsymbol{g}_2=\left(\frac{1}{a},0,0\right),\\&\boldsymbol{g}_3=\left(-\frac{1}{a},-\frac{\xi^1}{a}\sin\xi^3,\frac{\xi^1}{a}\cos\xi^3\right).\end{aligned} \tag{B.2.10}$$

$$\begin{aligned}&g_{11}=\frac{1}{a^2},\qquad g_{22}=\frac{1}{a^2},\qquad g_{23}=g_{32}=-\frac{1}{a^2},\\&g_{33}=\frac{1}{a^2}\left(1+(\xi^1)^2\right),\qquad g=\frac{(\xi^1)^2}{a^6},\\&g^{11}=a^2,\qquad g^{22}=a^2\left((\xi^1)^{-2}+1\right),\\&g^{23}=g^{32}=a^2(\xi^1)^{-2},\qquad g^{33}=a^2(\xi^1)^{-2},\end{aligned} \tag{B.2.11}$$

the not mentioned g_{ij} and g^{ij} are zero.

Because $g_{23}\neq 0$ it follows that this curvilinear system is not orthogonal. This is the reason that we only give the Laplacian of the scalar function Φ which is needed in section 6.1. We find from (B.1.13)

$$\Delta\Phi = a^2\left[\frac{1}{\rho}\frac{\partial}{\partial\rho}\rho\frac{\partial\Phi}{\partial\rho} + \left(1+\frac{1}{\rho^2}\right)\frac{\partial^2\Phi}{\partial\zeta^2} + \frac{2}{\rho^2}\frac{\partial^2\Phi}{\partial\zeta\partial\sigma} + \frac{1}{\rho^2}\frac{\partial^2\Phi}{\partial\sigma^2}\right]. \qquad \text{(B.2.12)}$$

References

1. Adams G.N., Propeller research at Canadair limited. Cal/Usaav-labs symposium Proc., Aerodynamics Problems associated with V stol aircraft, Vol. 1, 1966.
2. Andrews J.B., Cummings D., A design procedure for large hub propellers. Journal of Ship Research, Vol. 16, No. 3, 1972.
3. Batchelor G.K., An introduction to fluid dynamics. Cambridge University Press, 1974.
4. Belotserkovskii S.M., The theory of thin wings in subsonic flow. Plenum Press, 1967.
5. Betz A., Schrauben propeller mit geringstem Energieverlust. Kgl. Ges. d. Wiss., Nachrichten, Math.-Phys., Heft 2, 1919.
6. Borisenko A.I., Tarapov I.E., Vector and tensor analysis with applications. Prentice Hall, Inc., 1968.
7. Bridgman P.W., Dimensional analysis. Yale University Press, 1963.
8. Chopra M.G., Hydromechanics of lunate-tail swimming propulsion. Journal of Fluid Mechanics, Vol. 64, part 2, 1974.
9. Choquet G., Lectures on analysis. Vol. II, Benjamin, 1969.
10. Comstock J.P., Editor, Principles of naval architecture. The Society of Naval Architects and Marine Engineers, 1967.
11. Couchet G., Les profiles en aerodynamique instationnaire et la condition de Joukowski. Librairie Scientifique et Technique, Albart Blanchard, 1976.
12. Courant R., Hilbert D., Methods of mathematical physics, Vol. II. Interscience Publishers, 1966.
13. Cummings D.E., The Effect of propeller wake deformation on propeller design. International Shipbuilding Progress, Vol. 23, 1976.
14. Gelfand J.M., Schilov G.E., Verallgemeinerte Funktionen, Vol. I, V.E.B., Deutscher Verlag der Wissenschaften, 1962.
15. Gent W. van, Unsteady lifting surface theory for ship screws, derivation and numerical treatment of integral equations. Journal of Ship Research, Vol. 19, No. 4, 1975.
16. Gent W. van, Oossanen P. van, Influence of wake on propeller loading and cavitation. International Shipbuilding Progress, Vol. 20, 1973.
17. Goodman T.R., Momentum theory of propeller in shear flow. Journal of Ship Research, Vol. 23, No. 4, 1979.
18. Graaf R. de, On optimum fish tail propellers with two blades. Thesis, University of Groningen, The Netherlands, 1970.
19. Greenberg M.D., Non linear actuator disk theory. Zeitschrift für Flugwissenschaften, Bd. 20, Heft 3, 1972.
20. Grimm O., Propeller and vane wheel. Journal of Ship Research, Vol. 24, No. 4, 1980.
21. Gröbner W., Hofreiter N., Integral tafel I, II. Springer Verlag, 1973.
22. Hadler J.B., Contra rotating propeller propulsion, A state of the art report. Marine Technology, Vol. 6, No. 3, July 1969.
23. Hadler J.B., Morgan W.B., Meyers K.A., Advanced propeller propulsion for high powered single screw ships. The Society of Naval Architects and Marine Engineers, 1964.
24. Happel J., Brenner H., Low Reynolds number hydrodynamics. Prentice Hall, 1965.

25. Hess F., Boomerangs, Aerodynamics and Motion. Thesis, University of Groningen, The Netherlands, 1975.
26. Horlock J.H., Actuator disk theory. Mc Graw Hill Inc., 1978.
27. Isay W.H., Propellertheorie, hydrodynamische Probleme. Springer Verlag, 1964.
28. Isay W.H., Moderne Probleme der Propellertheorie. Springer Verlag, 1970.
29. Isay W.H., Kavitation. Schifffahrts-Verlag "Hansa", C. Schroedter und Co., 1981.
30. Jacobs W.R., Tsakonas S., Propeller induced velocity field due to thickness and loading effects. Journal of Ship Research, Vol. 19, No. 1, 1975.
31. Katz J., Wheiss D., Hydrodynamic propulsion by large amplitude oscillation of an airfoil with chordwise flexibility. Journal of Fluid Mechanics. Vol. 88, part 3, 1978.
32. Kellogg O.D., Foundations of potential theory. Verlag von Julius Springer, 1929.
33. Kelly H.R., Rentz A.W., Siekmann J., Experimental studies on the motion of a flexible hydrofoil. Journal of Fluid Mechanics, Vol. 19, part 1, 1964.
34. Kerwin J.E., The solution of propeller lifting surface problems by vortex lattice methods. M.I.T., Naval Architecture Department Report, 1961.
35. Kerwin J.E., Computer techniques for propeller blades section design. International Shipbuilding Progress, Vol. 20, No. 227, 1973.
36. Kerwin J.E., Leopold R., Propeller incidence due to blade thickness. Journal of Ship Research, Vol. 7, No. 2, 1963.
37. Klaren L., On the efficiency of a ducted non stationary actuator disk. Journal of Engineering Mathematics, Vol. 12, No. 3, 1978.
38. Klaren L., Sparenberg J.A., On optimum screw propellers with endplates, inhomogeneous inflow. Journal of Ship Research, Vol. 25, No. 4, 1981.
39. Kotschin N.J., Kibel J.A., Rose N.W., Theoretische Hydromechanik Bd I. Akademie Verlag, 1954.
40. Kruppa C., Practical aspects of high speed small propellers. International Shipbuilding Progress, Vol. 23, 1976.
41. Lamb H., Hydrodynamics. Dover Publications, 1945.
42. Ljusternik L.A., Sobolev W.I., Elemente der Funktionalanalysis. Akademie Verlag, 1960.
43. Manen J.D. van, Ergebnisse systematischer Versuche mit Propellern mit annäherend senkrecht stehende Achse. Jahrbuch der Schiffbautechnische Gesellschaft, Bd. 57, 1963.
44. Marchaj C.A., Sailing theory and practice. Dodd Mead and Company, 1964.
45. Milgram J.H., The analytic design of yacht sails. The Society of Naval Architects and Marine Engineers, 1968.
46. Mueller F., Recent developments in the design and application of the vertical axis propeller. The Society of Naval Architects and Marine Engineers, 1955.
47. Muskhelishvily N.I., Singular integral equations. P. Noordhoff N.V., 1953.
48. Ordway D.E., Sluyter M.M., Sonnerup, B.O.U., Three dimensional theory of ducted propellers. Therm. Advanced Research, Ithaca, New York, 1960.
49. Potze W., Sparenberg J.A., On optimum large amplitude sculling propulsion, finite span. International Shipbuilding Progress, Vol. 30, No. 351, 1983.
50. Schmidt G.H., Sparenberg J.A., On the edge singularity of an actuator disk with large constant normal load. Journal of Ship Research, Vol. 21, No. 2, 1977.
51. Schmidt G.H., Sparenberg J.A., On the linearized theory of hub cavity flow with swirl. Twelfth Symposium Naval Hydrodynamics, Office of Naval Research, 1978.
52. Sneddon I.N., The use of integral transforms. Mc. Graw-Hill Publishing Company L.T.D., 1974.
53. Sokolnikoff I.S., Tensor analysis. J. Wiley and Sons, Inc., 1960.
54. Sparenberg J.A., Application of lifting surface theory to ship screws. Koninklijke Nederlandse Akademie van Wetenschappen, Proc. series B, 62, No. 5, 1959.
55. Sparenberg J.A., On the existence of small-amplitude optimum hydrofoil propulsion. Journal of Ship Research, Vol. 22, No. 4, 1978.

56. Sparenberg J.A., Takens F., On the optimum finite-amplitude motion of a thrust-producing profile. Journal of Ship Research, Vol. 19, No. 2, 1975.
57. Sparenberg J.A., Thomas E.G.F., On the existence of small-amplitude optimum hydrofoil propulsion. Mathematical Methods in the Applied Sciences, 3, 1981.
58. Sparenberg J.A., Vries J. de, On sculling propulsion by two elastically coupled profiles. Journal of Ship Research, Vol. 27, No. 2, 1983.
59. Szantyr J., A computer programm for calculation of cavitation extent and excitation forces for a propeller operating in a non-uniform velocity field. International Shipbuilding Progress, Vol. 26, 1979.
60. Szeless A.G., Undulating plate type propeller: the two dimensional ideal case. Journal of Ship Research. Vol. 13, No. 3, 1969.
61. Taylor A.E., Functional analysis. John Wiley and Sons, Inc., 1958.
62. Uldrick J.P., Siekmann J., On swimming of a flexible plate of arbitrary finite thickness. Journal of Fluid Mechanics, Vol. 20, part 1, 1964.
63. Urbach H.P., On the existence of optimum large amplitude hydrofoil propulsion. To be published.
64. Urbach H.P., External force fields acting at an inviscid and incompressible fluid. To be published.
65. Vučinić A., Theoretical fundamental researches of dynamical blade and propeller loadings caused by unsteady inflow. Thesis, Rijeka, 1975.
66. Wald Q., Performance of a propeller in a wake and the interaction of propeller and hull. Journal of Ship Research, Vol. 9, No. 1, 1965.
67. Watson G.N., A treatise on the theory of Besselfunctions. Cambridge University Press, 1922.
68. Weissinger J., Linearisierte Profiltheorie bei ungleichförmiger Anströmung, I. Unendlich dünne Profile (Wirbel unf Wirbelbelagungen) Acta Mechanica 10, 1970.
69. Weissinger J., Linearisierte Profiltheorie bei ungleichförmiger Anströmung, Teil II: Schlanke Profile. Acta Mechanica 13, 1972.
70. Weissinger J., Maass D., Theory of the ducted propeller, a review. Seventh Symposium Naval Hydrodynamics, Office of Naval Research, 1968.
71. Wiersma A.K., On sailing to windward. Thesis, University of Groningen, The Netherlands, 1979.
72. Wu T.Y., Flow through a heavily loaded actuator disk. Schiffstechnik, Bd. 9, Heft 47, 1962.
73. Wu T.Y., Hydromechanics of swimming propulsion. Journal of Fluid Mechanics, Vol. 46, parts 2 and 3, 1971.
74. Wu T.Y., Hydromechanics of swimming of fishes and cetaceans. Advances in applied mechanics, Vol. 11, Academic Press, 1971.

Index